Investing in Science

Investing in Science

Social Cost-Benefit Analysis of Research Infrastructures

Massimo Florio

The MIT Press
Cambridge, Massachusetts
London, England

This book was set in Times New Roman by Westchester Publishing Services. Printed and bound in the United States of America.

Library of Congress Cataloging-in-Publication Data

Names: Florio, Massimo, author.
Title: Investing in science : social cost-benefit analysis of research infrastructures / Massimo Florio.
Description: Cambridge, MA : The MIT Press, [2019] | Includes bibliographical references and index.
Identifiers: LCCN 2019006862 | ISBN 9780262043199 (hardcover : alk. paper)
Subjects: LCSH: Science—Social aspects. | Science and state. | Science and the arts. | Discoveries in science—Social aspects.
Classification: LCC Q125 .F5845 2019 | DDC 338.9/26—dc23
LC record available at https://lccn.loc.gov/2019006862

10 9 8 7 6 5 4 3 2 1

Contents

List of Abbreviations

ADC	Analog-to-digital converter
AIRC	Italian Association for Research on Cancer
ALICE	Accelerators and Lasers In Combined Experiments
ALS	Advanced light source
APIs	Application programming interfaces
ARPANET	Advanced Research Projects Agency Network
ASI	Italian Space Agency
ASTERICS	Astronomy ESFRI and Research Infrastructure Cluster
ATLAS	A Toroidal LHC ApparatuS
BCG	Boston Consulting Group
BCR	Benefit-cost ratio
BRC	Bioinformatics resource center
CAD	Canadian dollars
CAFÉ	Clean Air for Europe
Caltech	California Institute of Technology
CBA	Cost-benefit analysis
CE	Cost-effectiveness
CERN	European Organization for Nuclear Research
CET	CERN Expenditure Tracking
CHF	Swiss franc
CML	Chronic myeloid leukemia
CMS	Compact Muon Solenoid
CNAO	National Centre of Oncological Hadrontherapy
COSMO SkyMed	Constellation of small Satellites for Mediterranean Basin Observation
CPI	Consumer price index
cryoEM	Cryogenic electron microscopy
CT	Computerized tomography
CV	Capital value
CVM	Contingent valuation method
DCF	Discounted cash flow
DoE	(US) Department of Energy
DSP	Data science platform
EARSC	European Association of Remote Sensing Companies

EBI	European Bioinformatics Institute
EBIT	Earnings before interest and taxes
EBITDA	Earnings before interest, taxes, depreciation and amortization
ECR	Early career researcher
EGA	European Genome-Phenome Archive
ELI	Extreme Light Infrastructure
EMBL	European Molecular Biology Laboratory
ENA	European Nucleotide Archive
ENPV	Economic net present value
ENVISAT	Earth Observation Environmental Satellite
EPO	European Patent Office
ESA	European Space Agency
ESFRI	European Strategy Forum on Research Infrastructures
ESRF	European Synchrotron Radiation Facility
EST	European Solar Telescope
ETS	Emissions trading system
EU	European Union
EU-SILC	European Community Statistics on Income and Living Conditions
EVRI	Environmental Valuation Reference Inventory
EXV	Existence value
FCC	Future Circular Collider
FDR	Financial discount rate
FFRDC	Federally Funded Research and Development Center
FIRR	Financial Internal Rate of Return
FMDV	Foot and mouth disease virus
FNPV	Financial net present value
FTEs	Full-time equivalent employees
FTIR	Fourier transform infrared spectroscopy
FY	Fiscal year
GATC	G, A, T, and C (guanine, adenine, thymine, cytosine)
GATK	Genome Analysis Toolkit
GDP	Gross domestic product
GERD	Gross domestic expenditure in R&D
GHG	Greenhouse gas
GPS	Global positioning system
GRI	Global research infrastructure
GTC	Gran Telescopio Canarias
HECToR	High-End Computing Terascale Resource
HEP	High-energy physics
HGP	Human Genome Project
HIV	Human immunodeficiency virus
HL	High-Luminosity
HMS	Her Majesty's Ship
HPA	Human Protein Atlas
HTML	Hypertext Markup Language

HTTP	Hypertext Transfer Protocol
HUGO	Human Genome Organisation
ICT	Information and communication technology
IMF	International Monetary Fund
IMPLAN	IMpact Analysis for PLANning
INFN	Instituto Nazionale di Fisica Nucleare
I-O	Input-output
IOP	Institute of Physics
IPR	Intellectual property rights
IRR	Internal rate of return
ISS	International Space Station
IT	Information technology
IV	Instrumental variable
ITER	International Thermonuclear Experimental Reactor
JASPERS	Joint Assistance to Support Projects in European Regions
JAXA	Japan Aerospace Exploration Agency
JPO	Japan Patent Office
KSC	Kennedy Space Center
KT	Knowledge transfer
LBLN	Lawrence Berkeley National Laboratory
LE	Light echo
LEP	Large Electron-Positron Collider
LHC	Large Hadron Collider
LIGO	Laser Interferometer Gravitational-Wave Observatory
LNA	Low noise amplifier
LSST	Large Synoptic Survey Telescope
MCCT	Multicenter clinical trial in medical research
MICE	Muon Ionisation Cooling Experiment
MIT	Massachusetts Institute of Technology
MPC	Marginal production cost
MRI	Magnetic resonance imaging
MSV	Marginal social value
NACE	National Association of Colleges and Employers
NASA	National Aeronautics and Space Administration
NCHGR	National Center for Human Genome Research
NEEDS	New Energy Externalities Development for Sustainability
NGI	National Graphene Institute
NH3	Ammonia
NIAID	National Institute of Allergy and Infectious Diseases
NIH	National Institutes of Health
NMR	Nuclear magnetic resonance
NMVOC	Nonmethane volatile organic compound
NOAA	National Oceanographic and Atmosphere Administration
NOx	Nitrogen oxide

NPL	Nonpatent literature
NPV	Economic net present value
NPV	Net present value
OECD	Organisation for Economic Co-operation and Development
OHL	Open hardware license
ONS	British Office of National Statstics
OSS	Open-source software
PCA	Principal component analysis
PDF	Probability density function
PhD	Doctor of philosophy
PPI	Public procurement for innovation
PPP	Purchasing power parity
PRL	*Physics Review Letter*
PRO	Public research organization
PV	Present value
QALY	Quality-adjusted life-year
QOV	Quasi-option value
RAMIRI	Realizing and Managing International Research Infrastructures
RHIC	Relativistic Heavy Ion Collider
RI	Research infrastructure
RIR	Research infrastructure resource
RO	Real option
ROS	Return on sale
ROSCOSMOS	Russian Federal Space Agency
RP	Revealed preference
RV	Replacement value
SAR	Synthetic aperture radar
SDR	Social discount rate
SFA	Swedish Forestry Agency
SIB	Swiss Institute of Bioinformatics
SKA	Square Kilometre Array
SL	Synchrotron light source
SMEs	Small and medium enterprises
SMU	Southern Methodist University
SNA	System of National Accounts
SPS	Super Proton Synchrotron
SRTP	Social rate of time preference
SSC	Superconducting Super Collider
STEM	Science, technology, engineering, and mathematics
STFC	Science and Technology Facilities Council
STScI	Space Telescope Science Institute
TB	Technology breakthrough
TCM	Travel cost method
UK	United Kingdom
UNIDO	United Nations Industrial Development Organization

URL	Uniform Resource Identifier
US	United States
USPTO	United States Patent and Trademark Office
USSR	Union of Soviet Socialist Republics
VOLY	Value of a life-year
VOSL	Value of a statistical life
WIPO	World Intellectual Property Organization
WLCG	Worldwide LHC Computing Grid
WTA	Willingness to accept
WTP	Willingness to pay
WWF	World Wildlife Fund
WWW	World Wide Web

List of Main Variables

AR	Benefits of applied research (all ar-types: $ar = (1, ..., n)$) to external users and other consumers at any t-time (benefits to patients in Chapter 10)
B_n	Nonuse benefits
B_u	Use benefits
c_j	Costs of the company $j = (1, ..., J)$
C_u	Use costs
$CERN_j$	Dummy variable that takes value 0 before the first order is received by the company j and 1 thereafter
CIT_i	Value of citations of papers of wave $i = (0, ..., n)$
CPI_c	Country's c consumer price index
CS	Consumer surplus
CU	Benefits to users of cultural goods
d_t	Financial discount factor at time $t = (1, ..., \mathcal{T})$
Dw_z	Increased earnings gained by former Research Infrastructure's students or former employees z since the time they left the Research Infrastructure
ε	Elasticity of the marginal utility of consumption
$\mathbb{E}$	Expectation operator
EBIT	Operating revenue and earnings before interest and taxes
EXP	Number of years of potential labor market experience
$expw_z$	Range of students' salary expectations taking integer values from 1–F.
EXT	Negative externalities
EXT_p	Externality of patents granted by the Research Infrastructure
EXV_0	Existence value
φ	Time when students and former employees leave the Research Infrastructure
FDR	Financial discount rate
g	General public categories $g = (1, ..., G)$
GDP_c	Gross domestic product in the firm's country c
gth	Expected growth rate of per capita consumption
h_s	Average hours of working time for the production of a scientific paper

HC	Benefits to human capital
i	Waves of scientific papers (in chapter 7 only: i = innovations)
j	Companies $j = (1, ..., J)$
K	Economic value of capital
L_o	Labor cost of administrative and technical staff
L_s	Labor cost of scientists
m	Multiplier of impact of a publication
M	Amount of money
$mcpf$	Marginal cost of public funds
MSV	Marginal social value (id est shadow price)
n_s	Number of scientists
NH	Number of years of schooling
NPV_{RI}	Net present value of the RI
NPV_u	Net present value of use-benefits
OP	Other operating costs
p	Set of patents over time $p = (1, ..., P)$
Π_j	Incremental shadow profits of companies
P_0	Value of the papers authored by the Research Infrastructure's scientists (wave 0)
P_i	Value of subsequent i-waves of papers produced by other scientists using the results of Research Infrastructure's scientists
$Patents_j$	Number of patents filed by the company j
$PRIV_p$	Private value of patents granted in the relevant technological field
PS	Producer surplus
$Productivity_j$	Sales per employee of the company j
Ptp	Pure time preference
pub	Total number of papers produced by scientists
PV_{B_u}	Present value of use benefits
PV_{C_u}	Present value of use costs
$PVPatent$	Present value of patents
PV_{pub}	Present value of scientific publications
q	Quantities
QOV_0	Quasi-option value
$R\&D_j$	Intangible fixed assets per employee of the company j
r_j	Revenues of the company j
Ref_i	Average number of references contained in scientific paper of wave i
Ref_p	Average number of references included in patents issued in the relevant technological field
$Revenues_j$	Yearly change of the operating revenues for the company j
s_t	Social discount factor at time t
s	Scientists
SC	Value of publications for scientists
SDR	Social discount rate

$Size'_j$	Vector, including information on the yearly change of assets and number of employees of the company j
T	Time horizon
t	Time
TE	Technological externalities to firms, including information and communications technology externalities and other technological learning
use	Average rate of usage of granted patents
w	Level of earnings
w_0	Level of earnings of an individual without any education or experience
w_s	Level of earnings per hour per scientist
WTP_g	Willingness to pay for cultural goods produced by the Research Infrastructure
X	In Chapter 8: Outreach activities
X_z	In Chapter 4: Vector of independent variables aiming at explaining the range of students' salary expectations
Y	Social cost of producing knowledge outputs (not just publications, but any other scientific communications)
y_s	Number of papers produced by each scientist
z	Students and former employees $(z = 1, ..., Z)$

Introduction: Beyond Big Science, the Research Infrastructure Paradigm

Large particle accelerators, outer space probes, radio telescopes, networks of submarine observatories, and genomic platforms are focal points of intellectual curiosity, creativity, and eventually scientific discovery managed in a new form: the research infrastructure (RI).

An RI can be defined as a collective scientific enterprise, designed and managed with the objective to create knowledge and related services for a number of different users. Dynamic and cosmopolitan scientific communities are structured around major RIs in new collaborative ways, reshuffling the intellectual boundaries across nations, universities, research institutes, and diverse disciplines.

These projects are often costly and ultimately funded by taxpayers. What are the benefits for the society of supporting these investments and their operations? This book suggests a framework to (at least partially) answer this question from the perspective of applied welfare economics.

Two working hypotheses are the starting points for this inquiry. The first one is that several contemporary major scientific projects, such as the Large Hadron Collider (LHC), no longer fit into the traditional "Big Science" model. This term was used for the first time by the physicist Alvin Weinberg (1961). He was the director of the Oak Ridge National Laboratory in Tennessee, a facility that was formerly part of the United States (US) atomic bomb Manhattan Project. According to him, huge rockets, particle accelerators, and high-flux energy reactors are the "monumental enterprises" Weinberg (1961, 161) of our days, as in past societies, the pyramids in Egypt or the cathedrals in the Middle Ages were "the extraordinary exertions by their rulers." He suspected that Big Science was ruining science and disrupting public finance, and that space travels and high-energy physics did not deserve a high priority relative to other more relevant and urgent human needs.

Two years later, the historian of science Derek de Solla Price published a seminal book, *Little Science, Big Science* (de Solla Price, 1963) and pointed out the exponential process of knowledge creation during and after World War II. He also suggested that this evolution was similar to the growth pattern of large corporations in the past century. In the posthumous augmented edition (de Solla Price, 1986), the title became *Little Science, Big Science ... and Beyond*. What is the "beyond"?

A possible answer is provided by the relatively new RI concept. According to the European Union Horizon 2020 work program for science (European Commission, 2017a, 4):

Research infrastructures are facilities, resources, and services that are used by the research communities to conduct research and foster innovation in their fields. Where relevant, they may be used beyond research, e.g. for education or public services. They include: major scientific equipment (or sets of instruments); knowledge-based resources such as collections, archives, or scientific data; e-infrastructures, such as data and computing systems and communication networks. ... Such infrastructures may be "single-sited," "virtual" or "distributed."

The key concept in this definition is an *infrastructure where knowledge is created by and shared with multiple users*. This suggests something more complex than a single laboratory, either large or small. In many fields, the word *infrastructure* per se implies a long-term investment aimed at the delivery of services to several users. The concept has evolved to include soft, intangible investments and institutions, along with fixed capital. The empirical estimation of the correlation between infrastructure investment and economic growth has attracted noticeable research attention in economics for decades. The role of different infrastructures in past and contemporary economies is discussed in the introductory chapters of Picot et al. (2015) and Cassis, De Luca, and Florio (2016).

The second working hypothesis in this study is that the evaluation of the socioeconomic impact of RIs can take advantage of some of their specific features. The most important one is that multiple stakeholders benefit from the organizational architecture of RIs, not just a closed group of insiders: coalitions of governments, funding agencies, scientific communities, students, higher-education institutions, companies, users of software and data, users of innovative products and services, and eventually the general public. RIs are thus much more open to social interactions than the traditional top-down Big Science model. Some benefits for these RI stakeholders are often empirically measurable against investments, operating expenditures, and other costs. Taking advantage of progress in the evaluation of environmental and cultural projects, a social cost-benefit analysis (CBA) of RIs can be implemented, with due caution.

CBA is an established methodology for the evaluation of public investment and consists in comparing the socioeconomic costs and benefits of an investment project. In this book, I use the standard toolkit of CBA, augmented with some new extensions, to suggest an evaluation frame of the socioeconomic impact of investment projects in science under the broad RI definition. I think this evaluation is deserved, and the CBA is appropriate, for two reasons: First, governments and citizens want to understand the social return of large-scale investments in science against the funds they need; and second, scientists who are increasingly facing skepticism about what they do should be able to defend their case in a more structured way than using generic arguments about the expected future social benefits of any research project.

0.1 Big Science versus Research Infrastructure

To illustrate the difference between the traditional Big Science model and the RI concept, two very large-scale scientific enterprises can be compared: on one hand, the Apollo Lunar Program, managed by the National Aeronautics and Space Administration (NASA) on behalf of the US government; and on the other hand, the particle physics laboratory of the European Organization for Nuclear Research (CERN). The Apollo Program is one of the archetypal Big Science examples. Another famous Big Science example is the Manhattan Project, which employed up to 130,000 people to produce and test the first nuclear weapons for the US government.[1]

CERN is instead the first example of a very large internationally sponsored RI,[2] without any military implications. Both projects, the Apollo Program and CERN, were built to be at the forefront of the research, but they were radically different in terms of their mission, governance, organizational perspective, and openness.

The goal of the Apollo Program (1961–1972) was famously stated by the special message that President John F. Kennedy addressed to Congress on May 15, 1961):[3]

If we are to win the battle that is now going on around the world between freedom and tyranny, the dramatic achievements in space which occurred in recent weeks should have made clear to us all, as did the Sputnik in 1957, the impact of this adventure on the minds of men everywhere, who are attempting to make a determination of which road they should take. ... I believe that this nation should commit itself to achieving the goal, before this decade is out, of landing a man on the moon and returning him safely to the earth. ... But in a very real sense, it will not be one man going to the moon—if we make this judgment affirmatively, it will be an entire nation.

Clearly, this was a *national* goal in the Cold War context. The mission was accomplished by Apollo 11 in 1969, at an estimated total cost of around $25 billion, by simply adding fiscal spending over fifteen years of the program. In fact, the total cost has been reestimated at around $110 billion in 2010 (Lafleur, 2010). Thousands of scientists were involved together with a large number of aerospace and defense contractors. At the end of the 1960s, according to some estimates, the program funded 400,000 jobs and involved 20,000 private firms (Scott and Jurek, 2014).

The core ingredients of the paradigm embodied in the Apollo Program (or even more clearly in the Manhattan Project) were: first, the close association of science, directly or indirectly, with the US Defense Department and the industrial complex of the major powers; second and consequently, the availability of generous government budgets arising from such an association, and hence, the ownership of the laboratories was firmly in the hands of national governments; third, secrecy about most of the research activities, censorship about the results, and selective exposure to apologetic press coverage; and fourth, the best minds had to be recruited, but only after their political loyalty was carefully assessed. As Hughes (2002, 13) stated, the "five Ms of Big Science" were "Money, manpower, machines, media and the military."

CERN started in Geneva in 1951, ten years after the Manhattan Project and ten years before the Apollo Program. Its first particle accelerator, the Synchrocyclotron, was built in 1957. CERN was conceived as a treaty among countries, not as a national project. It aimed to advance fundamental nuclear and particle physics in postwar Europe. Military applications were explicitly excluded from its scope. Its constituency was international (twelve Western European countries), and thus its funding was shared by its member-states and its ownership was diluted. Recruitment of scientists was not influenced by political loyalty considerations. Openness was the norm, and secrecy and censorship were virtually nonexistent. Since the earliest days, research priorities, organization, and investment decisions have been deeply influenced by the consensus of a global scientific community of particle physicists, even outside the boundary of the CERN staff or the government of its member-states. Moreover, and crucially, CERN was conceived as a platform: Its scientists and engineers design, build, and operate the particle accelerators complex, while experiments are designed and managed by international collaborations, each with its own organization and rules (Boisot et al., 2011).

Thus, two models of large-scale science coexisted in the second half of the twentieth century. In some major powers, particularly in the United States and the Union of Soviet Socialist Republics (USSR), but also in France, the United Kingdom (UK), and elsewhere, the government funding of Big Science was often directly linked to military or strategic national priorities.

In the biography of Zhores Alferov, a Soviet physicist and Nobel laureate, Josephson (2010, 193) writes: "Through the early 1980s, billions of roubles were available in the search for technological breakthroughs in the race for parity with the United States in the Cold War... research on missile defense, fusion, lasers, and communication. Money seemed to grow on trees or in snow banks....."

Aerospace research at NASA in the era of Wernher von Braun (the inventor of the V2 for Germany under Hitler, and later the key scientist on the Apollo Program) was never too disconnected from military and national security strategies, as President Kennedy's statement indicates. The same often can be said for some developments in the information and communication technologies (ICT) industry, such as research in telecommunication networks such as the Advanced Research Projects Agency Network (ARPANET), a packet-switching network funded in the late 1960s by the US Department of Defense, which was instrumental in the later development of the Internet.

Hence, the bottom-line argument for securing government funding to large-scale scientific projects was either their possible direct military implications or their national strategic importance. This is why famously, on April 17, 1969, Robert R. Wilson, the director of Fermilab (founded in 1967 as the US National Acceleration Laboratory in Illinois) felt that it was necessary to testify in front of the Joint Committee on Atomic Energy of the US Congress that the social value of building Fermilab's first particle accelerator was in a different league:[4]

Senator John Pastore: Is there anything connected with the hopes of this accelerator that in any way involves the security of the country?

Wilson: No sir, I don't believe so.

Pastore: Nothing at all?

Wilson: Nothing at all.

Pastore: It has no value in that respect?

Wilson: It has only to do with the respect with which we regard one another, the dignity of men, our love of culture. It has to do with: are we good painters, good sculptors, great poets? I mean all the things we really venerate in our country and are patriotic about. It has nothing to do directly with defending our country except to make it worth defending.

This statement is often cited by scientists because it marked a clear demarcation in the public discourse between the new rationale for funding large-scale science for the sake of knowledge per se, as opposed to the typical argument of national security. If the national security argument is invoked to justify public funding, then the social value of knowledge is not in the limelight. After all, knowledge related to national security is supposed to be protected rather than disseminated. If instead the linkage with national security is severed, then scientists should find other justifications to ask the taxpayers to support costly research projects. This is what Wilson was trying to say in his testimony.

Admittedly, the difference between these two models and rationales may be less clear in some cases. Hughes (2002) observes that when the US Atomic Energy Commission inherited the Manhattan Project sites (such as Los Alamos, Berkeley, Oak Ridge, and Argonne) after World War II, the potential relationship between basic physics research and military applications was still often used by physicists to justify government funding of particle accelerators for fundamental research. The US Department of Energy (DoE) budget (around $28 billion in 2018) is distributed among four major mission areas: (1) nuclear security, protecting the United States from nuclear threats, and propelling the nuclear navy; (2) basic scientific research; (3) energy innovation and security; and (4) environmental cleanup to meet obligations from the Manhattan Project, Cold War, and nuclear energy research.[5] The defense-related part of the budget (items 1 and 4) is about 75 percent of the total, while most of the basic research in physics and in energy[6] (items 2 and 3) is supported by the remaining share of government funds. At least in this case, for historical reasons, the two models of large-scale science are interwoven within the same institution. Some of the DoE-sponsored research programs are similar to what is done at CERN, in terms of international openness of the science model, wide consultation of the scientific community, and outreach. The DoE itself also contributes to experiments and to the upgrade of the LHC of CERN. However, other DoE programs, for obvious security reasons, are top secret.[7]

From a historical perspective, it is possible to adopt the Hughes (2002) view that the Manhattan Project (and pre–World War II science-based military research) were instrumental in shaping the concept of large-scale laboratories in physics, including CERN, at least in terms of organizational complexity and size of budgets. However, differences in missions and the fundamental features between the two models of science and their justification in the public discourse cannot be exaggerated.

0.2   Little Science

The traditional alternative to Big Science, with its legacy of national security, mission-oriented, and top-down approach, has been Little Science. Another physicist, Steven Weinberg (2012), reminds us that when Ernest Rutherford discovered the atomic nucleus (1911), his team consisted of just a postdoc researcher and an undergraduate. Some of the post–World War II breakthroughs, from the discovery of antibiotics to the invention of transistors, required only limited resources. Laboratories were relatively small, involving teams that rarely comprised more than a handful of scientists and doctoral students, but some of these people were part of large organizations (two notable instances are the Bell Laboratories in the case of the transistor and St. Mary's Hospital in London in the case of penicillin). The typical arrangement was (and still is in most universities and research institutes) that a group leader or principal investigator with his or her own funds, a dedicated team, a focused agenda, physical facilities, and supplies strictly needed to perform the experiments that he or she designs (sometimes shared with other teams). Eventually, articles published in scientific journals are authored by a relatively small number of team members. This is still the prevailing pattern in many fields, even within major universities and research institutions with large, cumulated budgets.

Life sciences were often regarded as the realm of Little Science. They often do not require large capital expenditures or very large teams, as in physics, although things are gradually changing. In their brief history of the role of collaborative work in life sciences, Vermeulen, Parker, and Penders (2013) observe a shift to large-scale collaboration in biology, albeit with a variety of patterns.

In some cases, it is not easy to ascertain to which organizational paradigm of science an organization belongs. The National Institutes of Health (NIH) is the most important US government agency for biomedical research. In recent years, on average, its yearly budget was around $30 billion, one-third more than NASA and around thirty times the yearly budget of CERN. The NIH hosts 1,200 principal investigators and around 4,000 postdoc staff. It is overall an amazing success story. It claims to have supported through its grants the research of 148 Nobel laureates to date.[8] Nevertheless, the NIH is certainly far from Big Science in the traditional paradigm, but neither is it a unique RI: it is an array of 27 research institutes, each of which includes many different labs and teams, several of them relatively small-scale.

The diversity of the projects is very wide and comprises research on cancer, the human immunodeficiency virus (HIV), eye disease, aging, alcoholism, deafness, and mental disorders, just to mention a few. Nevertheless, the NIH was instrumental in the implementation of the Human Genome Project, which has been often considered Big Science, and is a RI in the abovementioned sense. It currently operates or funds some very large-scale data centers where billions of bio data are archived and made accessible for research to third parties.

Most of the examples in this book are in high-energy physics, astronomy, material sciences, medical research with particle accelerators, and space science. However, several scientific fields are evolving in the direction of the RI paradigm, albeit more slowly and in a different way. A contact point across scientific domains is the emergence of the systematic digitalization of information, including in the life sciences and humanities, which is per se an ingredient of the RI paradigm. This issue is further discussed next.

0.3 Drivers of Change in Different Fields

New models of organization of the scientific research have gradually evolved, often within large research bodies such as CERN and NASA, or within the system of National Laboratories sponsored by the DoE (Westwick, 2003), but also within the networking role of the scientific societies in Germany—such as the Max Planck Society[9] or the Helmholtz Association[10]—or through the creation of consortia of smaller institutes supported by European Union Horizon 2020, just to mention some examples.

The core shift to the RI model was mainly driven by two determinants: first, the acknowledgment of the scientific community of the effectiveness and efficiency to create common open platforms, shared by a plurality of teams beyond national borders; and second, by dramatic changes in ICT. As has been mentioned, the multiuser platform is the essence of the RI concept, with far-reaching consequences in terms of funding, ownership, governance, organization model, stakeholder involvement, and openness to outsiders, including the general public. The digital revolution has made this concept very cost effective. Thus, the two drivers should be considered as mutually self-reinforcing. It was not by chance that the World Wide Web (WWW) was conceived at CERN as a communication tool among scientists (while ARPANET, with its association with US national security, was decommissioned in 1990).

While the drift toward the RI paradigm is particularly evident in physics, other sciences are not entirely lagging. Physics research requires costly, tangible assets, but when intangible assets are considered as well, it seems that life sciences are also adopting the RI paradigm in their specific way.

According to Hood and Rowen (2013), the Human Genome Project (HGP) has transformed biology through the adoption of a Big Science approach. The HGP preliminary research was sponsored in 1986 by the same DoE Health Office that studied the effects of radioactivity at Los Alamos (the place where the first atomic bomb was tested). The HGP was subsequently supported mainly by the NIH. It aimed at determining the sequence of the entire human genome over a period of fifteen years.[11] Launched in 1990, it was completed in 2003 after sequencing 3.2 billion base pairs—a unit of the deoxyribonucleic acid (DNA) double helix. It involved twenty core research institutes in six countries (and twelve other countries were associated with it in some form as well), more than 200 laboratories in the United States, and more than 1,000 scientists, including engineers, mathematicians, computer scientists, theoretical physicists, and biologists.

Large institutes, such as the Wellcome Trust Sanger, the Broad Institute of the Massachusetts Institute of Technology (MIT) and Harvard, the Genome Institute of Washington University, the Joint Genome Institute, and the Whole Genome Laboratory at the Baylor College of Medicine, are still providing human genome sequencing on a large scale and pushing technological developments in this field. In fact, new technologies to sequence and analyze DNA in general and to store and retrieve the data were at the core of the project (Chial, 2008). The HGP cost (while difficult to define precisely) is estimated more than $3 billion[12] or, in very crude terms, around $1 per base pair. The sequencing cost has decreased dramatically since then.

A second example of evolving organizations involves the large-scale, multicenter clinical trials (MCCTs) in medical research.[13] These facilities are increasingly designed and managed by several teams, and the experimental evidence is recorded in large digital data repositories. They are costly and risky. One of the many examples of this point is the unsuccessful trial of a drug for osteoporosis, odanacatib, completed in 2015 after twenty-two years, which involved 16,000 patients and 387 sites in forty countries. The total cost of such an MCTT is in the range of hundreds of millions of dollars (Rosenblatt, 2017). The data archives created by an MCTT are covered by the abovementioned definition of RI. The US National Library of Medicine maintains a database of over 260,000 such studies in 200 countries,[14] and it is itself an RI.

Finally, as a third example, Esparza and Yamada (2007, 702) discuss the proposal advanced in 2003 by some lead scientists in the field to create a Global HIV Vaccine Enterprise[15] in the form of a global network of institutes worldwide. What they say about the enterprise model for the project is of more general relevance for the RI paradigm:

The Enterprise proposes to complement individual investigator-driven research with a Big Science model to address this problem, taking full advantage of new scientific opportunities, creating new strategies for collaborative partnerships, and organizing a more coordinated global effort. To achieve its goal, the Enterprise proposes to (a) prioritize key scientific questions and rapidly direct resources to try to answer the most critical questions; (b) support product development, manufacturing, and testing of novel candidate vaccines; and (c) implement common standards and processes to maximize learning through the sharing of data and materials. The Enterprise model represents ... a new way for scientists to behave as a global community of problem-solvers.

Clearly, the term *Big Science* is used here in a very different way from Weinberg's "monumental enterprise" concept. No unique "ruler" would be the owner of this scientific enterprise and could claim glory for it. In the next decades, this pattern of large numbers of scientists creating knowledge by using a shared research platform or a well-integrated network will probably become the typical way in which the science frontier is designed and managed in several fields.

0.4 The Ingredients

The main ingredients of the RI paradigm may be identified as follows:

• *A process of identification of priorities within the scientific community*: While investment priorities in science linked to military and national ambitions were ultimately decided by governments, RI projects tend to be the expression of a consultation between funders and the relevant scientific community, often from an international perspective. In some cases, as in high-energy physics and other domains, the consultation process is formalized in a system of committees and review panels that agree on road maps and priority projects such as the European Strategy Forum on Research Infrastructures (ESFRI 2016a, 2018), discussed further later in this chapter.

• *International coalitions of funders*: Funding of RIs is often achieved by combining international resources such as government funds, financial contributions from charities, and other private investors. These funding arrangements make expenditure planning, cost accounting, and legal arrangements more complex than in traditional Big Science, which typically relies on transfers from a national budget. Thus, the RI paradigm of the twenty-first century is deeply cosmopolitan and eclectic. Even when a national government is the main financier and is interested in seeing national or local impacts, it usually is not the sole owner of the project.

• *Flexible accessibility to common resources by multiple users*: This involves a unique design of common facilities to support the work of large teams, or networks of smaller teams involved in a plurality of projects, at the same time or in sequence. The generic infrastructure is in fact addressed to multiple users, also in more traditional fields. For example, transport networks support various types of users, having different mobility needs, travel schedules, and willingness to pay (WTP) for the service. An electricity network supports a wide array of users, from business to residential, from low to high intensity of consumption, and with different needs in terms of quality of the service, such as voltage stability. This variability of demand should be managed precisely to avoid blackouts or other accidents. Congestion or underutilization should be avoided in any infrastructure, and reliability is imperative. In the RI framework, infrastructures can be single-sited, virtual, distributed, or mobile (such as satellites or oceanographic vessels), but in any case, their operations must allow a fine-tuned coordination of research projects, sometimes with very strict precision.

• *Shared governance*: Because the RI's core tangible and intangible assets, ranging from databases and software to scientific equipment, are shared by different communities of scientists, ownership is often diluted among partners. This feature requires special governance arrangements. A new managerial style is imperative. Large, traditional, hierarchically organized laboratories existed and will continue to exist. However, a large RI is also different from a consortium of scientific teams sharing just minimal services and logistics, as its funding involves multiple actors with decision-making arrangements tailored to hear such a plurality of voices (e.g., spokepersons are elected by scientists in experimental collaborations, and agreement on decisions requires formal and informal consultation).

• *Human capital incubator*: For students and young scientists, a large RI is a unique learning environment that goes beyond the scope of what a university can usually offer. The RI can be described as a collective intelligence environment.[16] Such an experiential learning process in large RIs may have long-term effects on the future careers of students and postdoctoral researchers in terms of employability and salaries in fields that are often outside academia or research. The universities themselves may take advantage of joining the RI's collective environment. Collaborative arrangements between several universities and RIs may change the investment strategy of the former, as the latter are more efficient than the traditional, smaller-scale laboratories in the academia, as well as being a more effective learning environment.

• *Technological hub*: New technologies needed for RIs often imply the close involvement of several firms, and in some cases of an international supply chain as well. Because many of the current and future large scientific projects have no military implications and are not always perceived as national strategic priorities beyond science, these supply chains can be transboundary, or constrained only by issues of fair returns to member-states in international ventures such as the European Space Agency (ESA). Hence, the existence of international supply chains of technological procurement of RIs is a feature that parallels (and occasionally anticipates) the evolution of the automotive or the aeronautics that have led the establishment of global value chains. Even relatively small high-tech firms can be involved in the procurement of several RIs or related research and development (R&D) activities in a number of countries at the same time. In other cases, RIs are embedded in regional or national strategies and need to pay more attention to territorial development, but also preserve scientific and technological excellence in procurement.

• *Big Data generators and managers*: Data gathering, analysis, and communication are closely associated with new RIs. The amount of information to be collected, stored, and considered in order to support scientific discoveries is critically determined by the progress of computer science and ICT in general. Furthermore, large externalities are created because solutions are often available free of charge to any potential user. The new RI is a digital information hub of unparalleled size in the history of science, and in some cases, it requires a completely new approach to issues such as data taking, filtering, storage, transmission, and usage. From this perspective, new RIs are part of the more general digital revolution.

• *Open science*: The scientific data and results of major RIs are highly visible and quickly disseminated through talks, online preprints, journals widely available to the scientific community, and increasingly open-access sources. The adoption of the open-science model, often required by funders, implies a communication process that is extremely fast compared with traditional academic procedures that lead to publications. An electronic preprint in some fields is available a few days after the experimental data are available (arXiv, managed by the Cornell Unversity Library, is a notable example). Progress in ICT allows data from the RI to be distributed online in real time or with a short delay to distant users. Large number of outsiders, including citizen-scientists, can then have open access to science and, to a certain extent, contribute to it.

• *Public involvement*: RI outreach, particularly through online social media sites, plays an important role in creating public support. Such outreach includes organized scientific tourism on site and online, as well as scientific divulgation through traditional means and social media. The RIs' webs and social media activities engage large numbers of outsider users. This situation creates virtual communities of citizens who are interested in what happens within science. Science fiction movies and documentary films are inspired by their activities. The Internet is flooded with news and comments arising from the most visible projects, such as the International Space Station (ISS) and the LHC. Thus, cultural goods are generated as a side effect of research.

While some of these features were not absent in the traditional Big Science model and are not common to all contemporary, large-scale scientific projects, they are strongly reinforcing within the new RI paradigm and are evolving fast. There is also some contagion in the Little Science world, which in some cases is adopting some of the features of the RI paradigm. For example, creating transboundary consortia and networks is a key requirement for applications in the EU Horizon 2020 program.

0.5 Examples

Some concrete examples may clarify the topic of this book. A few years ago, a science magazine for the general reader (*Popular Science*, 2011)[17] published a ranking of the ten most ambitious existing scientific projects, classified according to seven criteria: construction cost, operating budget, size of staff, physical size, scientific utility, utility to the average person, and what they call the "wow" factor, as evaluated by a jury of journalists. The winners of this competition were ranked as follows:

(1) The Earthscope (sponsored mainly by the US National Science Foundation), recording with 4,000 instruments several terabytes per year of geological data from North America

(2) The LHC (CERN, Geneva), running experiments on fundamental particle interactions in a 27-km circular underground tunnel

(3) The Spallation Neutron Source (Oak Ridge National Laboratory, sponsored by the US DoE), generating pulsed neutron beams to produce scattering data of atomic nuclei in various materials

(4) The ISS (jointly managed by the space agencies of the United States, European Union, Russia, Japan, and Canada), testing spacecraft technologies and human physiology in outer space

(5) The Advanced Light Source (Lawrence Berkeley National Laboratory, California, DoE), a synchrotron producing soft and hard X-rays and infrared photons, used for spectromicroscopy in material sciences, biology, chemistry, and other disciplines

(6) Juno (NASA), an orbiter to study the composition of the atmosphere and the magnetic fields of Jupiter

(7) The National Ignition Facility (Lawrence Livermore National laboratory, California, DoE), a high-energy laser facility used to study fusion reactions in deuterium and tritium

(8) The Very Large Array (National Radio Astronomy Observatory, New Mexico), a radiotelescope of twenty-seven antennas designed for the study of a range of topics, such as plasma ejections from supermassive black holes and gamma ray bursts

(9) Neptune Ocean Observatory (off the coast of British Columbia, a project of the University of Victoria), an underwater observatory network consisting of 530 miles of cables and 130 instruments for the study of chemical and physical conditions of the sea, local ecosystems, and effects of climate change

(10) The Relativistic Heavy Ion Collider (RHIC), sponsored by the Brookhaven National Laboratory and the DoE, studying the effects of collisions of gold ions and the creation of quark-gluon plasma

The total cumulated construction cost of these ten facilities was reported to be $19.626 billion (but my guess is that the real cost in current dollars is much higher). Another magazine for the general public, *Scientific American* (2015),[18] has published in visual form a map of the past, existing, or planned large-scale projects with total costs in excess of $1 billion: these include inter alia the HGP; the 100,000 Genomes Project (the number refers to the patients of the UK National Health Services whose full genome would be mapped); the Human Brain Project; the International Thermonuclear Experimental Reactor (ITER) project for nuclear fusion in southern France; the James Webb Space Telescope; the Mars Science Laboratory; the New Horizons Pluto Mission; the proposed Circular Electron Positron Collider in China (a 50–70-km-long underground facility, as opposed to the 27 km of the LHC tunnel and the 100 km of the proposed CERN Future Circular Collider).

These two lists are far from exhaustive, and these rankings are obviously subjective. I shall discuss elsewhere in this book the way that investment and total costs should be more precisely estimated and reported. However, these examples give an idea of the scope of hundreds of contemporary, large-scale facilities, most of them based on the RI paradigm. A more institutional perspective is offered by the RI road maps set by national and international bodies. These road maps are documents where a consultation process leads to suggested priorities for funding. This bottom-up process is the first feature of a contemporary RI in my earlier discussion.

ESFRI (2016b) provides a list of twenty-one new RI projects considered as a priority in Europe and twenty-nine already-implemented RIs (landmark projects).[19] Research fields include energy, environment, health and food, physical sciences and engineering, and society and culture. The construction costs, considering both existing and planned projects, range from the gargantuan EUR

1.843 billion of the European Spallation Source (production of pulsed neutrons for fundamental and applied physics) to the humble EUR 1.5 million of the European Clinical Research Network (support services for the preparation and management of MCCTs). The expected yearly operating budget of these projects ranges from EUR 234 million for the Facility for Antiproton and Ion Research to less than EUR 1 million for the Digital Research Infrastructure for the Arts and Humanities (a network of researchers on digital archives). The wide diversity of costs (further discussed in chapter 2 of this book) not only reflects the specifics of experimental, observation, or elaboration methods in different domains, but also the diverse architecture of the projects, with some RI based on hard, single-sited dedicated facilities, while others are soft networks of distributed resources.

Individual countries produce their own road maps.[20] Two examples include France and Australia. The French National Strategy on Research Infrastructures (2016)[21] lists 95 RIs, classified into four types: international organizations in which France participates, very large RIs (mostly included in the ESFRI list), the RIs, and future projects. This list of RIs covers a wide array of fields: biology, health, Earth system and environmental sciences, social sciences and humanities, energy, material sciences, astronomy and astrophysics, nuclear and high-energy physics, digital sciences, technologies and mathematics, and scientific and technical information.

The Australian 2016 National Research Infrastructure road map[22] has identified "nine focus areas that require ongoing support to ensure that Australia will be able to maintain its position as an emerging or established global leader: Digital Data and eResearch Platforms; Platforms for Humanities, Arts and Social Sciences; Characterisation; Advanced Fabrication and Manufacturing; Advanced Physics and Astronomy; Earth and Environmental Systems; Biosecurity; Complex Biology; Therapeutic Development." These examples point out the difficult political choices that governments have to face when they are required to fund such diverse projects in terms of domains, objectives, and scale.

0.6   Social Cost-Benefit Analysis

The promoters of the large-scale RI ultimately depend on government funding and, to a lesser extent, grants by charities or other donors. As national defense or strategic interests are no longer the overarching rationale to justify major research projects, the ability of the scientific community to convince governments and other public or private funders becomes crucial. There may be, however, a reasonable scientific case for a lot of competing projects. Public and private budgets must be allocated according to some kind of priority. Hence, taxpayers (or at least governments and lawmakers) want to know why they need to foot the bill of any project proposed by a specific scientific community. Is sending human explorers to Mars actually a worthy spending priority? Or probing the existence of supersymmetrical particles? Or tracking the genome of rare marine species? Because scientists' curiosity is virtually infinite, and because they cannot claim urgent national security reasons for their projects, which project should a government target for funding?

Governments are not willing to foot any bill. In the early 1990s, the Superconducting Super Collider (SSC), a large particle accelerator (87 km in circumference), was to be built in Waxahachie, Texas, with an initial budget of $4.4 billion. After having already spent around $2 billion and dug 23.5 km of underground tunnel and seventeen pits, the expected cost for the project completion rapidly surged to $11 billion, and the project was eventually abandoned by the US Congress in October 1993.[23]

The judge and prolific author Richard Posner, in the context of a discussion of the extremely unlikely but possibly catastrophic risks of some scientific projects (Posner, 2004, 251), wrote:

The experiments conducted in such facilities (the LHC or the RHIC[24]) cannot be quantified or even it seems estimated whether by the method of contingent valuation or otherwise. Those benefits are the incremental value of human welfare (in its broadest sense) of the experiments, and are unknown and cannot be presumed to be great. Because research accelerators are costly both to build and to operate, the net benefits of such accelerators may actually be negative quite apart from safety costs.

He then concludes that governments should not finance particle accelerators. Hughes (2002) argues that large-scale science is pathology, and that the demise by the US Congress of the SSC may have marked the end of megascience. Steven Weinberg reminds us of the political climate that killed the SSC (Weinberg, 2012, 5):

During the debate over the SSC, I was on the Larry King radio show with a congressman who opposed it. He said that he wasn't against spending on science, but that we had to set priorities. I explained that the SSC was going to help us learn the laws of nature, and I asked if that didn't deserve a high priority. I remember every word of his answer. It was "No."

Setting priorities in public expenditures unavoidably has a political dimension, and science policy is not an exception. The day before the SSC was aborted by a large majority, the same US Congress approved (by a thin majority of just one vote) the funding of the ISS, despite the expectations of "much greater cost overruns (up to USD 30 billion) than the SSC … because it was not centered in a particular state, because contracts and jobs were more evenly distributed across the US and because the NASA had significant foreign financial contributions (and better spin doctors)" (Hughes, 2002, 143).

As for any other public policy, being able to say something about the socioeconomic impact of investment decisions is important, even if it will not necessarily be the final argument in the political decision process. In this context, despite the specific features of science, something can be learned from more than one century of CBA tradition regarding government decision-making on major infrastructures.

RI investment mainly produces a good: namely, new scientific knowledge. Knowledge, in many fields, often has the peculiar characteristics of a public good. Its use is nonrival, and it is costly to exclude third parties. In fact, government funding of nondefense-related science is increasingly tied to a strict obligation to make knowledge available in the public domain. Some benefits of RI investments are immediately apparent; others remain unknown for a long time. Moreover, some benefits are more easily measurable because they involve market transactions such as technological procurement. Other benefits, such as the enjoyment of some cultural services, appear mostly outside the markets; however, they still have welfare effects that can be measured. As this book will show, the quantitative evaluation of some of the social benefits and costs of RIs is possible if an empiricist evaluation strategy is adopted. Citizens, or governments on their behalf, should be able to know the answers to simple questions, such as: What is the total social cost of this research project over its life cycle? To what extent can social benefits be forecast and valued? Who will enjoy such benefits? Are there measurable risks?

Of course, putting together what we can know about the social benefits and costs of a specific RI is a complement to, not a substitute for, its scientific case, which essentially implies a peer review process. Discoveries ex post will reveal if the project was successful. However, ex ante taxpayers, donors, or investors of private funds, are interested in the expected or achieved socioeconomic impacts of the projects that they are supporting, beyond the scientific impact. These impacts should be duly forecasted or reported as far as possible. The way that such evidence is gathered and data elaborated must be professional, impartial, and based in itself on a consistent evaluation framework.

Social CBA, duly adapted to a new topic, can offer such a framework, and its theory can guide the evaluation. Its analytical framework, after its distant origins in the nineteenth century at the École des Ponts in Paris, has been developed over the last several decades by international institutions such as the Organisation for Economic Co-operation and Development (OECD), the United Nations Industrial Development Organization (UNIDO), the World Bank, and the European Commission, as well as by several national governments, particularly those of the United Kingdom, United States, Canada, France, the Netherlands, and the Scandinavian countries, and by many public agencies and development banks. CBA methodology has been systematically applied to the appraisal of thousands of projects in various fields, initially in transportation, then in energy, telecommunications, health, education, and notably in environmental issues. In the past, several top economists, including Nobel laureates in economic sciences (particularly Kenneth Arrow, Peter Diamond, James Mirrlees, Paul Samuelson, Amartya Sen, and Joseph Stiglitz), have contributed to structuring the CBA theory and its application. This is still an active research field for hundreds of economists,[25] as without a solid, well-established theory, no effective project and policy evaluation is possible, in any field.

I will try to convey in this book the lessons I have been learning during my professional career about the way CBA of large-scale projects should be structured. Some of the reports available in the "gray" literature on the socioeconomic impact of RIs often tend to rely upon a mixture of qualitative assessments or a number of different indicators (sometimes a patchwork quilt), as well as on rhetorical arguments. This is often because of the untested assumption that an RI is too different from conventional infrastructure projects, and hence it cannot be analyzed with the well-established methods developed in other fields. Nevertheless, something can be tried.

Large-scale RIs can be subject to the scrutiny of social CBA, as far as possible, as a complement to the scientific case for it, before a decision is made, or several years later to learn the midterm or ex post impact of the investment decision. Some benefits can be measured and should be, in the way they are measured in other fields, with some adaptations. While the CBA will never be the core argument to justify the investment of taxpayer money in science (or in education and health), citizens (and governments, on their behalf) should be able to know the answer to simple questions about the net benefits and risks of any large-scale RI project. The answers must be honest, avoid double-counting and trivial mistakes in valuation, and must, as far as possible, carefully consider the uncertainty implicit in any forecast. The net benefits can be positive or negative. This can be known only after a serious analysis, not as a dogmatic prior view. The proof of the pudding is in the eating.

0.7 Structure of the Book

The core concept of a CBA of any infrastructure is to undertake the consistent, intertemporal accounting of social welfare effects using the available information. Chapter 1 suggests a simple framework for such accounting in the RI context. The model is straightforward: the social value of an RI is the algebraic sum of the benefits, net of costs, for certain social groups of users and for the general public of nonusers, where each value is discounted to a common base year. All these values are expressed by a common numeraire and in terms of the expectations computed by the probabilities attached to some critical variables. What makes this model for the RIs different from the analog for a high-speed railway project, for example, is that the producers, users, and nonusers involved are greater in number and more diverse than railway passengers and a railway firm's shareholders. Moreover, externalities and uncertainties are much more pervasive in scientific projects than in conventional infrastructures. However, they can be identified and, to a certain extent, evaluated.

Chapter 2 explains how to make these calculations on the cost side within a long-term intertemporal frame. From one perspective, an RI project's costs should be forecast ex ante and reported ex post in the same way as in the economic analysis of any infrastructure. Thus, the costs should include not just the capital expenditure, but also the discounted sum of operational and maintenance costs over the years from start-up until the decommissioning of the project or the end of its useful life. This approach has the unpleasant effect for proposers and funders that the cumulated costs are much higher than the initial grants asked from governments or other sponsors. However, the approach is a more complete and realistic way to report data.

The main benefits of an RI project should be analyzed in terms of the social agents involved. This analysis is undertaken systematically in the subsequent chapters of this book as follows: The benefits to scientists arising from publication (namely, the typical output of any research activity) are discussed in chapter 3. The interplay between universities and RI projects is also discussed. Early-career researchers (ECRs) in many RIs are students and postdoctoral researchers. How they benefit from their experience in RIs is discussed in chapter 4.

Large-scale research facilities require the involvement of firms under procurement contracts. In this context, there are often knowledge spillover effects from an RI to business. Such an externality can be gauged, and the method for doing this is described in chapter 5. Similar externalities positively affect the users of information technology (IT) because an RI produces or consumes Big Data and needs to invent solutions, which are then made available to outsiders. Examples of such effects, and ways to account for them, are presented in chapter 6. Although this often is not their main objective, large-scale RIs also directly or indirectly boost innovation, which is eventually embodied in goods and services. Chapter 7 discusses how, to a certain extent, this circumstance can be accounted for in a CBA, and provides practical examples such as hadron therapy research for cancer and the benefits of research on observations of the Earth by satellites.

Chapters 8 and 9 are devoted to the impact of RIs on the welfare of citizens who are not directly involved in any of the previously mentioned activities. Such citizens are not scientists, ECRs, suppliers of technology, professional users of IT products, or direct beneficiaries of innovations supported by discoveries; instead, they comprise an often large number of laypeople who use cultural services produced by RIs through outreach or produced by third parties. Chapter 8 considers these laypeople and suggests the approaches needed to measure the welfare effects of cultural services originated by outreach activities. Chapter 9 then explores the reasonable conjecture that the general public has social preferences for knowledge per se—in other words, a preference for the intrinsic value of scientific discovery. This preference may be shared to some extent by taxpayers in the form of a generic willingness to pay for scientific knowledge creation. This conjecture leads to testable empirical approaches based on contingent valuation experiments.

Finally, chapter 10 shows how the foregoing accounting of social benefits and costs can be expressed in the form of a stochastic net present value (NPV) or other summary indicators that include, as far as possible, intertemporal effects and probabilities of states of the world for the critical variables that affect the outcome. A number of caveats and suggestions for further research are also given.

Each chapter adopts a twofold approach. From one perspective, it suggests why certain features, such as knowledge spillovers, human capital externalities, and cultural effects, are typical of the new RI paradigm. From another perspective, it shows how such features can be accounted for in a CBA framework. Moreover, each chapter provides practical examples based on empirical evidence that has been collected in fieldwork in recent years by myself and my colleagues.

The case studies are based on a detailed CBA of the LHC at CERN, as well as on several examples of the empirical estimation of benefits of relatively smaller RIs, such as the National Centre of Oncological Hadrontherapy (CNAO), a center for particle therapy of cancer in Pavia (Italy); the ALBA synchrotron light source for material science and biology near Barcelona; the European Molecular Biology Laboratory–European Bioinformatics Institute (EMBL-EBI), in Hinxton, United Kingdom, an open-access database provider covering all fields of molecular biology; and the Copernicus Sentinels Satellites for Earth observation. Several other examples of RIs are provided in the discussion of social benefits and costs. Not all of the aforementioned examples share the new RI paradigm traits to the same extent; however, they are all broadly similar in terms of the empirical evaluation opportunities and challenges they pose.

Further, I have tried to consider the diverse backgrounds of potential readers of this book. In the last few years, I have been often invited to plenary talks at conferences of physicists and other scientists, and I have tried to offer self-contained presentations of my arguments and evidence. The book is firmly framed in applied welfare economics, but I hope it also will be read by scientists and decision-makers, and I have adopted a presentation style for readers who do not have familiarity with applied welfare economics and the formal theory of CBA. Some readers without previous exposure to CBA may prefer to read chapter 10 immediately after chapter 1 in order to grasp how the CBA model assumptions eventually lead to the NPV as an empirical concept. The "Further Reading" sections and some appendixes in these chapters are designed for those readers who want to go more in depth into the topic.

An important caveat: This is not a book on science policy, nor does it cover the history, politics, management, sociology of science, and economics of knowledge in general. There are several wonderful works already available on these topics, and several of them are selectively cited in this book. My contribution is more circumscribed. It is about what an applied economist can say (or often cannot say) on the quantitative estimation of the socioeconomic impact of investment in RIs as a specific way to organize science.

1 The Evaluation Framework: A Cost-Benefit Analysis Model

1.1 Overview

Knowledge creation has a special place in contemporary economic theory. The traditional neoclassical growth model, created by the MIT professor and Nobel laureate Robert Solow (1956), was admirably simple and elegant. Long-run output change was explained by a function of two inputs (fixed capital and labor force), plus an unexplained residual component (total factor productivity), which could be interpreted as being related to the exogenous technological progress. The empirical measurement of growth, in this setting, revealed a surprising puzzle: the residual was too big, as it accounted for more than 80 percent of growth. In the words of Acs, Estrin, Mickiewicz, and Szerb (2014, 2):

The Higgs boson, an elementary building block of modern physics, was first conceptualized in 1964 and its existence only confirmed in 2013. For economists, the inability to explain much of the cross-country variation in output by inputs, termed the Solow residual, has—like the Higgs boson—consumed the efforts of a generation of scientists.... To resolve this puzzle is important, because what lies behind the residual is presumed to be both a building block of the modern economy and essential to economic growth.

New growth theories have identified such ingredients: human capital (knowledge embodied in education and skills); high-quality institutions (good government, entrepreneurial capacity, and cultural attitudes); environmental sustainability; and ultimately, knowledge creation, which in turn would shift the technological level of the economy.[1]

The perspective of endogenous growth theory, a framework well established since the 1990s, has important policy implications. Governments can contribute to growth not just by policies supporting investments (e.g., by the provision of public infrastructure) and employment (e.g., by fiscal policy and labor market regulation), but also knowledge creation. Research by private firms produces an externality, an uncompensated benefit to a third party, as knowledge, once it is created, can easily spill over elsewhere. Not everything can be patented or kept secret, and the returns of research and development (R&D) expenditures by a firm may be captured by other firms. Hence, private investors are reluctant to incur such risk and would invest in R&D less than what would be socially optimal. Governments need to counteract this effect by funding research in areas where the private investors are not willing to invest.

To put the discussion about the socioeconomic impact of research infrastructures (RIs) in this broader context, one can consider how much governments and firms have invested in R&D in recent years. According to Organisation of Economic Co-operation and Development (OECD, 2017) data, in 2015 the United States spent more than $500 billion for R&D, China spent $409 billion, and the combined EU28 area spent $386 billion.[2] Most of this expenditure (about 70 percent) was supported by firms (including state-owned enterprises), while charities contributed 2.4 percent, and the rest was funded by governments.

Basic research, defined as knowledge creation without an identified application, accounted for around 17 percent of the total gross domestic expenditure on R&D (GERD) in the OECD countries in 2015; the share for applied research was 21 percent, whereas the remaining expenditure was invested in the creation of new products and processes: the development component of R&D.

Governments play a critical role in supporting research. In the OECD area, around $315 billion was funded from governments. As the total population in the OECD member-states was 1.281 billion in 2015, each citizen (all ages included) contributed, on average, around $246 per year to R&D through taxes. This contribution has been declining in some countries since the beginning of the Great Recession (2008).

Figure 1.1a shows the average GERD per capita, including both public and private contributions, for a selected sample of developed economies in the last twenty years. While there has been an overall increase over time, figure 1.1b highlights how the government contribution to R&D has been declining in recent years compared to other sources of funding. For example, in Europe, it was about 40 percent in 1995, while it was 32 percent in 2015. In the meantime, the number (and the share in the workforce) of researchers has increased (OECD, 2017). Government support per each scientist or researcher has been stagnant or decreasing since 2010 in the United States (but not elsewhere, such as in Germany; see figure 1.1c).

According to the current director of the Institute of Advanced Study in Princeton (Dijkgraaf, 2017, 33), the federal R&D budget in the United States

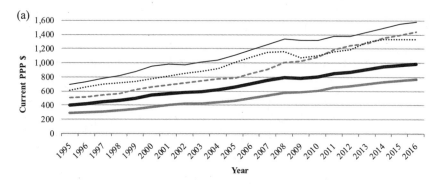

Figure 1.1a
GERD per capita.

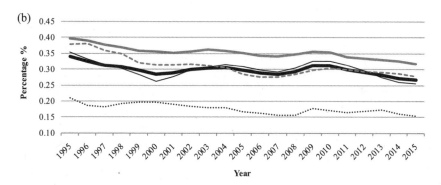

Figure 1.1b
GERD funded by government.

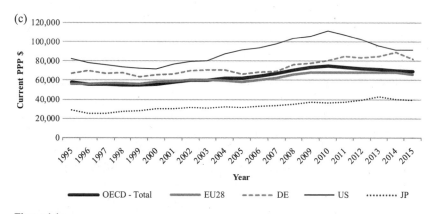

Figure 1.1c
GERD funded by government per researcher. *Source*: Own elaboration on OECD data, http://www
.oecd.org/sti/msti.htm.

Legend: EU28: European Union (28 countries), US: United States, DE: Germany, JP: Japan.

was 2.4 percent of gross domestic product (GDP) in 1964 (in the midst of the space race and the Cold War climate), while it was 0.8 percent of GDP in recent years, still half of it for defense. For example, in the last decade, the budget of the US National Institutes of Health has been cut in real terms. Not surprisingly, competition for public funds is more and more stringent, in the United States and elsewhere.

Even before the recent recession and budgetary problems, government funding of science, and particularly of large-scale scientific projects, has been criticized. In an article of *Nature*, Macilwain (2010, 682) writes: "Beneath the rhetoric ..., there is considerable unease that the economic benefits of science spending are being oversold." He also reports the opinions of several experts about the poor quality of the assessments of individual scientific projects' social benefits.

We may know that governments should support science, but which projects should take priority, given their costs and unknown benefits? Typically, the decision to fund highly expensive RIs is advocated by a coalition of scientists, often supported by peer reviews or other expert opinions to convince policy-makers about the importance of a new project. This process can be described as lobbying for science, a form of advocacy. Riordan, Hobbeson, and Colb (2015, X) describe, in the case of the Superconducting Super Collider (SSC), how particle physicists had to build "uncomfortable compromises and ... unfamiliar alliances with the US Congress, the Department of Energy (DOE), Texas politicians and business leaders, and firms from the US military-industrial complex." There is nothing particularly surprising or bad about this. Lobbying is historically a feature of any major public investment decision-making process, such as major transport, energy, and water projects. This process has been described as the creation of an infrastructure elite (Cassis, De Luca, and Florio, 2016), leading to a variable composition and dynamics of coalition-building over time and across countries to make large-scale investment decisions.

It is in this context that the economic analysis of infrastructure projects has been proposed, since its remote origins at the French École National des Ponts et Chaussées (Dupuit, 1844), as a way to counterbalance, as much as possible, flawed arguments in investment decisions. Social cost-benefit analysis (henceforth simply CBA) can then be seen as part of an intellectual attempt since the nineteenth century to introduce a dose of economic rationality in public policies, leading to the broad framework of the economics of welfare by the Cambridge University economist, A. C. Pigou (1912), and by subsequent developments. *Rationality* here means looking for efficient solutions to increase the well-being of society at large against vested interests.

Jules Dupuit (1844) asked the following question: Suppose that somebody proposes to build a bridge and no tolls are levied? This project would not be supported by private investors because there are costs and no revenues; hence, the bridge would produce financial losses. Does this necessarily mean that the project is worthless? In fact, the bridge produces a valuable service to society, which can be measured in terms of the willingness to pay by the users. This willingness to pay is proportional to the value of time saved compared to the no-bridge scenario.

The marginal social value (MSV) of time is the value of saving a small amount of time in the perspective of the society. For example, instead of traveling, a citizen may do something else that has a value, such as earning additional income by working or enjoying more time for leisure, which also has a value. Moreover, saving time in transport may decrease the production costs of goods and services, which is also a benefit for producers and consumers.

In general, CBA wants to assess whether benefits from a project (to whomever they accrue) exceed costs (from whomever they are incurred), thereby showing whether the project represents a net benefit to the whole society. The key strength of this approach is that it produces information in a quantitative form about the project's net contribution to society, summarized as simple numerical indicators such as the net present value (NPV) of a project (presented later in this chapter).

Only recently has CBA started being applied to investment projects in R&D. The European Commission (2014) included, for the first time, a chapter on the evaluation of research infrastructures in the fifth edition of its reference CBA guide for funding major projects under the European Union Structural and Investment Funds (Florio, Morretta, and Willak, 2018). More recently, the European Commission (2017a) mentioned that the preparatory phase of the new European Strategy Forum on Research Infrastructures (ESFRI) projects (the priority RI road map in the European Union) should include a CBA, and it explicitly refers to the abovementioned European Commission (2014). A report by ESFRI (2016b, 19) mentions that "Cost Benefit Analysis methods, well established in sectors like transportation or energy, have recently been applied to research infrastructures."

The methodological issues related to the design of a CBA framework for large-scale RIs require the development of a conceptual model based on the first principles of applied welfare economics, with several adaptations to meet the needs of project evaluation in this area. We need a model to answer questions such as: What is the value to society of building and operating a radio-telescope, a digital archive of marine life, or a gravitational wave detector, if the knowledge created is given for free to any user? In financial terms, these projects incur losses. What are the socioeconomic benefits? The situation is similar to the Dupuit bridge without a toll, where nobody directly pays for the time saved and still the service provided has a social value. An RI is also a *bridge*, a bridge between a state of the world where we do not know something and one where we can discover it. As we shall see, however, the answer for an RI is more complex than for a project serving the needs of one type of user, such as commuters between two cities, because of the peculiar RI feature of generating impacts on multiple stakeholders.

The chapter's structure is as follows: Section 1.2 introduces the conceptual framework of CBA for RIs; section 1.3 presents the net present value (NPV) concept; section 1.4 presents a taxonomy of the main types of social benefits; section 1.5 presents a simple CBA model; section 1.6 explains how the NPV test works; and section 1.7 suggests further reading on CBA and other evaluation approaches on the socioeconomic impact of science and R&D.

1.2 Benefits and Costs as Social Values

While CBA was first applied in transport, and later in water infrastructure (e.g., US Green Book, 1950), it was also extensively used in energy, telecommunications, other services, and policy evaluation, including the economic impact of regulations. In the 1980s, some authors argued that investment projects in sectors such as education or health could not be evaluated by using CBA techniques.[3] However, since then, progress in empirical methods and theory has happened so fast that CBA in education, health, and particularly in the environment and regulatory impacts in general is widespread nowadays.[4] However, until recent times, investment in science has remained outside the scope of CBA methods. Why was there such a lag in spite of the abovementioned interest of governments in evaluating the impact of science projects and policies?

In the introduction, I mentioned some features of RIs. While some are peculiar to them, several aspects are similar to those of other categories of infrastructures. For example, the first critical ingredient of any infrastructure is the capital intensity at the initial stage of the project cycle (Gramlich, 1994). This is particularly true in large-scale RIs, which use some of the most expensive machines ever built (except those for military purposes: for example, a contemporary US air carrier costs few billions more than the Large Hadron Collider, or LHC).[5]

Virtual infrastructures, such as social surveys and clinical trial data archives, are more labor than (fixed) capital intensive. For example, ESFRI (2016b) considers some electronic surveys, such as the European Social Survey,[6] as RIs. In these cases, capital takes the form of intangible fixed assets accumulated over time. This does not change the framework: assets, either tangible or intangible, must be cumulated in several initial years before services can be delivered in later years.

The second ingredient of any infrastructure is the long time horizon involved on both the cost side and the benefit side. For example, the CERN accelerators built in the late 1950s (Proton Synchrotron) and in the 1970s (Super Proton Synchrotron, or SPS) are still used today as injectors of proton beams in the LHC. The conceptual design report of the Future Circular Collider (FCC) was published in 2019,[7] for a project whose construction, if approved, would start no earlier than 2040. The Human Genome Project (HGP) was funded in 1988 and (almost) completed fifteen years later in 2003.[8] The time horizon of the RI is not necessarily longer, however, and it is often shorter than some traditional infrastructures, such as roads, railways, or dams.[9] This point is further discussed in chapter 10.

Third, traditional economic infrastructures are often associated with externalities and spillover effects: part of the economic benefits of infrastructure is usually not appropriated by its owner. I shall show that this is a core feature of RIs as well. As previously mentioned, an externality is a benefit that accrues to third parties without compensation.[10] While this frequently happens with knowledge-based goods, it is not uncommon for the benefits of health and education projects or of preserving a beautiful landscape actually to spill over beyond, respectively, patients, students, or landowners.

Fourth, there is no explicit market for most of or all the services delivered by the RI, and there is very limited competition (Irvin and Martin, 1984). Sometimes, in principle, the same research question could be answered by more than one competing RI[11] (see Baggott, 2012). An example is the parallel effort of the publicly funded HGP and Celera Genomics, a private corporation (Venter et al., 2001), but this can hardly be described as market competition. This situation of limited competition also frequently happens with traditional infrastructures, which operate in a context of natural monopoly because it is costly to duplicate the core facilities, such as electrical grids or rail tracks.

Hence, high capital costs in the initial period, long time horizons, externalities, and lack of proper competitive markets are all RI features that are common to other infrastructures and are typical CBA issues. Thus, what is needed is a model that can deal with both the generic and the specific features of the RI and can handle both facilities for pure and applied research, including some large-scale university laboratories.

As mentioned in the introduction, most RIs are single-sited, including particle colliders, telescopes, electronic microscopy facilities, research nuclear reactors, and some supercomputers. However, there are also examples of geographically distributed facilities, such as grid computing systems or atmospheric measurement stations, which are located in different areas and record data that then are centrally studied.[12] In such cases, there may be network externalities to consider in the project's assessment. According to the OECD (2014a), a *distributed infrastructure* is a network or multinational association of geographically separated organizational entities that jointly operate a set of independent research facilities, such as the European Very Large Baseline Interferometry Network, which is a collaboration of the major radio astronomical institutes of Europe, Asia, and Africa. Some RIs are mobile, as oceanographic vessels and satellites. Clearly, these other features should be considered and, in some cases, may deserve special treatment.

The core concept of CBA is that the socioeconomic impact of a project is represented by the difference over time in the benefits to different agents and the costs of producing such benefits. All of these are expressed in an appropriate accounting unit.

Benefits and costs are in essence (*quantities*) * (*prices*) = *values*, and the net socioeconomic benefit of any project is represented by the net present value (NPV), expressed as the difference between benefits and costs represented in terms of present values (PVs), to avoid just adding values that occur now to values that are related to possibly distant events. Values that occur at various times can be expressed in PV terms by a process of discounting, as explained in the appendix of this chapter. Accounting prices in this context are not necessarily market prices because for many goods, such prices do not exist or do not provide information about the social welfare effects. Most environmental, cultural, and, as we shall see, scientific goods and services are such that markets do not give them an empirically observable price; yet they have social value. The basic principle of CBA is that when market prices do not exist or are inappropriate to value inputs and outputs, or benefits and costs in general, such prices must be replaced by empirical estimations of MSVs: namely, *shadow prices.*

In practice, and without loss of generality, shadow prices for project evaluation can be estimated by an agent's marginal willingness to pay (WTP) for a good, by the marginal production costs, or by a combination. The marginal WTP is how much you would be happy to pay to buy an additional small unit of a good. The marginal production cost is the value of the resources needed to produce such an additional unit.

The picture is completed with three further well-established ideas in applied welfare economics. A project can produce externalities as side effects. These may be negative (e.g., radioactivity) or positive (e.g., improvements to the landscape). Typically, there are no market prices for such effects; thus, CBA is particularly helpful in project evaluation when such effects are important. As we shall see, in the case of RIs, most of the measurable benefits are externalities.

Finally, as previously mentioned, because there is uncertainty in forecasting future benefits and costs, all the variables have to be considered as stochastic. A project's NPV should be expressed as an expected value from a probability distribution, conditional to the distribution of some other variables. The notations $\mathbb{E}(NPV)$ and $\mathbb{E}(PV)$ are used when the mean value of a distribution is explicitly considered. When needed (particularly in chapter 2), to distinguish between socioeconomic concepts and financial concepts, the following notation will be used: economic net present value (ENPV) and financial net present value (FNPV). When it is not strictly needed, as in the next discussion, the notation NPV means $\mathbb{E}(ENPV)$, the expected net economic PV.[13]

1.3   The NPV, and How Not to (Mis)understand Its Meaning

Economists try to have a commonly agreed meaning for each term that they use, as scientists do in their various fields. However, jargon differs across disciplines. This book has been written with both economists and scientists or, in general, readers who is not familiar with CBA, in mind. Thus, I need to proceed step by step with the following definitions and elaborate on them.

The bottom line of a CBA of any project is the empirical estimation of NPV. This is often reported as a single number, but in fact it is an *empirical function* of several economywide parameters (such as GDP growth, per capita income, and unemployment) and of project-specific variables (such as procurement expenditure, labor cost, and number of doctoral students). In a fully developed project evaluation, there may be hundreds of variables and parameters that determine the NPV, and the important point is not a single value, but the underlying model.

Net in the NPV function means *Benefits minus Costs*. Hence, in whatever ways that benefits or costs are measured and expressed in a common unit of account, such a difference can be positive, zero, or negative. A negative NPV of, say, an oceanographic vessel project does not per se warrant a negative funding decision. NPV < 0 is a shortcut empirical assessment of what we know and what we do not know, in terms of the expected socioeconomic impact of a project: no more, no less. I am not saying that the HMS *Beagle* expedition (1831–1836), with the young Charles Darwin on board, should not have been funded (hence missing the Galapagos and the theory of evolution of species) if the NPV were expected to be negative according to some Admiralty economist. Neither am I saying that the Mars mission budget of the National Aeronautics and Space Administration (NASA) should not be approved by Congress if somebody suggested that the estimated uncertain socioeconomic benefits are less than the expected costs. What I am saying is that the empirical estimation of the NPV is an informationally constrained measure of our knowledge in terms of the real resources needed and expected to be created by a project.

Second, *Present* in the NPV concept means that we need to deal with the fact that we cannot compare the values of today with values occurring twenty years ago or fifty years in the future. Whatever the values of the costs and benefits of a scientific project are, the knowledge created will last several years and will possibly have an impact over decades; hence, everything should be expressed at time $t = 0$, a conventional starting point. The appendix explains how to do this, along with a discussion of specific implications for the RI.

Third, to any economist, a *Value* is not a philosophical or ethical construct. It does not have the same meaning as in the book by Henri Poincaré (1905), *La Valeur de la Science*, or in claims that science is value free, or value in mathematics, or many other meanings of this ubiquitous word. In economics, the term *value* is the product of price (either a market price or a shadow price) *times* quantity. By convention, a *cost* is recorded as a *negative* quantity times its price, and this is also mentioned as an *input* value. A good produced is a *positive* quantity, or *output*. Not surprisingly perhaps, this is where some of the literature on the qualitative evaluation of RIs uses concepts in a very different way from economics.

For example, to any economist (and accountant), the labor (scientific, technical, or unskilled) needed to build and operate a telescope generates a cost (the number of full-time equivalent employees *times* wages) because labor is an economic resource needed as an input to bring about what the facility in question wants to produce (i.e., observation of the sky) as an output. In turn, the price of scientific labor is what it costs to society to employ it. It is usually expressed in monetary wages paid to the scientists, measured in a given currency (in a base year). In some cases, an actual wage may be corrected by a shadow wage when the MSV (discussed further later in this chapter) of scientific work is less (or more) than the observed wage (e.g., because of an imperfect market for scientists), but this is still a cost.

MSV, in turn, is the change in social welfare generated by the production or consumption of a good (technically, it is the first derivative around the optimum of a social welfare function—that is, of the rule we use to aggregate the well-being of agents in a society). Only in a loose discourse may one say that creating jobs for astronomers by public funding is a valuable thing to do (supposedly better than creating jobs for, say, opera singers?). This proposition would have no more economic meaning than saying that it is desirable to create new opportunities to consume energy. What may have a socioeconomic value is the *product* of the effort of astronomers (and of opera singers), not their employment per se.

Once inputs and outputs associated with the RI (including externalities such as uncompensated benefits and costs to third parties) have been identified and their baseline quantitative estimates have been valued in monetary terms, the resulting NPV should be read as saying that the *measurable* net benefit of an oceanographic vessel is, for example, $100 million, or $100 million in excess of costs. This has nothing to do, of course, with the profitability of this oceanographic venture, which usually, in fact, will create losses in financial accounting terms. Profitable scientific or technological projects are usually supported by private investors and there is no government money directly involved; hence, there usually is no need to estimate social benefits and costs.

Moreover, *"dollars" in the NPV are not conventional dollars*. The use of the same unit of account of a day-by-day monetary transaction is a convention in applied welfare economics, but, for example, there are no real dollars attached to the public good value concept of scientific discovery, as presented in the introduction. In a sense, CBA dollars are virtual money. One may use any other numeraire, such as quality-adjusted life-years (QALYs) of a statistical human being;[14] for instance, saying that the NPV of a genomic platform is worth 1 million QALYs means that its benefits (including health benefits) are greater than its costs, if all these are measured in terms of life years.

Understanding this point is crucial. No meaningful, comparable, comprehensive socioeconomic impact is possible if all the measurable benefits and costs cannot be converted into values through a common accounting metric, whatever it is. Scientists know that they could not study the thermodynamic properties and processes of a system if they were unable to measure energy in different forms by a common unit. An RI project is a system that exchanges certain quantities of inputs and outputs with the rest of society, a larger system; and the NPV is a balance of such exchanges, given a unit of account. In a precise sense, prices used in CBA are proxies of MSVs, as they aim at reflecting the unit change of the social welfare associated with the use or production of a good. The empirical estimation of such MSVs or shadow prices may be imprecise, but if one misunderstands this concept, one cannot really interpret what the NPV of an RI means.

Slightly restating what has been said here, one can write:

$$\mathbb{E}(NPV_{RI}) = \sum_{q=1}^{Q} \mathbb{E}(MSV_q)\,\mathbb{E}(q),$$

which should be read as follows: The expected ($\mathbb{E}$) NPV of a research infrastructure is the sum—*at time* = 0 *equivalent values*—of all the products of quantities (of inputs, negative quantities; and outputs, positive quantities $q = 1, 2 \ldots Q$), each valued by its estimated MSV and conditional on the probability distributions attached to each MSV_q and q element. Consequently, $NPV_{RI} = Change\ of\ Social\ Welfare$. The NPV is the net contribution of the RI to the aggregate welfare of society, where everything is expressed in the same unit of account.

GDP is a particularly crude way to measure social welfare changes empirically *per year*, as it is based on some simplistic accounting conventions (e.g., the value of depleted natural resources or environmental damages is not subtracted from the GDP). Welfare economists prefer other concepts (social welfare functions),[15] but if you believe that GDP is an adequate measure for the well-being of a society, $NPV_{RI} > 0$ means that if the project is implemented, the intertemporal value of GDP virtually increases by exactly the same amount of the NPV, consistently measured.

As previously mentioned, the NPV is expressed in terms of the unit of account used for measuring the MSV (whether money, time, or any other element). Hence NPV is not a pure, dimensionless number, and sometimes it may be helpful to communicate the results of the CBA simply as the benefit-cost ratio (BCR). More precisely, the BCR is the ratio between equivalent benefits and costs at $t = 0$. A BCR > 1 indicates that the project has a positive NPV. Indeed, the BCR uses the same information as the NPV, but presents it in a slightly different manner. However, the BCR can be misleading when ranking projects, a process that produces mutually exclusive alternatives because it does not account for the scale of the projects.[16] Moreover, it is not robust to the way that benefits and costs are defined (e.g., if time saved is considered a benefit, it increases the numerator, but if it is considered a negative cost, it decreases the denominator, and the ratio is not the same in the two accounting conventions).[17]

Another way to collapse all the CBA of a project into a unique, dimensionless number is the internal rate of return, discussed later in this chapter in the appendix.

Before introducing the CBA model for the research infrastructure, I discuss qualitatively the way that benefits may enter into such a model. The latter point is in fact the core issue; the remaining issues are standard for those who are familiar with CBA.

1.4   Taxonomy of Benefits

A customary partition of economic agents in the applied welfare economics literature is as follows: firms, consumers, employees, taxpayers. Firms are ultimately owned by individual shareholders and have an objective of profit maximization;[18] consumers want to maximize their utility;[19] employees want to maximize their income in return for a given amount of effort; and taxpayers adjust their decisions as a consequence of the existing fiscal constraints to minimize the burden of taxation. Obviously, most taxpayers are also employees, some of them are shareholders, and all are consumers; and there are other combinations of roles and possible disaggregations of these simplified social categories (see figure 1.2). Thus, it is natural to look at research infrastructures as projects (i.e., changes in the world that affect each type of agent in different ways). This is an important point because tracking the benefits in terms of agents is a structured evaluation strategy to avoid double-counting or picking up redundant indicators, which have no precise meaning in terms of social impact.

For RI project evaluation, it is convenient to divide the social intertemporal benefits into two broad classes. First, there are *use* benefits, which accrue to different categories of direct and indirect users of infrastructure services. Second, there are *nonuse* benefits, which reflect the social value of the discovery potential of RIs, regardless of predictable ex ante actual use. A possible taxonomy of benefits deriving from an RI project includes six main measurable use-benefits:

• The value for scientists of increased knowledge outputs; namely, publications and other forms of specialized communication that disseminate new theories, results, methods, and concepts. RIs, including distributed RIs such as networks of data repositories, increase the productivity of scientists in terms of efficiency of the management of information. Scientists are simultaneously producers and users of such output. They can be employed directly by the body managing the RI or by universities and other institutions. Publications and citations have measurable value. A confusion to avoid is between the average publication value per se (which is similar to the price of a book or of an article) and the value of knowledge embodied in the publication, which is the value of the long-term effects of such knowledge.

• Human capital formation of PhD students and postdoctoral researchers involved in experiential learning at RIs. Typically, they have a university affiliation, and there are spillovers from the RI to different university departments and learning mechanisms that ultimately increase job opportunities and income for early-stage researchers.

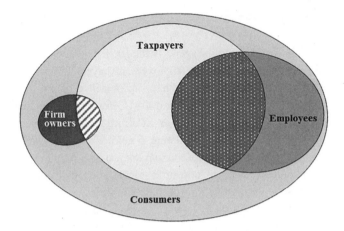

Figure 1.2
Partition of economic agents in applied welfare economics. The illustrative Venn diagram shows sets of agents, and possible intersections. For example, some consumers, but not all of them, are also taxpayers and employees. *Source*: Author.

• Technological and other types of learning and spillover for firms involved in procurement contracts for the construction and operation of the RI. An externality arises when the RI staff, which is intrinsically motivated by nonprofit objectives, helps the suppliers of technology to solve new problems arising from procurement contracts. Such solutions may create innovation opportunities for the firms and additional sales in other markets.

• Two core benefits are related to information technology (IT), which often involve broad communities of users, far beyond the scope of scientific research. An externality arises when the RI managing body does not protect inventions, including new software. Another externality arises when data sets are open access, and virtually donated to any potential user.

• Services directly provided by some RIs to third parties, such as medical research for patients and testing of materials for firms. Indirect benefits to users of science-based innovations (when strict causality can be proved) later arise, usually from market mechanisms.

• The cultural goods created by outreach, which accrue to specific groups of citizens through tourism at RI facilities, traveling exhibitions, media exposure, websites, and other facilities.

Finally, there are two nonuse benefits. The first is the benefit of the future use of the knowledge created. This often remains hidden in our ignorance and cannot be given a value ex ante (while, in some cases, it can be given a value ex post). The second is the intrinsic value of knowledge as a public good: the perceived generic social value of a discovery is correlated to the WTP for knowledge per se by the median taxpayer. The difference between these two ideas is further explained in chapter 9.

Subsequent chapters justify in some detail the reasons why these broad types of RI benefits should often be expected to be the core benefits in an ex ante project evaluation, albeit in variable proportions, according to the specific project to be appraised.

Having broadly defined the taxonomy of the RI benefits, for each benefit, there are two crucial steps to implementing a social CBA. The first step is forecasting the benefit in quantitative terms; the second is valuing it through a shadow price, which expresses the social value of a marginal change in the availability of the good (namely, MSV).

The estimation of shadow prices is the main conceptual difficulty involved with the calculation of the NPV. Drèze and Stern (1987, 1990), drawing on rich, earlier literature, prove that in some cases, the shadow price of a good can coincide with its long-run marginal production cost (MPC), such as the social cost of increasing the production of that good by an additional unit, holding the production level of all other goods constant. An alternative approach is to consider the WTP stated by the project users or indirectly revealed through specific techniques. This approach is particularly appropriate to determine a monetary value for nonmarket goods. In some circumstances, the MSV of a good can also be obtained by a combination of the long-run MPC and WTP.[20]

The main reason why market prices are unlikely to represent a relevant signal for the decision-makers of RI projects is because the most relevant goods produced by the RI are public goods. In economics, a public good, according to a famous definition by Paul Samuelson (1954), is a good for which exclusion of anybody from its consumption is impossible or too costly and there is no rivalry in consumption among consumers. Joseph Stiglitz (1999) suggests that knowledge is a global public good. The knowledge by somebody that gravitational waves have been detected by the Laser Interferometer Gravitational-Wave Observatory (LIGO) does not prevent the use of this information by anybody else in the world. The open model of science typically associated with RIs would make it contradictory to exclude somebody from such knowledge, in contrast to the esotericism of the ancient alchemists or the top-secrecy of their heirs: the Manhattan Project's nuclear scientists. Other goods produced are in the form of externalities, benefits to third parties without compensation.

In general, the price system does not work efficiently for RIs, and this is the reason for using social CBA at shadow prices in this context.

Other benefits may be considered as well, but often by qualitative methods only, such as the contribution to peaceful relations among nations arising from international scientific collaborations (Schopper, 2016). The cafeteria of the European Organization for Nuclear Research (CERN) is an amazing melting pot, where a cosmopolitan community meets and creates a social capital of interpersonal relations beyond religion, ethnic, and national boundaries. Tangible assets are also surprisingly cosmopolitan. During a shutdown of the LHC, I was shown a detector component for which some mechanical parts were provided from Iran and the electronics from the United States. However, as an economist, I do not know how to measure the effect of good personal relations between Iranian and US scientists and how they could have an impact on the social cost of the adversarial relations between these two countries. I also do not know how to measure the contribution of the International Space Station (ISS), a mainly US/Russian joint venture that is open to the participation of several other countries, to the economic value of peace. Unfortunately, the unknown incremental effect of something on something else that is also unknown is not measurable. Thus, because our understanding of these other effects is practically negligible, I will leave them aside, even if other social scientists may try—with different methods of analysis—to assess them in a qualitative way.

This is also why I think that a CBA of Big Science projects motivated by military priorities, such as the Manhattan Project (and several other contemporary ventures), is technically impossible for an applied welfare economist to perform. While some side effects of the traditional model may be similar to those discussed here for the RI (e.g., learning effects for contractors), the main objective of a science project for the military is not knowledge creation per se; rather, it has to do with war and power. Hence, for such an investment, I could not make the crucial assumption about the nonnegative social value of the knowledge created. Suppose that instead of the Manhattan Project (which many think was justified in the context of World War II), we had to evaluate the socioeconomic benefits of the parallel Nazi German nuclear weapon project,[21] which motivated the famous 1939 letter on the atomic bomb by Albert Einstein to President Franklin D. Roosevelt in the first place.[22] The ultimate outcomes of the two rival research projects would have been very different for the world. War and peace, important as they are, should be left outside a quantitative model of socioeconomic impacts of science, for good reason.

1.5   A Simple CBA Model for RIs

Given the previous discussion, from an ex ante perspective, the forecast of the economic NPV of RIs over the time horizon T is defined as the expected intertemporal difference between benefits and costs valued at shadow prices, which in turn are defined as the MSV of goods. The time horizon is assumed. It seems reasonable to assume a long, but finite, time horizon for the benefits of an RI, given the obsolescence process of the value of knowledge over time. The residual value of these effects can be included in the final year of the analysis. A social discount rate (SDR) is used to translate future values into present ones, as covered in more detail in the appendix. All the effects to be considered are expressed in incremental terms in a counterfactual scenario—a state of the world without the specific RI project under evaluation. This issue is further discussed in chapter 2.

As previously mentioned, the NPV of RIs can be broken into two parts: the NPV (NPV_u) of use-benefits B_u and costs C_u and the nonuse value of knowledge created, or discovery (B_n). These terms will be used interchangeably (u and n represent a use-value and a nonuse-value, respectively):

$$NPV_{RI} = NPV_u + B_n = (PV_{B_u} - PV_{C_u}) + B_n. \tag{1.1}$$

The PV of use-benefits PV_{B_u} is the sum of benefits to users of the RI services, including the value of publications for scientists (SC); benefits to staff, particularly students, arising from human capital accumulation (HC); benefits to firms that we define as technological externalities (TE), including information and communications technology (ICT) externalities; benefits of applied research to external users or other consumers (AR); and benefits to users of cultural goods (CU). Nonuse benefits (B_n) refer to the future possible economic effects of any discovery (quasi-option value) and the intrinsic value of discovery per se, a public good. The PV of costs PV_{C_u} is the sum of the economic PV of capital (K), the labor cost of scientists (L_s), the labor cost of other administrative and technical staff (L_o), other operating costs (OP), and negative externalities, if any (EXT):

$$NPV_{RI} = [SC + HC + TE + AR + CU] + B_n - [K + L_s + L_o + OP + EXT]. \tag{1.2}$$

The discounting process, discussed here and later in this chapter, is represented by the T terms: $s_t = 1/(1 + SDR)^t$, such that, for example, $K = \sum_{t=0}^{T} s_t K_t$, and so on for each benefit and cost.

Scientists are both producers and consumers of knowledge outputs generated by the RI. The direct MSV of such knowledge output SC (i.e., the output per se, mainly publications and preprints without considering the wider effects of its content) is measured by the opportunity cost of producing and using a publication. This is represented by the value of the time that scientists spend on these activities. Taking the *marginal cost* of a service as an empirical proxy of its *benefit* or MSV is standard practice in CBA when the marginal WTP for such goods is not available. The idea has been accepted since the Little and Mirrlees (1974) manual was published, proposing shadow pricing rules for traded and nontraded goods. As the market value of scientific publications and the WTP by scientists for being published, read, and cited is not available,[23] taking the marginal cost is justified, and probably conservative, at least if scientific work is on average efficient. To clarify this intuition, the output of public services in the System of National Accounts[24] is based on their production cost. For example, in GDP statistics, the value of the provision of public health and education, law and order, and environmental protection is based on the production cost accounting.

Each publication has a chain effect in the literature. Hence, an operational shortcut to estimate the social value of publications consists of computing the sum of the PV (in terms of value of time) of papers authored by the RI's scientists (P_{0t}) and the value of subsequent flows ($i = 1, \ldots n$) of papers produced by other scientists using the results of the RI's scientists (P_{it}), with each term divided by the number of references they contain (Ref_{it}): $\left(\dfrac{P_{it}}{Ref_{it}} \right)$. The sum of these terms is a proxy of the subsequent waves of papers citing the RI literature output. Eventually the value of citations each paper receives is a proxy of the social recognition that the scientific community acknowledges to the paper (CIT_{it} with $i = 0, \ldots n$):

$$SC = \sum\nolimits_{t=0}^{T} s_t \cdot P_{0t} + \sum\nolimits_{i=1}^{n} \sum\nolimits_{t=1}^{T} \frac{s_t \cdot P_{it}}{Ref_{it}} + \sum\nolimits_{i=0}^{n} \sum\nolimits_{t=0}^{T} s_t \cdot CIT_{it}. \tag{1.3}$$

The details, variants, and implications of this approach are discussed in chapter 3. Again, this is not the value of knowledge, but only a relatively small fraction of it, embodied in the value of writing, publishing, reading, and citing papers: that is, the day-by-day business of scientists.

Human capital accumulation *HC* is valued as the increased earnings (*Dw*) gained by the former RI's students and former employees (*z*) since the time (φ) they leave the RI project, against a suitable counterfactual scenario:

$$HC = \sum\nolimits_{z=1}^{Z} \sum\nolimits_{t=\varphi}^{T} s_t \cdot Dw_{zt}. \tag{1.4}$$

The increased earnings are a premium—a salary and employability effect of the skills acquired, reputation and networking effects, and other benefits to students and young researchers in general (postdocs, junior scientists). The premium is derived from the value added to their curricula from their experience in the RI environment. The details of this approach and empirical evidence for it are presented in chapter 4.

The PV of technological spillovers TE is given by the discounted incremental social profits Π_{jt} by companies (j) of the RI's supply chain or other economic agents who have benefited from a learning externality:

$$TE = \sum_{j=1}^{J} \sum_{t=0}^{T} s_t \cdot \Pi_{jt}. \tag{1.5}$$

The learning mechanism is crucial in this context. The knowledge created in the procurement relationship between firms and the RI leads to R&D and innovations, generating productivity and profitability changes in the long term, well after the initial contact. The same applies to developers of open-source, free software and databases released by the RI in the public domain. Social profits are usually different from private profits because, in principle, they are gross of taxes and interest, and because of the possible wedge created between market and shadow prices. These issues and empirical evidence are presented in chapters 5 and 6.

The PV of benefits produced by (mainly applied) research infrastructures on other external users and the economic value of services provided by the RI is

$$AR = \sum_{ar=1}^{n} \sum_{t=0}^{T} s_t \cdot AR_{art}, \tag{1.6}$$

where AR_{art} is the annual benefit for external users and other consumers of ($ar = 1, \ldots n$) type of activities. These services are project specific (e.g., health services of particle accelerators for hadrontherapy research), and each of them is ultimately valued by users to its WTP. This also applies to any benefits embodied in innovations caused by the RI activities, provided that it is possible ex ante to forecast such benefits, or ex post to disentangle the impact of the RI from other determinants of innovation. These issues are discussed and illustrated with examples in chapter 7.

To conclude the presentation of users' benefits, outreach activities CU carried out by the RI directly and indirectly produce cultural goods for different categories of the general public (g). The goods produced can be valued by estimating the WTP_{gt} (or other proxies of it) for such activities:

$$CU = \sum_{g=1}^{G} \sum_{t=0}^{T} s_t \cdot WTP_{gt}. \tag{1.7}$$

In fact, the vector ($g = 1, 2 \ldots G$) includes a plurality of users, such as general-public visitors to scientific facilities, exhibitions in science museums and elsewhere, visitors to websites, and users of traditional and new (social) media. The WTP for the cultural goods is different across these groups and requires different estimation strategies, which are presented in chapter 8.

Finally, the term B_n captures the sum of two types of nonuse values related to the research discoveries: their quasi-option value (QOV_0) and the existence value (EXV_0) of discovery, both of which are evaluated at the present time (these are undiscounted effects because they can be evaluated only at the present time, when the evaluation is performed):

$$B_n = QOV_0 + EXV_0. \tag{1.8}$$

The first term is a quasi-option value of potential discoveries. It can be defined as the social value of information under the uncertainty and irreversibility of some decisions—an interesting idea that harks back to a seminal paper by Arrow and Fisher (1974) that placed it in the context of the management of natural resources. In a nutshell, if a resource has both a certain current value being irreversibily used (A) and an uncertain future value in an alternative use (B), which would be impossible to know when decision (A) is made, there may be a benefit to waiting and acquiring more information.

In chapter 9, I argue that for an RI, waiting (i.e., delaying a potential discovery) in some cases may create both social costs and benefits that are too uncertain to be predicted, but one can assume they balance. Hence, given that QOV_0 is, in general, intrinsically uncertain and therefore not measurable, it is simply assumed to be nonnegative and conservatively set to zero, and thus it can be disregarded. This assumption has the important and perhaps surprising consequence that the value of unpredictable applications of discoveries must be left to future evaluations and provisionally left aside. Perhaps one day we shall know what to do with the discovery of a distant galaxy, a strange elementary particle, or a genetic mutation of an extinct species, but it seems prudent to admit that—to the best of our current understanding—knowing these facts does no harm in general (there may be exceptions, though, as discussed in chapter 9).

However, curiosity is a social attitude that supports knowledge creation per se. One may think that scientists alone are curious about nature, but this is more a matter of intensity than of a fundamental divide between various social groups. Scientists are people who have transformed curiosity into a profession. They are a relatively small group—perhaps around 0.1 percent of the world population[25]—but hundreds of millions, if not billions, of citizens, including those in primary schools or in their very old age, are occasionally interested to know something about the universe, its laws, and strange features. Government funding of science is also a reflection of these widespread cultural attitudes, not just a forecast of economic growth effects.

This is similar to the concept of existence value in environmental economics, a nonuse value of natural resources that citizens want to be protected, whether or not they plan to use them in the future. The term EXV_0 in equation (1.8) represents the social value of knowledge as a pure public good and can be proxied by stated or revealed WTP for potential discovery. While the intrinsic or existence value concept is not new,[26] its application to RI's evaluation is novel, and I shall discuss it later in detail in chapter 9, with an empirical application.

Turning to RI costs in equation (1.2), their PV can be expressed as

$$PV_{C_u} = \sum_{t=0}^{T} s_t \cdot (K_t + L_{st} + L_{ot} + OP_t + \text{EXT}_t), \tag{1.9}$$

where K_t is the annual capital expenditure; L_{st} and L_{ot} scientific labor and administrative/technical labor costs, respectively; OP_t other operating costs; and EXT_t the value of negative externalities, if any. Shadow prices should replace market prices in determining the social costs if market prices are poor representations of the MSV of a good. If the marginal cost of scientists' labor input in publications is taken as a proxy of the value of knowledge outputs produced by scientists, then the terms including L_{st} in equation (1.9) and P_{0t} in equation (1.3) cancel each other out (under the reasonable assumption of linearity of the cost function).

In summary, the CBA model for pure and applied research infrastructures turns into the following equation:

$$
\begin{aligned}
NPV_{RI} = & \left[\left(\sum_{i=1}^{n} \sum_{t=1}^{T} \frac{s_t \cdot P_{it}}{Ref_{it}} + \sum_{i=0}^{n} \sum_{t=0}^{T} s_t \cdot CIT_{it} \right) \right. \\
& + \left(\sum_{z=1}^{Z} \sum_{t=\varphi}^{T} s_t \cdot Dw_{zt} \right) + \left(\sum_{j=1}^{J} \sum_{t=0}^{T} s_t \cdot \Pi_{jt} \right) \\
& + \left(\sum_{ar=1}^{n} \sum_{t=0}^{T} s_t \cdot AR_{art} \right) + \left(\sum_{g=1}^{G} \sum_{t=0}^{T} s_t \cdot WTP_{gt} \right) \left. \right] \\
& + (EXV_0) - \left[\sum_{t=0}^{T} s_t \cdot (K_t + L_{ot} + OP_t + EXT_t) \right].
\end{aligned}
\tag{1.10}
$$

In fact, this simple model is just a starting point, and variants according to the type of RI will be discussed chapter by chapter in this book.

1.6 The NPV Test

As a first step one can estimate the net present use-value of the RI, NPV_u. This is the sum of the terms in the right side of equation (1.1) except the last one (the nonuse effect). The NPV_u test could produce three possible baseline results:

- The net present use-value of the research infrastructure NPV_u is greater than zero (i.e., $PV_{B_u} > PV_{C_u}$; hence, $NPV_u > 0$).
- The net present use-value of the research infrastructure is equal or close to zero net of the unknown nonuse effects, $NPV_u \cong 0$.
- The net present use-value of the research infrastructure is negative net of the unknown nonuse effects, $NPV_u < 0$.

In the first two cases, the RI passes the ex ante CBA test if the evaluator guesses that the uncertain B_n would be at least nil, so that the total NPV_{RI} is not expected to be negative (within a range of associated probabilities). In other words, when the use-benefits of the RI are at least equal to the costs of producing them, in principle there is no further need to try to estimate B_n, so long as it can be excluded that nonuse effects are nonnegative. This is clearly a considerable computational advantage. The pure public good value of discovery, EXV_0, is still an unmeasured externality of the project, but society gains (or at least does not lose) by having the RI just because of the welfare effects on users, net of costs. For many RIs in applied research, the CBA test should be passed on these grounds.

In the third case, which may be typical of fundamental research, the RI project passes the CBA test if B_n is positive and large enough to compensate for the possibly negative net use-effects. In this situation, we cannot avoid an estimation of the WTP for the pure value of the (ex ante potential) discovery as a public good. The nature of such an estimation is discussed in chapter 9. The CBA model is offered as a starting point for empirical testing and further analysis of appropriate RI projects. It is just a preliminary sketch intended to provide a broad framework.

The presentation strategy is simple: I start with the estimation of costs in chapter 2; then I discuss the benefits in the subsequent chapters; and finally, I sum up everything in chapter 10 with a discussion of the way that the benefit net of costs can be presented.

1.7   Further Reading

Some sections of this chapter draw on Florio and Sirtori (2016), who discuss earlier research. This section very selectively cites some references on three topics: (1) the economic literature on the role of knowledge and R&D, (2) some contributions from other fields, and (3) applied welfare economics of infrastructure projects in general.

1.7.1 Innovation Economics

The economic literature on the impact of R&D, technology progress, and innovation is a major stream, or array of streams, that has been flourishing over five decades. Del Bo (2016) identified 871 and 1,221 papers, respectively, in the Scopus and Web of Science databases (which include only refereed journals), published between 1973 and 2014 on the subject of the rate of return for R&D. The publication trend clearly increased after the 1990s. Citations boomed, with 14,984 and 19,216 articles in Scopus and Web of Science, respectively, cited in the last twenty years (since 1998).

Griliches (1958) and Mansfield, Rapoport, Romeo, Wagner, and Beardsley (1977) are among the most-cited early papers on the private and social returns of R&D. The basic idea is to augment the standard production function (in some cases, a cost function) with one or more variables correlated with R&D, and to study the marginal effect on productivity at different levels of aggregation. The macroeconomic theory counterpart of this research trend is the shift from the Solow (1956) and Swan (1956) exogenous growth models to the endogenous growth theory. A first intuition was provided by the seminal papers by Arrow (1962) and Romer (1990), both of which focus on the role of human capital. Aghion, Akcigit, and Howitt (2015) review the Schumpeterian growth models, where innovation plays a key role. Barro and Sala-i-Martin (2004) summarize the story in an advanced textbook; see also Griliches (1998).

The divergence between private and social returns of R&D is summarized by Hall (1996) and updated by Hall, Mairesse, and Mohnen (2009). Crépon, Duguet, and Mairesse (1998) started an entire subfield of literature on the econometric study of simultaneous equations of the relationship between R&D, patents, and productivity. The econometric literature on patents is now a flourishing subfield within the economics of innovation.

These papers, however, do not focus on large-scale research infrastructure, but rather on firm-level R&D and spillovers. Monjon and Waelbroeck (2003) present a good example of the study of spillovers from universities to firms, an empirical topic well developed in the literature.

A comprehensive presentation of various research topics is provided by the *Handbook of the Economics of Innovation* (Hall and Rosenberg, 2010).

1.7.2 Earlier Literature on the Benefits of Big Science

This body of economic literature, in spite of the diversity of assumptions and often-conflicting views among economists, is usually based on models that, at least in principle, can be given a formal structure and tested empirically according to certain standards and methods. In contrast, the contributions from other fields in the social sciences tend to be more based on qualitative approaches, or even when based on quantitative data, are more heterogeneous in terms of the way hypotheses are formulated and tested.

Autio (2014), in a survey paper on Big Science for the UK government, proposes a taxonomy of scholarly traditions. Outside economics, he identifies approaches from sociology (covering topics such as the rate of increase of scientists, social context and science, and collaborative arrangements); assessment and scientometrics (including benefits from spinoffs, citations, and construction requirements); history (patterns of Big Science and trends in science evolution); policy (management of international collaborations and knowledge creation effects); project management (complexity and learning); and innovation studies (case studies, mechanisms, and industry engagement). He concludes: "In summary, while the big-science mode of doing science appears to be increasingly important, and while there exists a fair number of case studies detailing individual impact mechanisms, much of the received literature is fragmented and lacks a coherent underlying framework" (Autio, 2014, 12).

Salter and Martin (2001) and Martin and Tang (2007) try to conceptualize the literature on the benefits of research. They focus more on the methods and identify three approaches: econometric studies, surveys, and case studies. Martin and Tang (2007, 6) also describe seven channels through which the benefits of research materialize:

Channel 1: increase in the stock of useful knowledge;

Channel 2: supply of skilled graduates and researchers;

Channel 3: creation of new scientific instrumentation and methodologies;

Channel 4: development of networks and stimulation of social interaction;

Channel 5: enhancement of problem-solving capacity;

Channel 6: creation of new firms;

Channel 7: provision of social knowledge.

Other reviews about the social benefits of RIs tend to mix very different approaches and list social benefits at very different levels. See, for example, SQW (2008); Science and Technology Facilities Council (STFC, 2010b); COST Office (2010); OECD (2014b); Simmonds, Kraemer-Mbula, Horvath, Stroyan, and Zuijdam (2013); Technopolis (2011); and European Space Agency (ESA 2012).

Outside economics, there are many papers based on multiple indicators, mostly in the gray literature.[27] Some of these are reports for funding agencies or governments such as the Belgian Science Policy Office (2013); Almirall,

Moix Bergadà, and Queraltó Ros (2008), on spatial data infrastructure in Catalonia; ESFRI (2013); European Commission (2012) on the assessment of the socioeconomic impact of software and internet services; European Commission (2015), on the Horizon 2020 projects; Lee (2018), on the Korean contribution to the International Thermonuclear Experimental Reactor (ITER) nuclear fusion research project; OECD (2015b); PAERIP (2013); BIS (2010), on the role of research councils in the United Kingdom; and Simmonds et al. (2013), on Big Science and innovation. Also see Technopolis (2015); Hallonsten, Benner, and Holmberg (2004); Hallin (2012), on Canada; KPMG (2014), on Australia; and Technopolis (2010b), on impacts of European research and technology organizations.

Several papers are on CERN or other particle physics facilities, such as Åberg and Bengtson (2015); Arenius and Boisot (2011); Autio, Bianchi-Streit, and Hameri (2003); Kamer (2005), on the Brookhaven National Laboratory; Le Goff, Heuer, Koutchouk, Stapnes, and Stavrev (2011); Lebrun and Taylor (2017); OECD (2014b), on case studies at CERN; Schmied (1977); Martin and Irvine (1984a, 1984b); and the Lawrence Berkeley National Laboratory (LBNL, 2001).

In other fields, see such sources as BBMRI-ERIC,[28] on the European research infrastructures for biological and medical Sciences; OECD (2014b), on international distributed research infrastructures; Technopolis (2010a, 2014), on molecular biology; BBSRC (2015), on the bioeconomy; and the Tauri Group (2013), on NASA.

1.7.3   Applied Welfare Economics

As this book focuses on a CBA framework, readers who want to have a more complete picture of theory, methods, and applications may read the manuals by Boardman, Greenberg, Vining, and Weimer (2018); De Rus (2010); and Brent (2006, 2017). The latter is a very concise and clear presentation with worked examples. These manuals are mostly framed in a partial equilibrium approach, a setting that considers market equilibria one by one, implicitly disregarding the mutual simultaneous interactions of supply and demand of goods in different markets. In a nutshell, the difference between partial and general equilibrium tradition in CBA concerns the way that welfare effects on economic agents are considered, either in one market and then adding effects on other markets in the former case, or simultaneously in several markets and on public finance in the latter case. Farrow and Rose (2018) explain the implications of these two approaches.

The general equilibrium approach to CBA is summarized by Drèze and Stern (1987), and in a shorter paper by Drèze and Stern (1990). These are rather technical presentations, but Florio (2014) provides a simplified discussion of the topic. In this framework, shadow prices capture the opportunity costs of goods and services and are defined as the MSVs of goods when the economy is at its (constrained) optimum. There are other general equilibrium strategies for this estimation, such as starting from a position of initial disequilibrium (Johansson and Kriström, 2018) and using observed prices with several corrections to estimate cost-benefit rules.

CBA has been developed extensively in both theory and practice. Applications (very selectively cited) include such sectors as transport (Quinet, 2007); energy (Von der Fehr et al., 2013); environment (Hanley and Spash, 1993; Pearce, Atkinson, and Mourato, 2006; Atkinson et al., 2018); culture and education (Loomis and Walsh, 1997; Báez and Herrero, 2012; PricewaterhouseCoopers, 2007; Hough, 1994; Woodhall, 1992; Psacharopoulos, 1987); health (Birch and Donaldson, 1987; McIntosh, Clarke, Frew, and Louviere, 2010), and many others, including impact analysis of any kind of regulations and policies, particularly in the United States. To see the wide scope of contemporary CBA, I would suggest browsing the *Journal of Benefit-Cost Analysis*, a scholarly journal sponsored by the Society for Benefit-Cost Analysis (for curious reasons, CBA became BCA after crossing the Atlantic).

Other references on CBA of RIs are mentioned in the "Further Reading" section of chapter 10.

1.A.1   Appendix: Discounting

Suppose that you own a sum of money M at time $t=0$, M_0. Then you invest it at an interest rate i, a pure number, to be paid after one year. The amount of money at time 1 (i.e., after one year from now) would be $M_1 = M_0(1+i)$. If you go ahead in the process by additional years, you end up in the year T with the amount $M_T = M_0(1+i)^T$. Equivalently, you can say that $M_0 = M_T/(1+i)^T$. Hence, any amount of money (or, in fact, of any good that is used as a numeraire) available at a given time M_t can be expressed at its $t=0$ equivalent. You can define the NPV function as

$$NPV(q,\ MSV,\ SDR,\ t) = \sum_{t=0}^{T}(B-C)_t(1+SDR)^{-t}. \tag{1A.1}$$

The expectation operator has been omitted, and $B = (B_1,\ B_2 \ldots B_v)$ or $C = (C_1,\ C_2 \ldots C_R)$ are vectors including V and R benefit and cost types, respectively. Hence, the NPV is a nonlinear function of the vector of quantities q, shadow prices MSV, time t, and the social discount rate SDR (defined here and later to simplify, assumed to be constant). If you look back from time $t = 0$, the equation is such that a past value is augmented by $(1 + SDR)^t$ because $t < 0$, while if you look forward, it is decreased by $(1 + SDR)^{-t}$. The case studies that follow, LHC and the National Centre of Oncological Hadrontherapy (CNAO), a particle accelerator in Pavia (Italy) for research on cancer therapy, respectively, use both perspectives (forward and backward), as their $t = 0$ was midtime in the history of the evaluated RI.

In the financial analysis of an RI, the relevant interest rate is based on information given by the financial markets. It is the opportunity cost of capital for an investor, which is generally estimated, in the case of government being the main funder of the RI, by looking at the interest paid on sovereign bonds (note: it is *not* the interest rate paid by the RI on loans).

However, in a social CBA, we are interested in the opportunity cost of capital for society, not for any investor. This is the MSV of a unit of capital to the economy, which may diverge from the interest rate in financial markets because such markets are imperfect. What is the most appropriate social discount rate in the RI context?

There are many possible empirical strategies (see the "Further Reading" section earlier in this chapter), but a popular one is based on the Ramsey equation (invented by the Cambridge University mathematician Frank Ramsey), expressed as the social rate of time preference (SRTP):

$$SRTP = ptp + \varepsilon \cdot gth, \tag{1A.2}$$

where ptp is the pure time preference, ε is the elasticity of the marginal utility of consumption, and gth is the expected growth rate of per capita consumption in future.

The effect of discounting can be nonnegligible for long-term RIs. With a 0.03 social discount rate, such as the one recommended by the European Commission (2014) for more developed European countries, 1 EUR at $t = 2016$ would be worth EUR 1.96 at $t = 1993$ (the starting year of the LHC) and 0.50 at $t = 2038$ (the end of its major upgrade, the High-Luminosity LHC). It is important to stress that this effect has nothing to do with inflation or exchange rates; rather, it is an effect of the social preference for the present consumption against the past. Even if the pure time preference (ptp) is equal to 0, as Ramsey himself suggested (based on an argument of intergenerational neutrality), the expected growth rate is never zero in the long term, and the parameter ε is often empirically estimated as greater than 1, as it expresses the social aversion to inequality. The richer we expect the next generations to be compared to the current ones, the more we care about the present poor generations compared to the future richer ones; therefore, the greater the SDR will be.

In most CBA practices, a constant social discount rate is used, which implies an exponential discounting process of a project's flows.

Benefits of knowledge occurring far in the future are discounted more than the costs of investments, which instead typically occur in the initial years of the time horizon. This approach unavoidably underrates the welfare of future generations. One possible way to address the issue raised by a constant rate is to adopt a sufficiently low discount rate. One example is the *Stern Review: The Economics of Climate Change* (HM Treasury, 2006). This influential report adopted the social rate of time preference (SRTP) approach and assumed a near-zero pure time preference (equal to 0.1 percent), a figure that is assumed to consider the risk of extinction of the human race. The elasticity of the marginal utility of consumption is set to 1, and the expected rate of future per capita growth is assumed to be equal to 1.3 percent per annum, in accordance with historical data. When these values are entered into the SRTP formula, they result in a low rate of 1.4 percent in real terms. Criticism was raised by Dasgupta (2007, 2008); Weitzman (2007), and others. However, see Stern (2015) for a rebuttal of this criticism. The *Stern Review* is relevant to our discussion because some RIs are directly related to climate change research (e.g., Earth observation satellites, nuclear fusion, and superconducting electricity transmission), but also because some projects may have effects in the very distant future.

A declining discount rate, following a hyperbolic (or decreasing over time) discounting function (Laibson, 1997), characterized by relatively high discount rates in the short term and relatively low rates over long horizons, has also been advocated and adopted by some governments in their official guidelines, based on the argument of increasing uncertainty (see the "Further Reading" section). Such governments include those of the United Kingdom (HM Treasury, 2018) and France, following Quinet (2013).

We can now discuss the concept of economic internal rate of return (IRR) of an investment project in this context. This is a dimensionless number, such that the benefits and the costs are intertemporally equal if discounted as follows:

$$NPV = \sum_{t=0}^{T} [(B - C)_t][(1 + IRR)]^{-t} = 0. \qquad (1A.3)$$

In such a case, the CBA test passes if $IRR > SDR$.

The three indicators, NPV, BCR, and IRR, provide the same information in principle. However, in practice, they can differ in some specific cases. Thus, computing the three of them makes it possible to examine project performance in a slightly different way. In particular, for a project with positive socio-economic performance, there is usually a convergent message: (1) the NPV (expressed in monetary terms) is higher than zero, and the higher the NPV, the larger the social benefits achieved, net of costs, and negative externalities; (2) the IRR is comfortably higher than the adopted social discount rate; and (3) the BCR has a value higher than 1. There are some caveats in dealing with these indicators. For mathematical reasons, the IRR (which usually cannot be derived by solving the NPV equation with analytical methods and requires a computational approach) should be used with caution (Florio, 2014).

2 Costs and Financial Sustainability of Research Infrastructures

2.1 Overview

How much does a synchrotron light source, an oceanographic vessel, or a mission to Mars actually cost taxpayers? The total costs of research infrastructure (RI) projects are publicly reported in different ways, often quite confusingly. According to the European Space Agency (ESA), the estimated cost to develop, assemble, and run the International Space Station (ISS) over a period of ten years was EUR 100 billion;[1] on the US part, the ISS cost was estimated by the NASA Office of the Inspector General as follows: "The United States has invested almost $78 billion in the International Space Station … over the last 21 years, and going forward, NASA plans to spend between $3 and $4 billion annually to maintain and operate the Station, including transportation for crew and cargo" (NASA 2015, 2). Further, independent estimates, such as Lafleur (2010), cite a figure of around $150 billion over thirty years as of 2015:

The ISS has been described as the most expensive single item ever constructed. In 2010 the cost was expected to be $150 billion. This includes NASA's budget of $58.7 billion (inflation-unadjusted) for the station from 1985 to 2015 ($72.4 billion in 2010 dollars), Russia's $12 billion, Europe's $5 billion, Japan's $5 billion, Canada's $2 billion, and the cost of 36 shuttle flights to build the station, estimated at $1.4 billion each, or $50.4 billion in total.

The wide variability of how these figures are reported is evident when one considers differences concerning which items are included in total costs and which are excluded, which currency and exchange rate is adopted year by year, how inflation is accounted for, how values occurring in the future or in the past are expressed in present terms (which is not the same as accounting for inflation and exchange rate fluctuations), and whether only the main funder of the project is considered (NASA) or other contributors as well (the ESA and the space agencies of Russia, Canada, and Japan).

A standard way to present the total costs would also be helpful for making comparisons across different RIs. Table 2.1 reports the costs of selected priority RI projects and landmarks in Europe from the European Strategy Forum on Research Infrastructures (ESFRI, 2016b, 16–17).

While a certain degree of standardization has been attempted by ESFRI (e.g., all the figures are given in euros), there are still several issues concerning how to interpret the data on capital values, preparation costs, and yearly operational expenditures. As these expenditures occurred in the past or are predicted to occur at different times, the comparability of costs across existing and future RIs in Europe is limited without some agreed-upon accounting conventions.[2] In fact, some expert readers within the various scientific communities may find that the cost statistics in these tables are different from what they may have seen elsewhere.

This chapter discusses how a rigorous cost accounting of RI should be performed from different perspectives (financial analysis, economic analysis, domestic versus international government funding, etc.). After all, before assessing whether the social benefits are greater than the costs, one should be able to comprehensively report the latter.

Similar to other investment projects in any field, the costs of RIs can be broadly classified into two categories: capital and operating expenditure. The former includes the costs for the acquisition of durable assets, such as construction costs, acquisition of experimental equipment, and intangible fixed assets such as the value of research and development (R&D). Some costs may be incurred by third parties when there are in-kind contributions. Among operating expenditures running year by year, there are labor costs (including the costs of scientific personnel and other administrative and technical staff), materials, energy and other utilities, communication, and maintenance. Decommissioning costs should also be considered (e.g., when there is radioactive waste to be disposed of after the shutdown of a facility).

From a socioeconomic perspective, there may also be negative externalities during construction and operation, to be valued in appropriate ways. For example, ESA (2013) estimated the space debris from defunct human-made objects orbiting around the Earth to be around 29,000 pieces at sizes larger than 10 cm, or 670,000 (larger than 1 cm), or 170 million (larger than 1 mm). This is a form of pollution of space that may create problems for the operation of telecommunications and Earth observation satellites—thus, a negative externality.

Thus, several corrections are needed to go from financial to social cost accounting. These issues are typical of any infrastructure. There is a clear understanding, based on decades of experience, of how to count the social

Table 2.1
Selected ESFRI projects and landmarks

Project Name	TYPE		YEAR		COST (M€/Y)			COUNTRIES	
	SEC	TYP	YCO	YOP	CAV	CON	OAB	CC	PCS
Multipurpose hYbrid Reactor for High-tech Applications (MYRRHA) http://myrrha.sckcen.be	ENE	SS	2019–2034	2024*	1,500	NA	100	BE	16
European Carbon Dioxide Capture and Storage Laboratory Infrastructure (ECCSEL) http://www.eccsel.org	ENE	D	2014–2030	2016	1,000	80–120	1**	NO	9
European SOLAR Research Infrastructure for Concentrated Solar Power (EU-SOLARIS) http://www.eusolaris.eu	ENE	D	2018–2022	2020*	120	120	3–4	ES	9
Aerosols, Clouds, and Trace gases Research Infrastructure (ACTRIS) http://www.actris.eu	ENV	D	2019–2021	2025*	450	190	50	FI	11
International Centre for Advanced Studies on River-Sea Systems (DANUBIUS-RI) http://www.danubius-ri.eu	ENV	D	2012–2022	2022*	300	222	28	RO	18
Infrastructure for Analysis and Experimentation on Ecosystems (AnaEE) http://www.anaee.com	H&F	D	2014–2018	2018*	135.5	200	2–3**	FR	9
European Solar Telescope (EST) http://www.est-east.eu	PSE	SS	2019–2025	2026*	NA	200	9	ES	13

(continued)

Table 2.1 (continued)

Project Name	TYPE		YEAR		COST (M€/Y)			COUNTRIES	
	SEC	TYP	YCO	YOP	CAV	CON	OAB	CC	PCS
Cherenkov Telescope Array (CTA) https://portal.cta-observatory.org	PSE	D	2017–2023	2023*	400	297	20	DE	33
Jules Horowitz Reactor (JHR) http://www.cad.cea.fr/rjh	ENE	SS	2009–2019	2020*	1,000	–	NA	FR	10
Partnership for Advanced Computing in Europe (PRACE) http://www.prace-ri.eu	e-RI	D	2011–2015	2010	500	–	120	PRACE–AISBL, DE, ES, FR, IT	28
European Advanced Translational Research Infrastructure in Medicine (EATRIS ERIC) http://www.eatris.eu	H&F	D	2011–2013	2013	500	–	2.5	NL	11
European Spallation Source (ERIC) http://www.europeanspallationsource.se	PSE	SS	2013–2025	2025*	1,843	–	140	DK, SE	15
European XFEL http://www.xfel.eu	PSE	SS	2009–2017	2017*	1,490	–	115	***	12
High-Luminosity Large Hadron Collider (HL-LHC) http://home.cern	PSE	SS	2017–2025	2026*	1,370	–	100	CERN (CH)	26
Facility for Antiproton and Ion Research (FAIR) http://www.fair-center.de	PSE	SS	2012–2022	2022*	1,262	–	234	DE	10

Project Name	TYPE		YEAR		COST (M€/Y)			COUNTRIES	
	SEC	TYP	YCO	YOP	CAV	CON	OAB	CC	PCS
European Extremely Large Telescope (E-ELT) http://www.eso.org/public/teles-instr/e-elt/	PSE	SS	2014–2024	2024*	1,000	–	40	ESO	17
Extreme Light Infrastructure (ELI) http://www.eli-laser.eu/	PSE	D	2011–2017	2018*	850	–	90	ELI-DC, CZ, HU, RO	7
Square Kilometre Array (SKA) http://www.skatelescope.org	PSE	D	2018–2023	2020*	650	–	75	UK (ZA and AU)	16
European Social Survey (ESS ERIC) http://www.europeansocialsurvey.org	S&C	D	2010–2012	2013	NA	–	6	UK	23
Survey of Health, Ageing, and Retirement in Europe (SHARE ERIC) http://www.share-project.org/	S&C	D	2010–2012	2011	110	–	12	DE	22
Digital Research Infrastructure for the Arts and Humanities (DARIAH ERIC) http://www.dariah.eu	S&C	D	2014–1018	2019*	4.3	–	0.6	FR	18

Source: Author's elaboration on ESFRI (2016a).

Notes: * expected; **for centralized services; ***M€/Y: EUR (millions) per year.

Legend: e-RI: Electronic Infrastructure SEC: Sector: ENE: Energy; ENV: Environment; H&F: Health and Food; PSE: Physical Sciences and Engineering; S&C: Social and Cultural Innovation; TYP: Type: D=Distributed, SS =Single-Site; NA (not available); YCO: Construction Phase; YOP: Operation Year; CAV: Capital Value (EUR millions); CON: Construction Cost (EUR millions); OAB: Operational Annual Budget (EUR millions); CC: Coordinating Country; PCS: Number of Participant Countries—AU: Australia, AZ: Azerbaijan, BE: Belgium, CH: Switzerland, CZ: Czech Republic, DE: Germany, DK: Denmark, ELI-DC: ELI Delivery Consortium International Association, ES: Spain, ESO: European Southern Observatory, FI: Finland, FR: France, HU: Hungary, IT: Italy, NL: Netherlands, NO: Norway, PRACE: Partnership for Advanced Computing in Europe, RO: Romania, SE: Sweden, UK: United Kingdom, ZA: South Africa.

costs of a power plant or of high-speed rail, including negative externalities, even if mistakes can always be made. However, because RIs are often designed to perform a range of experiments or scientific activities, and projects generally consist of several interrelated but relatively independent components, such as with networks or distributed RIs, several specific problems arise when appraising their costs.

First, delimiting the RI project borders and, in turn, its financial and social costs, requires some preliminary analysis. The specific feature of an RI project calls for the aggregation and apportionment of several items under various contractual arrangements. For example, which past or retrospective costs should be considered as "sunk"[3] and not included in the computation of project costs? The Large Hadron Collider (LHC) at the European Organization for Nuclear Research (CERN) currently uses the tunnel excavated for its predecessor, the Large Electron-Positron Collider (LEP). The LEP costs should not be considered part of the LHC costs, though, as the new accelerator did not have to pay to use the previous tunnel, but only had to invest in certain improvements and other civil engineering works.

In general, the analysis should focus on one specific project, while the hosting organization is in fact managing a portfolio of projects. If different RIs are hosted at the same site, with some functional interrelation (e.g., energy generation equipment or data centers), their cost must be apportioned to different projects according to transparent criteria.

Second, typically, RIs are unable to cover their costs by selling services and products. A funding gap arises, and the way in which such a gap and financial sustainability are accounted for deserves some attention. For example, the way in which some cash inflows from research grants are accounted for has some consequences for the forecast of the project's long-term financial sustainability. This issue is related to the overall financial sustainability of the project.

Third, some social costs, particularly unskilled labor in regions affected by high levels of unemployment, and some inputs when markets are distorted, require the use of shadow prices. Therefore, the social costs of an RI may be different from its financial costs, and in any case, it is important to avoid confusion between the two perspectives of financial and social accounting.

This chapter elaborates on these topics and proposes a consistent framework for tabulating the costs of the RI, providing some examples. Section 2.2 discusses how to identify a project's borders and how to solve aggregation issues of RI components belonging to different owners, often from different countries; section 2.3 considers the financial costs of an RI and why these may differ from its social costs; section 2.4 discusses the project revenues, if any; section 2.5 covers the funding models and financial sustainability; section 2.6

describes the intertemporal accounting; and section 2.7 considers the financial performance. Finally, section 2.8 presents a case study: How much does the most powerful particle accelerator in the world actually cost?

2.2   Project Borders and Units of Analysis

A necessary step before carrying out the financial and economic evaluation of an RI is to identify clearly the object of the analysis: the project. This is an investment and its operation: a new, "greenfield" RI or an upgrade of an existing one ("brownfield"). The LHC is a brownfield RI, for example, while the Square Kilometre Array (SKA) radio telescope is South Africa is a greenfield RI. In principle, such evaluation should focus on all the components that are logically connected to the attainment of the RI's objective.[4] In other words, the border of the project should include any cost components that, at time $t = 0$, appear as incurred in the past or to be incurred in the future, and for that specific project and not for something else.

Delimiting the borders of an RI project may require some careful analysis when the project is designed to perform a range of experiments or activities simultaneously, or at different times, or when it consists of several interrelated but relatively independent components. For example, the operation of the LHC requires the combination of an upstream accelerator system and four main detectors, each managed by an international collaboration of scientists and laboratory staff in various combinations. The ISS project includes an assembly of components, including several ground facilities of the National Aeronautics and Space Administration (NASA), the Russian Federal Space Agency (ROSCOSMOS), ESA, the Canadian Space Agency, and the Japan Aerospace Exploration Agency (JAXA).

Any accelerator without at least one detector is worthless, as experimental data would not be available. Detectors without accelerators cannot observe anything. Thus, while CERN and the collaboration of scientists managing the experiments are different legal entities with different budgets, accounting for the costs of only one part of the RI would be like considering tracks without trains in a railway project or vice versa (not a recommended practice). The same applies to the ISS, where ground facilities, including control rooms, are of the essence.

Similarly, for distributed facilities, a project's analyst should ascertain whether synergies and functional relationships among the facility's components are such that they justify the entire infrastructure's assessment as a single unit of analysis.[5] One example is the Laser Interferometer Gravitational-Wave Observatory (LIGO) project, which consists of two separate interferometers in the United States, located in Hanford, Washington, and Livingston, Louisiana, operating in unison and in collaboration with the Virgo interferometer (near Pisa, Italy) to detect gravitational waves. The design and construction of LIGO was carried out by the LIGO Laboratory's team of scientists, engineers, and staff at the California Institute of Technology (Caltech) and the Massachusetts Institute of Technology (MIT), and collaborators from over eighty scientific institutions worldwide. In this case, there is a unique international collaboration and the RI is described[6] as a unique national facility, while in fact it consists of two facilities, geographically and functionally separated.

In principle, each of these interferometers can operate independent of the other, even if they take advantage of being operated in unison to confirm the observations. Thus, one may consider that LIGO comprises the combination of the two detectors but, in principle, each one can be evaluated separately. Instead, as mentioned, it would be wrong to analyze the LHC cost without including its main detector costs from the perspective of cost-benefit analysis (CBA).

There are also differences of perspective between ex ante and ex post evaluations. In the appraisal phase, the unit of analysis typically focuses on the financing decision. The body managing the RI applies for funding to an agency and focuses on covering the costs of certain facilities or operations related to a research program. Thus, for administrative reasons, Project Step 1 may be funded in one fiscal year and Project Step 2 in another fiscal year, but in fact, there is only one RI project in functional terms, whatever the funding schedule. The analysis of costs should integrate these expenditures.

In other cases, however, there are actually two subsequent projects, each building on the previous one, and they must be analyzed separately. This is the case with the High-Luminosity LHC, a major upgrade of the LHC, worth nearly 1 billion Swiss francs (CFH) in construction costs. Another example is the SKA, whose Phase 1 (Garret et al., 2010) is a subset of Phase 2, with a rather complex implementation sequence.[7]

In these examples, the analysis of costs should consider High-Luminosity LHC or SKA Phase 2 as new projects relative to LHC and SKA Phase 1, respectively.

2.3   Financial versus Economic Costs

After having identified the project through its borders, the next step is to determine which are the relevant prices to apply in order to give a value to its activities on the cost side. In standard economic theory, a price is associated with different factors of production (tangible capital, labor, human capital, purchased materials and services, knowledge, and other intangible assets). More specifically, each resource used in a project is associated with an equilibrium economic value through a price. Thus, the labor cost of scientific and technical personnel, for example, is the product of the numbers of full-time equivalents and their salaries. This market value, in monetary terms, is taken into account in the financial analysis. Instead, in the economic analysis, each good has an opportunity cost, which is the marginal social value of the next best economic alternative foregone for society as whole due to the chosen use of such a good. The two prices, as mentioned in the previous chapter, can diverge.

From this perspective, studying the financial performance of an RI has two roles: It is important per se to analyze the financial sustainability of the project, and it is preparatory to the economic appraisal, which assesses the project's worthiness to society (at the global, country, and regional levels, as appropriate). More specifically, the financial analysis is useful to determine the costs and revenues (if any) arising from the project over the reference period, and to verify whether the projected cash flow ensures the RI's adequate operation and its financial sustainability in the long term. In what follows, I first focus on the financial costs and then turn to the economic costs.

Costs are defined as cash outflows directly paid by the managing bodies of the RI to design, build, operate, maintain, and update the project. Although cost disaggregation is project-specific, common categories of financial costs, including investments that are generally related to RI tangible and intangible assets, include the following items: conceptual design, R&D expenditures, planning and technical design, land acquisition, construction of buildings and facilities, construction of plants and machinery, machinery and equipment purchases, utilities consumed during the construction phase (e.g., energy and waste disposal), other start-up costs, license acquisitions, and replacement costs. From such a perspective, scientific and technical personnel costs strictly necessary for the R&D before the construction of the RI are investment costs or capital expenditures, as they contribute to the accumulation of intangible fixed assets. The operating costs include scientific, technical, and administrative personnel expenditures to run the RI after its construction; ordinary maintenance; materials for the operation and repair of assets; utility consumption; services purchased from third parties; rental of machinery; environmental protection measures; general management; administration and quality control costs; royalties paid for the use of patented products or processes; promotional campaigns and other outreach expenditures; and decommissioning.

In fact, the weights of these generic classes of costs are different across scientific domains (with physics, astrophysics, and space exploration usually the most capital intensive). However, one important point is that in this context, only cash outflows should be considered without any depreciation, which is customarily used in standard accounting, particularly for corporate tax purposes. But there are two reasons to ignore depreciation. First, maintenance expenditures, if any, must be recorded year by year on a cash-flow basis. Any obsolescence of the assets needed for research (e.g., computers) is eventually taken into account by the fact that the assets' residual value at the end of the project life cycle is lower than at start-up.

The second reason to avoid the use of depreciation in accounting for the RI cost is that usually there is no way to assess the loss of value of a unique telescope or a probe against a market value, for the simple reason that such market value does not exist for major RIs, and in some cases, the decommissioning time is uncertain. Thus, it is not possible to write off part of the value of the assets according to a specified depreciation formula. From this perspective, an RI is similar to an environmental good such as a national park or a cultural good such as a fine arts museum, for which there is no way to estimate a depreciation rate. If, after some years, a particle accelerator is worthless, its residual value in cash terms will be zero, and this is all that we want to know.

Interest rates on loans and other funding arrangements should be dealt with in the ways suggested. For example, the European Commission (2014) does not consider any conceptual difference between an RI and any other infrastructure project from the perspective of financial arrangements.

I turn now to the social costs of an RI project. The core question is: Do the observed prices, considered in a preliminary, comprehensive financial analysis of an RI, accurately represent the marginal social value of certain inputs? I would argue that, in general, five types of RI costs may need a correction in economic terms: land acquisition or use of certain facilities; unskilled labor cost in locations affected by high levels of permanent unemployment; some utilities; some environmental externalities; and in-kind contributions. Next, I briefly discuss each of these.

First, many RIs are provided land or some facilities owned by public-sector entities, either for free or at a lower price than market rate. When this is the case, there is a hidden subsidy—a cost to society that does not appear in the financial accounting. In some cases, the correction may be significant for land-intensive RI or buildings in highly valued urban contexts. In other cases, there may be a pecuniary externality on real estate in the area.

Second, the price of some utilities, particularly electricity, occasionally may deviate from the long-run marginal costs because of regulated tariffs, and this should be a possible correction to introduce to the financial accounts. This issue may be particularly important for RIs located in less-developed countries, where electricity tariffs do not reflect the actual cost (including the cost of pollution).

Additionally, the geography of the RI is unbalanced in favor of the most developed areas, where there are clear advantages in terms of preexisting facilities, opportunities to attract scientists and technical staff, proximity to universities, good transport connections, reliable utilities, and other elements. In these developed local contexts, the labor market is usually competitive even for unskilled labor, and the unemployment rate is relatively low. As a consequence, Del Bo, Fiorio, and Florio (2011) suggest using actual wages as the most sensible proxy of shadow wages in such regions.

In other words, the social value of labor would be (often but not always) the same in economic analysis as in financial analysis. However, if an RI is built in a lagging region, where the labor market is affected by structural imbalances, the salaries paid to unskilled personnel, particularly during the construction phase, may be higher than the regional shadow wage. Public procurement regulations are often strict about compliance with wage legislation. In such a case, an appropriate conversion factor would virtually decrease the personnel cost in the CBA of an RI, relative to the financial analysis. This is implicitly a benefit arising from employment created by the RI, ultimately reducing unemployment. The correction, however, does not usually apply to scientific personnel and other skilled employees for whom international mobility is high, and also does not apply, as mentioned, to unskilled personnel in developed contexts. For more on this topic, see the discussion in section 3.7 of chapter 3.

Hence, in general, one would not expect the role of such corrections to be particularly important, even if occasionally it should be duly considered, particularly for unskilled personnel of RI in lagging regions. For example, the ESFRI Roadmap 2016 contains projects located in some Central–Eastern European countries, such as the Extreme Light Infrastructure (ELI), a facility for experiments on high-energy light-matter interactions. ELI is currently being implemented on three sites in the Czech Republic, Hungary, and Romania.

In these countries, given the labor market conditions, both the actual and shadow wages are lower than in Western Europe (the ELI headquarters are located in Brussels). For example, shadow wages are 70 percent of the market wages for Hungary, 79 percent for Romania, and 83 percent for the Czech Republic, according to calculations by Del Bo, Fiorio, and Florio (2011). Other examples of ESFRI projects located in countries where the shadow wage correction may apply is the European Solar Telescope (EST) at a site in the Canary Islands (Spain); one of the facilities of the KM3 Neutrino Telescope 2.0 in Sicily (southern Italy); the Danubius Centre for Advanced Studies on River-Sea Systems (Romania); and the EU-Solaris RI for Concentrated Solar Power (Seville, Spain).

Finally, there may be social costs related to externalities such as traffic congestion, particularly (but not only) during the construction and decommissioning phases, and some emissions of pollutants, including radioactivity. For most of these externalities, there are well-established methods in CBA practice on how to value the externalities (for more information, see the "Further Reading" section at the end of this chapter). In some cases, however, there is not yet a specific study on estimating the externalities of some scientific projects. One example is the abovementioned pollution by fragments of satellites and other spacecraft waste orbiting the Earth.

In principle, one may need to estimate the extra costs for the existing and new space projects from the risk of impacts with such fragments. See, for example, the accident concerning Sentinel-1-A, an ESA Earth observation satellite, which was hit by a high-speed (albeit luckily very small) piece of debris that made it suddenly change its orbit and damaged its solar panel.[8] Many of these events occur, and frequently the ground-station operators need to watch out for the impact risks and move the satellites to a different orbit to avoid damage from collisions.

Finally, several RIs receive in-kind contributions in the form of donations of equipment or of work paid by third parties. Such costs, while they do not enter into the financial accounting of the RI management body, must be integrated to provide a more complete picture of the social costs.

Overall, these five types of correction (land, utilities, labor, pollution, and in-kind contributions) do not seem to suggest that a considerable difference would affect the RI costs in general in financial versus economic analysis, provided that, as mentioned, the first one is comprehensive, which is usually not the case. However, for some RI projects, the difference may be significant and worthy of careful analysis.

2.4 Financial Revenues

Weighed against the costs of a project are its social benefits, most of which do not generate cash inflows for the RI management body. Nevertheless, some RI projects may earn revenues from services included either in their core activities or in side activities. These operating revenues, not to be confused with other sources of funding, such as grants or transfers from the government budget, are defined as cash inflows directly paid by some project users for receiving certain services or goods. Revenues can vary significantly from one project to another in relation to the specific type of service delivered by the RI. In most (but not all) cases, such RI revenues are negligible or modest compared with costs. Hence, I suggest usually considering such revenues just as a way to decrease the total net costs and the funding requirements of the project.

Possible revenues of an RI can include a wide variety of items, such as the proceeds from the sale of certain materials or manufactured products, including experimental equipment; fees for consultancy services; royalties gained from patents; income from industrial research contracts and precommercial procurement contracts; research contracts involving transfer of ownership of a specific research output; fees to access laboratories and to use research equipment paid by external researchers and businesses; rent of space; equity return to investors if the RI managing body is a shareholder of a spin-off; access fees for individual students or collective agreements with some universities and institutes; revenues from organizations using the research outputs (e.g., on behalf of patients receiving innovative treatments, government agencies in various fields); and revenues from outreach activities to the broader public (e.g., bookshop sales, entrance tickets).

Generic transfers to an RI managing body, loans from banks or financial institutions acting as intermediaries of government agencies, regular or exceptional donations from such agencies, and donations from charitable entities, philanthropic organizations, or individuals cannot be considered as operating revenues because they are not payments for specific services. According to the CBA Guide of the European Commission (2014), public research contracts or contributions, granted through either competitive or noncompetitive arrangements, should be considered as operating revenues when these payments are made against a service directly rendered by the project's managing body to the funding entity. Grants or transfers should instead be accounted for as cash inflows for the verification of the project's financial sustainability. The concept of an RI's financial sustainability in this framework is discussed next.

2.5 Funding Model and Financial Sustainability

Having mentioned financial costs and revenues, I now briefly discuss the concept of financial sustainability of an RI. An investment project is financially sustainable when the financial inflows (including both operating revenues and any other sources of financing) are able to cover the financial outflows (including investment costs, operating costs, reimbursements and interests on loans, taxes, and other disbursements) year by year. Hence, if the cumulated net cash flow is negative for even one year, the project is not financially sustainable. In this case, the project promoter is expected to demonstrate the capacity to raise additional sources of financing to cover the costs in each year of the time horizon.

In the RI context, a number of factors influence the long-term sustainability, which is only partially related to a sustainable funding profile. Thus, financial sustainability should be considered together with other sustainability criteria. According to the EIROforum (2015), an association of eight major scientific organizations including CERN, ESA, European Molecular Biology Laboratory (EMBL), European Southern Observatory (ESO), European Synchrotron Radiation Facility (ESRF), EUROfusion, Institut Laue-Langevin (ILL), and European X-Ray Free-Electron Laser Facility (European XFEL),[9] the following five criteria are key to ensuring the long-term sustainability of an RI:

- Relevance to its scientific community
- Sustainable governance model and legal framework
- Attraction of a critical mass of scientific expertise
- Positive long-term socioeconomic impact
- A financially sustainable funding model

The necessary ongoing investments needed for optimal operations should be guaranteed to ensure that the RI can continuously carry out its cutting-edge research activities.

These EIROforum criteria combine very different aspects, some of which will be discussed later in this book. Here, my concern is exclusively related to the last point—the funding model, which is often the most urgent concern of RI managers. However, a funding model that ensures financial sustainability needs the other criteria to be met. Without the ability to build a reputation in the scientific community, to attract talent, to adopt the right agenda and governance, and to show some socioeconomic benefits, it will be difficult to attract adequate funds in the long term.

Having said this, the funding model and the financial planning are also important per se; one of the most critical issues in the management of an RI is that many scientific programs are multiyear endeavors, in some cases taking decades, while governments and donors tend to adopt funding decisions year by year. Just to mention an example, the US Department of Energy (DoE), which is responsible for the US National Laboratories and for the National Nuclear Security Administration and other programs, needs to submit a Congressional Budget Request every fiscal year (FY). This process is fraught with some uncertainty and fluctuation.

The latest DoE Request opens with the following sentence: "The Department of Energy (DOE) requests $28.0 billion for FY 2018, a reduction of $1.6 billion from the FY 2016 Enacted level of $29.6 billion" (DoE 2017b, 1). Out of the total, the specific component targeted to the Office of Science is $4.5 billion, which was $874 million below the 2016 approved budget figure. In fact, under such an overall cut, there is some reshuffling, with an increase for advanced scientific computing and yet a decrease for basic energy sciences, biological and environmental research, fusion research, high-energy physics, and nuclear physics, in variable proportions. To put these figures in perspective, the total budget of CERN, in 2016, in current terms was around $1.150 billion.[10] Thus, the total DoE budgetary proposed cut of 2018 is equivalent to the shutdown of a CERN as an order of magnitude.

In certain circumstances, the mismatch between multiyear research programs and yearly budgetary appropriations can be disruptive (the Superconducting Super Collider (SSC) demise by the US Congress is probably the most extreme case). From this perspective, it would be better, if possible, for RI managers to revert for funding either to own capital endowments, such as large US universities and some foundations, or to convince funding agencies to frame the disbursements as a multiyear plan.

Another critical issue is the combination of regular funding of the overall RI budget and funding of a specific program. There is no golden rule for assessing the optimal combination of these mechanisms, and each managing body should find out how to survive when funding mechanisms are designed to support expenditure programs that have shorter time horizons than scientific programs. See more on this in the "Further Reading" section.

2.6 Intertemporal Accounting

Costs and revenues of any RI occur over several years. How much money, spent in possibly distant periods of time, should be accounted for? The typical spending profile and cost distribution over time of various categories of the RI infrastructure often show spikes.[11] The total spending pattern shows a relatively large investment peak during construction, a quasi-flat spending period during operation, and then a new peak for decommissioning. For a major upgrade during operation, such as the High-Luminosity LHC, another investment peak would have been visible somewhere, perhaps at two-thirds of the project cycle. Such upgrades are often planned well in advance, in anticipation of new technologies that were not available at the time of the initial design but were available at the start-up of the project.

Figure 2.1 shows the project time profile of the LHC, which includes a major upgrade occurring thirty years after the start of construction and more than fifteen years after the first run of the collider. The distribution of costs during construction shows that civil engineering and technical hardware costs often represent the vast majority of spending during the investment

Table 2.2
CNAO estimated investment and operating costs (2002–2031)

	Total CNAO Cost Nondiscounted (a)	Total CNAO Cost Discounted (a)
Past investment cost, 2002–2013	159.3	188.8
Future investment cost (including research) 2014–2031	28.6	23.8
Operating costs (2011–2031)	315.8	248.5
TOTAL	511.7	465.9

Source: Pancotti et al. (2015), based on CNAO data. 2013 constant price. Total amount includes decommissioning costs. Investment costs include any research costs.
Note: (a) EUR million.

phase. Salaries, consumables, and utilities are the main expenditures during the operation period. However, it is difficult to generalize, and there may be RI projects with different time profiles.

The design and the R&D phases of an RI can be very long.[12] As opposed to what is often reported in the official budgetary figures and the press, costs occurring in one year cannot simply be added to costs occurring in another year—perhaps a distant one, given the long time horizon of the RI. Rather, there are two corrections: First, each nominal value should be deflated or inflated by an appropriate price index to avoid changes over time in the purchasing power of the numeraire not being accounted for—see, for example, the huge difference between the estimates of the Lunar Apollo Program, mentioned in the Introduction, when expressed in nominal or real terms, according to the year used as the accounting base. Second, each future cash flow should be discounted by an appropriate financial discount factor, typically taking the form $d_t = 1/(1 + FDR)^t$.

The total economic costs are computed in the same way, using a social discount rate instead of a financial discount rate, because the opportunity cost of capital for the society usually differs from the private cost. The effect of discounting past and future value can be seen in table 2.2, which shows the difference between not discounted and discounted total costs, respectively, in the case of the National Centre of Oncological Hadrontherapy (CNAO), EUR 511.7 and 465.9 million.

2.7   Financial Performance Indicators

The financial performance of an RI project indicates its integrated net costs over time, regardless of the sources of financing (loans, private equity, or grants). The financial return on an investment can be calculated using two simple performance indicators: the financial net present value (FNPV) and the financial internal rate of return (FIRR). The former is expressed in monetary terms and is the discounted sum of the net financial flows for the entire time horizon. The latter is defined as the financial discount rate (FDR), which produces FNPV = 0. The FDR reflects the opportunity cost of capital from the perspective of the RI funder and is used to discount financial flows to estimate an investment's profitability indicators. The financial discount is valued as the loss of income from an alternative investment with a similar risk profile, and it is estimated by considering the return on an appropriate portfolio of financial assets lost from the best alternative investment, such as the real return on government bonds or the long-term real interest rate on commercial loans. A project with a *positive financial performance* is associated with a positive FNPV, meaning that the total discounted inflows exceed the total discounted outflows. Under a certain technical condition,[13] an FIRR higher than the reference FDR provides the same information. Conversely, a project with a *negative financial performance* is associated with a negative FNPV (and usually with an FIRR lower than the FDR), a situation typical of RIs.

Usually, an RI, particularly for basic research, has a negative financial performance. This is not surprising, and it is exactly why government funding is needed. In the European Union (EU), the abovementioned financial indicators are used in the context of EU Cohesion Policy to set the appropriate volume of public support to be committed to social welfare-improving projects (Florio, Morretta, and Willak, 2018). For example, the European Commission allows cofinancing through grants only if the proposed major project is not financially profitable; namely, the FNPV is negative and the FIRR is lower than the reference FDR for the analysis.[14] Table 2.3 presents, as an illustration of the concept, the financial performance indicators of a sample of RIs, cofinanced by the European Commission, during 2007–2013. These figures should be interpreted as indicators of the funding gap to be covered by government or third-party grants or loans.

Not all research projects are loss-makers. In the United States and elsewhere, scientific projects with potential future profitability are regularly assessed by venture capitalists. In some cases, RIs are involved in a transitioning process from the public to the private sector or the venture capital arena. Clarifying the wide difference between social CBA and the role of project analysis for venture capital investment may be useful. To screen investment opportunities, venture capitalists use a broad range of accounting and nonaccounting information from sources such as business proposals, contracts with other venture capitalists, interviews with entrepreneurs, interviews with potential investors,

Table 2.3
Examples of financial performance indicators of a sample of R&D projects (2007–2013)

Country	No. of Projects	Time Horizon (a)	FIRR (b)	FNPV (c)	Investment Cost (c)
Czech Republic	1	15	−30.0	−107,100	104,200
France	2	26	−0.2	−77,388	88,440
Hungary	2	21	0.0	−309,700	229,350
Italy	1	15	−	−39,142	83,000
Lithuania	2	15	−12.4	−65,805	76,818
Poland	8	16	−8.4	−68,202	79,018
Slovenia	1	15	0.0	−76,691	111,100
Spain	2	25	−8.8	−230,250	225,864
United Kingdom	5	21	−2.4	−9,762	65,608
Total	24	19	−6.6	−90,985	104,143

Source: Project applications for cofunding by the European Commission, author's elaboration on data provided by the Directorate General for Regional and Urban Policy. Sector classification taken and reelaborated from the Commission Implementing Regulation (EU) No. 215/2014 of March 7, 2014, at http://eur-lex.europa.eu/legal-content/EN/TXT/?uri=CELEX:32014R0 215, see Florio, Morretta, and Willak (2018).

Notes: Data do not include R&D investment in firms. (a) Average number of years; (b) average 2007–2013; (c) average EUR thousands; (d) average value across countries; (e) average value across countries.

and statistical services.[15] The collection and analysis of this information, the due diligence process, is needed to gain a thorough understanding of all business aspects, such as the viability of the product or service;[16] potential for sustained growth of the company; quality of the management team for efficient control and operation of the project; a balance between risk and expected profits; and investment criteria related to strategies in the context of financial markets. Additionally, the screening involves a variety of valuation techniques to determine the profitability of venture capitalists' investments, ranging from standard valuation methods based on discounted cash flow analysis[17] or the multiple earnings and the value of a company's assets, to the most innovative approaches based on option pricing theory (e.g. Seppä and Laamenen, 2001).[18]

However, whatever approach is adopted, the objective of venture capitalists, in principle, is to seek out value for their shareholders. The objective of social CBA is instead to seek out value for society. Hence, the role of the financial analysis of projects is quite different in the two cases, even if some of the techniques (such as the discounted cash flow method) may look similar. Some parameters may be different in any case; for example, the appropriate financial interest rate and the risk premium for a business analyst looking at a research project on behalf of a venture capitalist may be different from the corresponding variables in the perspective of CBA. An example is the role of taxes: For a venture capitalist, the relevant return on equity is net-of-tax, while the return gross-of-tax is more informative from the government perspective. I shall come back to this issue in chapter 5, when discussing the impact of RIs on firms and how to consider the profits arising from learning spillovers and other externalities typical of capital-intensive scientific projects.

2.8 Case Study: How Much Does a Collider Cost?

To show the complexity, but also the feasibility, of a comprehensive intertemporal cost accounting of large RIs, this section presents the assessment of the LHC's costs, heavily drawing on the work of Florio, Forte, and Sirtori (2016). As the LHC will often be cited as a case study in the rest of the book, box 2.1 presents a summary of its main features.

The LHC case is particularly challenging for CBA for several reasons. First, it is a very large infrastructure in terms of several variables: number of people involved, physical size, and total cost. Also, it has an especially complicated structure due to the intricate interplay of accelerator and detectors in the experimental collaborations between the host laboratory (CERN) and its participating institutions, with a large number of countries and different kinds of organizations involved (universities, research laboratories, and national academies). This poses difficulties in cost apportionment and aggregation when attempting to estimate the total economic costs.

Second, the life span (both past and future) of the facility is quite long: it requires both retrospective evaluation and appraisal techniques because capital costs for the LHC began to be incurred in 1993 and the generation of both operating costs and benefits is expected to continue for a few years into the future. Figure 2.1 shows the time pattern of costs, both in discounted (SDR = 0.03) and undiscounted values. To interpret the figure, two initial remarks are important: the total LHC cost includes the capital and personnel expenditures of the main experimental Collaborations except scientific personnel (see box 2.1), which do not fully appear in the CERN account, given their autonomous status; and the discounted and undiscounted values overlap exactly only at $t_0 = 2013$, the base year. This has the consequence that past expenditure (1993–2012) is greater than the undiscounted one (because one EUR in the past year $tpast = 1, 2, \ldots$ is discounted by $1/(1 + SDR)^{-tpast}$, while in future years (2014–2025): tfut = 1, 2 ... ; that is, the expenditure is discounted by $1/(1 + SDR)^{tfut}$, and consequently, is lower than the undiscounted one.

Box 2.1
The LHC in a Nutshell

The LHC is currently the largest particle accelerator in the world. A *particle accelerator* is a device in which particles (protons and atomic nuclei, in the case of the LHC) are accelerated and made to collide with a target or with each other, with the goal of studying the structure of matter. Particles are accelerated by subjecting them to electric fields and are collimated into focused beams by magnetic fields. Particle beams travel in a pipe in which a vacuum has been established and are made to collide in experimental areas in which the debris from the collisions is accurately measured by devices called *detectors,* which allow an accurate reconstruction of what has happened during the collision.

The main goal of the LHC is to study the precise nature of the forces that govern fundamental interactions at the shortest distances that are currently accessible, which requires the colliding particles to hit each other at the highest possible energy.

In operation since 2009, the LHC reached its first goal with the discovery, in 2012, of the *Higgs boson*—at the time, the only major missing piece of information in the existing theory of fundamental interactions. Current research involves both investigating the properties of the newly discovered Higgs boson and searches for deviations from the current theory, which is believed to be incomplete, and is foreseen to continue for at least about another decade.

The LHC was built by CERN, and the construction work lasted from 1993 to 2008. The LHC is the largest element in a chain of machines that accelerate particles to increasingly higher energies: the CERN accelerator complex. The accelerator complex is developed, maintained, and operated by CERN. This facility is exploited by the experimental collaborations that perform experiments in the areas where collisions occur. Each experiment is based on a detector, designed, built, and operated by a collaboration that involves both the participation of CERN and of scientists from a number of institutions (universities and research laboratories) from several countries. Four main experimental Collaborations exploit LHC collisions; the two largest ones, ATLAS and Compact Muon Solenoid (CMS) both involve several thousand scientists from several hundred institutions in almost fifty countries. The corresponding detectors are roughly the size of a ten-story building. When observing particle collisions, the four experiments produce about 1 GB of data per second, which are either analyzed inside by LHC Collaborations or sent to a number of other computer centers around the world, connected through the worldwide LHC computer grid.

Source: Adapted from Florio, Forte, and Sirtori (2016).

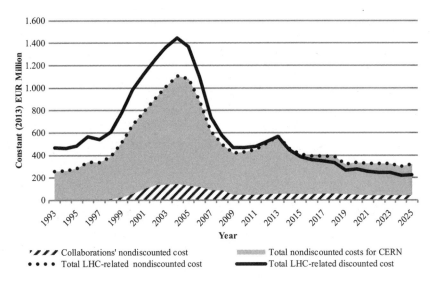

Figure 2.1

Time distribution of LHC costs (discounted and nondiscounted). *Source*: Author elaboration on CERN data, from Florio, Forte, and Sirtori (2016).

To be more specific, a consolidated LHC cost accounting should include past and future capital and operational expenditures, both of CERN and experimental collaborations, for building, upgrading, and operating the machine and conducting experiments, including in-kind contributions. Three categories of costs have been considered: construction capital costs, upgrade capital costs,[19] and operating costs. Florio, Forte, and Sirtori (2016) have estimated CERN costs from the beginning of its activity (1993) up to 2025, while for different collaborations (having different reporting systems), they have reconstructed costs using their financial reports, supplemented by some forecasts for years after 2013. As mentioned, integrated past flows have been capitalized to $t_0 = 2013$ with a 0.03 SDR (in line with European Commission 2014). Future costs have been discounted to 2013 EUR values by a 0.03 social discount rate as well; see the abovementioned formulas.

Budgetary allocations from CERN to LHC have been recovered from data communicated by the CERN Resource Planning Department to the research team, drawing on the CERN Expenditure Tracking (CET) system. These original data cover all CERN program and subprogram expenditure, in Swiss francs, from January 1, 1993, to December 31, 2013.

The CERN programs include accelerators, administration, central expenses, infrastructure, outreach, pension funds, research, and services. The cost for each program is disaggregated in various subprograms; for instance, under accelerators, there are nineteen subprograms, such as the Super Proton Syn-chrotron (SPS) Complex, LHC, LEP, and general R&D. In turn, each of these items shows expenditures on materials, personnel, and other financial costs, broken into recurrent and nonrecurrent expenditures. Bank charges and inter-ests are excluded, in line with European Commission (2014), which suggests performing the CBA without considering the sources of funding. Expenditures that can be attributed to the LHC rather than other CERN activities were identified. To do so, in many cases, it was necessary to estimate an apportion-ment share to the LHC project of the expenditure for each item in the CERN programs and subprograms.

To double-check these accounting data, the team interviewed the CERN staff in various departments.[20] For example, 80 percent of the outreach costs were attributed to LHC, as opposed to only 50 percent of the generic R&D of the Accelerators program. Some overheads are not recorded in internal reporting, as these are related to specific accelerators or programs. The authors have attributed 10 percent of CERN administrative costs, central expenses, and administrative and technical personnel costs to the LHC.[21] Scientific per-sonnel costs of CERN were identified from the reports of CERN Personnel Statistics, which are released each year. Every year, the share of this part of the personnel was between 19 percent in 1993 and 32 percent in 2013. This share of costs is assumed to balance with the contribution of CERN scientists to the direct value of the LHC publications, similar to what is assumed for non-CERN scientists in the collaborations. I discuss this assumption in detail in the next chapter, where the valuation of scientific output (publications and other forms) is presented. See the previous chapter for an initial explanation.

To these, direct CERN costs and past in-kind contributions from member-and nonmember-states have been added,[22] while forecasts of in-kind contribu-tions in future were not available. Many RI projects may revert to this form of in-kind support, such as the donation of scientific equipment, or other legal arrangements. In fact, this is a form of funding that usually does not appear in the budgetary cost of the project on the managing body side. Consolidated accounts of all the partners are often not available, but in terms of economic costs, comprehensive reporting would be needed. The 2014–2025 forecast of CERN expenditures has been communicated to the team by CERN staff.[23]

As previously mentioned, the cost implications of new investment projects, such as the LHC High-Luminosity Project and the LHC Upgrade Phase 2, were not included, as they will run mostly after the time horizon considered. This is an example of the fact that major upgrades of an RI in most cases can be considered as new projects (with the past expenditures categorized as sunk costs).

For the expenditures of the experimental collaborations using the LHC, the analysis focused on the four main experiments (ATLAS, see box 2.2, CMS, ALICE, and LHCb), as the remaining ones (LHCf, FELIX, FP420, HV-QF, MOEDAL, TOTEM) were comparatively quite small in terms of capital and operating costs; hence the social benefits of these experiments were also excluded from the computation of the net present value (NPV). Focusing on the four main experiments, the data sources have been the resource coordinators of each Collaboration.[24] Forecasts of future expenditures of the Collaborations have been based on the same sources. When only cumulative data at a certain year were available, the yearly distribution has been interpolated linearly. In the same way, some missing yearly data for the LHCb Collaboration have been interpolated by the team. To avoid double-counting, the CERN budgetary contributions to the Collaborations have been excluded from their total expenditures. Similar to what is assumed for CERN, the scientific personnel cost of the collaborations (salaries paid by their respective institutes and universities) has been taken as balancing the marginal cost valuation of the scientific publications attributed to each experiment and excluded from the grand total of cost. This is, of course, an important simplification because a major Collaboration such as ATLAS has a very complex personnel structure (see box 2.2).

Estimating the cost of the scientific work of thousands of scientists, belonging to so many institutions, would be possible in principle, but it would require a myriad of data on salaries and time allocated to the collaboration in the various countries and institutions. The balancing of the value of the publication output and the scientific personnel cost is a practical shortcut, which will be discussed in detail in the next chapter.

The overall trend of cumulated LHC-related CERN and Collaboration expenditures is shown in figure 2.1. The information up to 2013 was considered as given (within the limit of measurement error), while the data from 2014–2025 were treated as stochastic, with normal distribution and mean equal to EUR 1.97 billion and a standard deviation compatible with mean ±50 percent as asymptotic values. This range is based on in-depth interviews with experts at CERN and analysis of the most optimistic and pessimistic future cost scenarios.[25] Chapter 10 explains the probabilistic approach, more in general.

In summary, after including the value of in-kind contributions, Florio, Forte, and Sirtori (2016) reconstructed the time distribution of LHC costs to 2025, while CERN costs unrelated to LHC and costs for future upgrades have been excluded, as their benefits will occur beyond the time horizon. Scientific personnel costs are not included. The final estimate for the expected mean value

Box 2.2
The ATLAS Collaboration

ATLAS comprises around 3,000 scientific authors from about 182 institutions around the world, representing thirty-eight countries from all the populated continents. It is one of the largest collaborative efforts ever attempted in science. Around 1,200 doctoral students are involved in detector development and data collection and analysis. The collaboration depends on the efforts of countless engineers, technicians, and administrative staff.

ATLAS elects its leadership and has an organizational structure that allows teams to self-manage and members to be directly involved in decision-making processes. Scientists usually work in small groups, choosing the research areas and data that interest them most. Any output from the collaboration is shared by all members and is subject to rigorous review and fact-checking processes before results are made public.

Source: The Collaboration (https://atlas.cern/discover/collaboration), accessed on April 17, 2018.

---

of the total cost of the LHC over thirty-three years (1993–2025) is EUR 13.5 billion, net of scientific personnel cost and including capital and operating expenditures. Here and elsewhere in the case study sections, the mean value always refers to the outcome of the Monte Carlo process after 10,000 draws (see the final chapter of this book for details). No revenues were deducted from the present values of costs because they are negligible, given CERN's policy to offer free services to visitors, firms, users of its software, and others: an approach that is a feature of government-funded basic research.

How large is the LHC total cost, including the experiments? First of all, it is much higher than the amount that is usually reported in the press and in several publications about the LHC's costs. For example, a CERN (2017b) brochure for the general public, *FAQ: LHC The Guide*,[26] reported a total cost of CHF 4.332 billion, including the LHC machine and areas, CERN share to detectors and detector areas, and to LHC computing. This figure may be true in budgetary terms but is incomplete from a CBA perspective. This is because only CERN's capital expenditure costs are reported, not the operating expenditures. Moreover, the costs of equipment needed by the experiments are not added to the reported total cost; hence, most personnel costs and other variable costs are not considered, including a substantial electricity bill. There is also an inconsistency when the costs of something that happened in 1993 are added to the costs of something which is forecast for 2025.

The media sometimes make their own kind of cost estimates. For example, Knapp (2012), in an article in *Forbes* magazine, did some calculations and concluded that the total LHC cost was $13.25 billion. Nevertheless, he was optimistic on the benefits: "When you combine the potential for advancing computer technology, medical imaging, and scientific breakthroughs, $13.25 billion seems like a bargain. Especially when you consider the fact that on top of all of that, the Large Hadron Collider and its associated experiments are bringing us that much closer to understanding the mysteries of the universe."

To see if this was actually a bargain, readers may need to be patient and wait until the end of the book (section 10.8 in chapter 10).

2.9   Further Reading

ESFRI (2017) deals with the long-term sustainability of the RI and recommends: "Research Infrastructures must develop, right from the start of the planning phase and prior to the roadmapping exercise, a comprehensive business plan covering all stages of their lifecycle including upgrading and decommissioning" (p. 55). This is not always the case, and RIs often are launched without such multiyear, long-term funding commitments. EIROforum (2015, 4) is clear about the fact:

…Considering therefore that all RIs need adequate resources to stay at the frontline—both human and capital resources—these facilities can only be successful if they operate with a favorable long-term perspective, i.e. one that enables long-term financial planning with appropriate guarantees by the funding countries or other shareholders. This implies the existence of regulations ensuring a long-term commitment from the funders. Crucially, the long-term financial planning must cover operations costs, including the necessary maintenance and instrumentation upgrade costs.

See also European Commission (2016a). Funding issues are discussed by Maessen, Krupavičius, and Migueis (2016).

A detailed presentation of a project's financial versus economic analysis (the latter is synonymous with *CBA* in this context) is given by the European Commission (2014). Florio, Morretta, and Willak (2018) discuss the history of this guide (five editions) and of infrastructure project evaluation by the European Commission. The guide includes several case studies and examples in sectors such as transport, environment, energy, telecommunications, and others. The European Investment Bank (EIB, 2013), the world's most important multilateral borrower and lender by lending volume and capital, presents its approach, which is largely consistent with European Commission (2014). A number of sectors are considered, including education and research, power

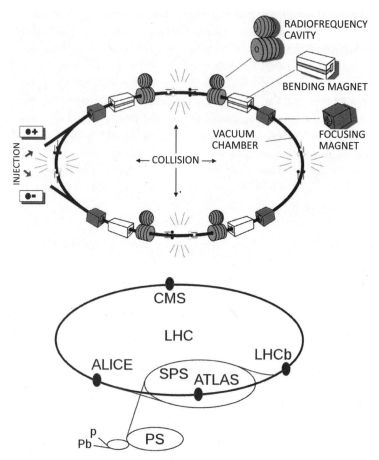

Figure 2.2

The LHC (near Geneva, Switzerland, and crossing the France border) is located in a 27-kilometer tunnel, where two high-energy proton beams are accelerated at close to the speed of light while superconductive magnets keep them on track. Two beams travel in opposite directions in separate ultravacuum tubes and made to collide. The effects of the collisions are observed by detectors at four locations around the accelerator ring. The top diagram shows the main components of the LHC. The bottom image shows the main detectors (ALICE, ATLAS, CMS, and LHCb) and the upstream accelerator system: SPS (Super Proton Synchrotron), PS (Proton Synchrotron); p is the source of protons, and Pb is the source of lead ions. *Source*: Courtesy of ©CERN and Professor Lucio Rossi.[27]

generation, renewable generation, health, private-sector research development and innovation, research on software, RIs, and telecommunications.

While the sector coverage is uneven, reading both the European Commission (2014) and EIB (2013) reports offers a good starting point to understand the difference between financial and economic analysis in practice. See also Campbell and Brown (2003) and Sell (1991). On specific points discussed in this chapter, see Jones, Moura, and Domingos (2014), on residual value and time horizons; Von der Fehr et al. (2013), on the project borders (with an example in the energy infrastructure); and Keef and Roush (2001), with a survey of discounted cash flow methods. The *Manual for the Preparation of the Industrial Feasibility Studies* by the United Nations Industrial Development ment Organization (Behrens and Hawranek 1991) is still a good introduction to project financial appraisal (beyond industry), particularly in chapter 10.

More specifically about RI costs, some definitions are provided by ESFRI (2016a), giving guidelines for the applications of RI projects to be included in the 2018 road map[28] which, similar to the previous editions, sets a de facto strategic priority list for European RIs, some of them located outside Europe and considered Global Research Infrastructures (GRIs).

In this context, the document provides several accounting definitions adapted to the RIs:

• Capital value (CV) is defined as the total asset value, or equivalently, the value of the investment, assuming that the depreciation is compensated by a constant upgrade; hence, the CV is stable and equivalent to the replacement value (RV). For distributed RIs, the CV is the sum of the asset values of the Central Hub and the Nodes with an apportionment formula: "if a CV-RV is defined for a given node (national node, institutional node) and if the Node is contributing X% to the distributed RI, then this X% of the node's CV-RV is added to the total CV-RV of the distributed RI" (ESFRI 2016a, 9). Surveys, data banks, and sample collections should be considered assets in a similar way.

• The RI design costs are defined as all the in-kind and cash expenditures needed to go from the conceptual design to a feasibility study and proposal. These expenditures include specific budgets obtained by funding agencies, as well as the labor costs of scientists and technical and managerial staff involved in prototype design and actual development.

• The preparation costs include past and future expenditures for the preparation and interim phase of an RI. These also include in-kind and cash third-party contributions.

• Implementation costs of the RI include hiring personnel; acquisition of land, buildings, and other assets on site; construction, legal, and coordination costs of the RI with the communities of scientific users; and expenditures for data management infrastructure, commissioning, and preoperation costs.

• Finally, the average annual operation costs should comprehensively cover the annual costs of running the RI, operating users' access and delivering scientific services as described by the project. They include all RI costs, such as personnel, power, rents/mortgages, taxes, maintenance and continuous upgrading, user support, and in-house scientific programme.

The ESFRI data reported in table 2.1 should reflect such definitions, which are based on a number of shortcuts and may differ from the framework of cost analysis described in this chapter or by the European Commission (2014). First, the capital value is not a discounted cash flow (DCF) method, if investment costs occurring in different years are summed. The assumption that the upgrading cost substitutes for depreciation is less precise than the residual value forecast. The replacement value concept is also inconsistent with the DCF method. The same applies to the remaining items, particularly to the abovementioned "average annual operation cost," which may be informative, but with the obvious limitations of an average over a relatively long time period.

Finally, I would cite a form of cost-effectiveness (CE) analysis performed by Del Bo (2016). She used data from Riportal[29] on costs and number of users of the RI, defined as the sum of internal and external users, excluding permanent scientific/engineering staff operating the RI.

The average cumulative total cost *per user* is EUR 0.087 million, with high variability across fields, with the highest CE ratios in engineering and energy, while the lowest values are in information and communication technologies, mathematics, and humanities and behavioral sciences. In a simple econometric analysis, the CE ratio is the dependent variable, while the explanatory variables are duration of the RI project and the size of its staff, after controlling for the country where the RI is located and its scientific field. My own interpretation of these results is that, while the analysis is obviously very preliminary, it points to economies of scale, which are perhaps the drivers underlying the trend for bigger RIs, open to wide international collaborations, and having a "natural monopoly" effect.

A very recent report (The Royal Society, 2018) on 135 out of the estimated 400 RIs in the United Kingdom shows some interesting facts: 69 percent were international in scope; 32 percent of the staff came from overseas; 93 percent involved business in their research; government funding was 47 percent, with the remainder being provided by universities, business, charities, and the European Union; in fact, 84 percent received EU funds (this was before Brexit) which may imply a loss of EU funding to UK facilities; models of ownership (which also have implications for funding) came to 46 percent with a specific RI legal entity; 30 percent national legal entities; 17 percent intergovernmental organizations; and 7 percent collaborative projects (such as H2020 consortia). The average annual operating cost of this sample was between 23.1 million pounds (physical sciences and engineering) and 31.4 million pounds (ecosystems and earth sciences), with intermediate figures for life sciences and energy. The average yearly operating cost for RI in the social sciences is much lower (3.3 million pounds). Over 500 RIs were surveyed by Stahlecker and Kroll (2013), with data on their costs and funding. On life science RI funding needs, see ERA Instruments particularly.[30]

On unit costs for RIs providing services to third parties, see The Norwegian Association of Higher Education Institutions (2014), which provides a full cost model for computing the cost per hour of the use of a research infrastructure resource (RIR). For an application to digital infrastructures, see Uninett Sigma2 (2016).

3 Benefits to Scientists: Producing and Using Knowledge Output

3.1 Overview

Research infrastructures (RIs) are enterprises producing a good: namely, scientific knowledge. They do not displace other modes of organizing science, such as research carried out by small teams, or even by individuals. Large-scale RIs, however, are redefining the frontier of the social, economic, and technological arrangements of science in several fields. In this chapter, I first discuss how RIs innovate the organization of knowledge production compared to other institutions; then I introduce the empirical approach to the evaluation of the direct benefits to scientific communities of the immediate output: publications (and similar products).

From a historical perspective, universities, scientific societies, royal academies, and some cultural and even religious organizations have evolved as core institutions, specializing in knowledge creation. Moreover, economic organizations have contributed to knowledge and innovation for centuries as a side-product of their development. Since the Industrial Revolution, corporate research and development (R&D) has been instrumental in sustaining innovations, but often only after basic research supported by governments created new market opportunities. In her popular book, Mazzucato (2015) presents several examples of the interplay between government-funded research and corporate R&D.

According to Foray (2004), most knowledge is created in two forms: discovery and invention. The former arises from the revelation of facts existing in nature, and the latter from research aiming at improving production activities. Universities have been traditionally considered as the core locus of scientific knowledge creation in the form of cumulative discovery work, building and transmitting scientific paradigms. Economic organizations, instead, usually specialize in invention activities, particularly since the Industrial Revolution. However, these two loci of knowledge creation are not entirely distant.

Since the nineteenth century, universities have evolved in different organizational models. Kerr (1963) contrasts the elitist and the German models
of university. The elitist model was envisaged by the "eminent Victorian,"[1]
Cardinal John Henry Newman, according to whom (Kerr, 1963, 2) a university is "the high protecting power of all knowledge and science, of fact and
principle, of inquiry and discovery, of experiment and speculation; it maps
out the territory of intellect. ... Knowledge is capable to be its own end. Such
is the constitution of the human mind, that any kind of knowledge, if it really
be such, is its own reward."

Newman's essay, "Idea of a University," was published in 1854, and his
model was Oxford. On the other hand, according to Kerr (1963), the German
university emphasized the link between higher education and society: knowledge had to be useful to a progressive economy. This, however, was not the
end of the story. According to Kerr, the contemporary evolution of university
is a "multiversity," a large-scale and inconsistent array of various communities
(of scientists and humanists, of teachers and students, of administrators and
staff, etc.) under the umbrella of a common academic brand. The core assets
are reputation, social networks, and influence, and the scientific work is part
of a broader mission.

As has been mentioned, the other place where knowledge is created in
our economies, mainly in the form of invention, is the firm. Contemporary
growth greatly relies upon the ability of economic organizations to create new
products and processes, through either formal R&D or informal processes of
experimentation, a topic explored in detail by the flourishing field of economics of innovation, but that goes back to Karl Marx and Joseph Schumpeter's
analysis of capitalism.

The state supports both universities and firms in their role of knowledge
creators. Therefore, why has RI, which is so different from both universities
and firms, also emerged in the last decades with the support of governments,
despite the waning of military implications? The answer may be found in a specific demand by scientists, initially in physics and astronomy and then in several
other domains, to overcome symmetric weaknesses belonging to universities
and firms. The former cannot sustain the burden of large-scale projects that
ultimately would destroy the complex balance of power among the various
communities hosted by the academic brand. Under the multiversity model, while
some projects may be a flagship for the scientific reputation of an academic
organization (and achieve Nobel prizes), they cannot become so dominant in
terms of capital and personnel that they crowd out all other projects.

A contemporary university can be seen as a manager of a diversified portfolio of intellectual projects, with long-term return objectives; hence, it must

be risk averse and invest in a plurality of fields. For instance, the University of California manages a large RI, such as the Lawrence Berkeley National Laboratory (LBNL) on behalf of the Office of Science, US Department of Energy (DoE), which funds most of its budget. However, more than 3,200 LBNL scientists, engineers, and support staff constitute a separate organization with respect to the university, with different arrangements.[2] In fact, for most universities, the internal managerial practice and governance of a large RI, such as the LBNL, would be difficult to accommodate in their own budgets. The universities are indeed participating in different ways with such organizations, but with special arrangements.

Firms (usually) cannot create large-scale RIs for a different reason: they focus on a portfolio of business investments and expect a financial return from it. The RI open model of science, its diluted ownership, and ultimately the bottom-up decision process by the international scientific community is incompatible with the firms' profit objectives, except perhaps for some government-owned enterprises in Europe with long-term R&D missions (such as EDF, the former electricity supply monopolist in France, with 2,000 R&D staff), or in other special cases (such AT&T[3] Bell Laboratories before the liberalization of the telecommunications market in the United States). Private investors may support charities out of the profits of firms, and in turn, charities may support research, including RIs, but this is mainly a funding arrangement.

Scientists have invented RIs as new mechanisms of knowledge production, different from the laboratories of both academic and economic organizations, simply because the former and the latter often did not provide the right environment and scale of operations for the ambitious projects designed by the research communities. While some leading personalities, scientific entrepreneurs, may have influenced the RI project trajectories, and some political circumstances may have been favorable to their foundation and continuation, RIs are intrinsically bottom-up collective projects grounded in the consensus of the scientific community. As the scientists are the promoters, the designers, the users, and ultimately the evaluators of the RI, it seems convenient to start the analysis of the social benefits of the RIs, focusing on this specific group.

For scientists, particularly academics, the main benefit of being part of an RI project is the opportunity to access and process new experimental or observational data and methods that are not available in other ways. New data contribute to the creation of knowledge, and ultimately to the production of scientific output released in the public domain in some form. The peculiarity of this process is that the demand for new knowledge (and hence the implicit willingness to pay for it) are driven by scientists who often are simultaneously users and producers of the knowledge output. This fact implies that when scientists spend time on a research project, they have an opportunity cost from not working on an alternative project. If this opportunity cost is assumed to equal the average scientist's compensation, then a reasonable proxy of the value of "statistical" scientific output is its marginal production cost, expressed in terms of the value of the scientist's time. Hence, after all, there is some merit in Newman's view that "knowledge, if it really be such, is its own reward," even if in a less idealistic perspective. In a precise, quantitatively measurable sense, scientific work *on average* pays for itself.

In this chapter, I elaborate on this idea. While some of the knowledge created in an RI is tacit, such as tricks that scientists use in the laboratory or in data processing (Foray, 2004), with regard to most professional researchers, communication with peers in the scientific community is mostly a formalized process. While new information generated at the RI is initially stored in computer memory[4] or in other technology, and obviously in the brains of the scientists as well, then it produces a stream of specialized literature. Scientific results based on the information generated by research and its subsequent elaboration must eventually be embodied in some sort of communication vehicle, such as talks supported by slides, technical reports, conference proceedings, preprints or working papers, articles in peer-reviewed scientific journals, and research monographs. RIs are directly and indirectly data factories supporting scientists' publications and other research output.

As suggested by earlier studies[5] and, more recently, by some cost-benefit analysis (CBA) guidelines (Clarke et al., 2013; European Commission, 2014), one simple, empirical measure of research output is given (albeit very imperfectly) by counting publications (including preprints, conference abstracts, etc.) and other productivity indicators. I use the generic term *knowledge output* and *publication* interchangeably to refer to these products. Their influence on the scientific community is then usually estimated by using the number of citations as a proxy. The statistical information about them is increasingly available for scientometric analysis, and the decay of publications' number over time is an empirical measure of the obsolescence of a scientific project, similar to the obsolescence of patents.

However, this is not enough from a CBA perspective. The key questions are: How can knowledge output be valued as a social benefit? What is the marginal social value of a scientific paper? In this chapter, I consider the RI simply as an organization that produces a flow of knowledge output over time, and I suggest how to estimate this value in monetary terms using the concept of the opportunity cost of scientists' time. It turns out that the preferred valuation approach is looking at the marginal cost of publications, augmented by the economic value of their measurable impact in the scientific literature (i.e., influence).

This approach has two consequences. First, the scientific staff cost of the RI, including the users' salaries, on average exactly balances with the monetary value of the publications. Benefits and costs cancel out because benefits are conservatively estimated with costs. Second, the proximate net benefit of the knowledge output production (in a narrow meaning) is only the value of its influence in the scientific community outside the RI scientific users. This idea was already presented very briefly in the previous chapter, but it is elaborated on here in more detail.

An important consequence of valuing the scientific output by its marginal cost, which is mainly the labor cost of researchers, is that to a certain extent, scientific work pays for itself. This may seem surprising, but it is not a unique situation.

A similar case is that of a self-employed subsistence farmer where the output's marginal benefit, which is obviously the unit value of produced food, net of any other costs (e.g., fertilizers, etc.), is exactly the value of the marginal labor input. The marginal social value of food and of labor must balance in this case (even if there is no market price for the food). Another example is Facebook (and other social media): Producers of content are the same users, and the benefits and costs of the content production would cancel out if the implicit willingness to pay for being part of Facebook is equal to the value of the time spent in producing and using the site's content. There are several other examples of such coproduction mechanisms by users of certain services (Ostrom, 1996).

It transpires that because the costs and benefits of producing knowledge output, *on average*, mutually sum to zero, the net benefit of publications (excluding the ultimate effects of their content) is estimated conservatively as their net impact or influence on the scientific community outside the RI personnel. This repays a small fraction of the overall RI costs because the scientific community is small (differently from Facebook and similar platforms with hundreds of millions of users). "Publish or perish" may be the rule for academics, but the societal benefits of costly investment in science should mainly lie elsewhere.

The structure of the chapter is as follows: after this overview, section 3.2 discusses the difference between the broader social value of knowledge and the narrower benefits to scientists of producing and using knowledge outputs, the key concepts for the empirical strategy. Next, section 3.3 explains why the marginal cost of production is the preferred approach to valuing the benefits of publications and other products. Section 3.4 mentions some scientometric evidence across disciplines; section 3.5 extends the valuation of benefits to citations and research data; section 3.6 is a step-by-step tutorial, and section 3.7 concludes with a case study.

3.2 Demand and Supply of Knowledge Output

In the context of the evaluation of the socioeconomic impact of RIs, it is important to avoid a possible confusion between the social value of *knowledge output* (i.e., publications and others) and the social value of *knowledge per se*, embodied in such publications. The former is usually predictable, while the latter is often immeasurable. For example, the social value of producing and selling a book or a specific publishing project is unrelated, or only weakly related, to the social value of the book's cultural content and its wider effects. The written communication of ideas in any form (e.g., in a book) influences society through their understanding and further elaboration by readers; however, little is known about this cultural process, either in the humanities or in science. A paper by Peter Higgs, introducing the theory of a massive boson in a short paragraph, was published in 1964, about the same time as other papers, written by other physicists, which are now acknowledged as leading to a similar theoretical prediction. It took nearly fifty years to confirm this intuition at the Large Hadron Collider (LHC) experimentally. Nobody currently knows if and when the theoretical hypothesis about the Higgs boson, made decades ago, its recent experimental discovery, and further precision measurements in the future at the High-Luminosity LHC will lead to any practical applications, or even to changes in theories of fundamental interactions. Hence, we currently ignore the wider social value of the content of the publications announcing these achievements.

In other cases, we know ex post the applications of some discoveries. More than a century after groundbreaking, celebrated articles by Albert Einstein on special relativity in 1905 and general relativity in 1916, practical applications of these theories are now widespread. In any global positioning system (GPS) device, which has many direct and indirect uses, there are relativistic corrections without which the system would not work[6]. However, this ex post information does not help to evaluate the social impact of a specific paper by Einstein. Tracking a causal chain from the paper concepts to their practical application after 100 years seems difficult because much more than Einstein's theory lies behind the GPS.

It may seem prudent to suggest that *on average*, there is a chance that any new knowledge embodied in a publication will have an economic impact someday in the future. This seems a reasonable assumption. One also must admit, however, that there is a chance that such impact may never materialize. This chapter focuses only on the narrow value of a stream of professional publications or other output to the scientific community, before anything can be said on their broader social impact (if any). What is this narrow value?

With regard to the scientists using RI data, the direct benefit of publications is related to their impact or measurable influence (I will use the terms interchangeably) on the scientific community. Reputation and career depend to a great extent upon such influence. However, this benefit refers to a specific group, and it is only a small part of the overall balance of RIs' social impact. This is unsurprising because, first, a specific scientific community in any field or subfield is small relative to other social groups. Second, as previously mentioned, the ex ante value of knowledge to society per se, embodied in publications, is unknown.

While one can measure the economic effects of a publication on the career of a scientist in principle, this is a second-round effect that is difficult to estimate in practice, except for early career researchers (ECRs). To them, having been able to publish a paper in a blue-ribbon journal may make a remarkable difference. For established scientists, the effect of publications on their career is cumulative and nonexclusive, as better job opportunities depend upon several other criteria, such as the ability to attract funds, managerial qualities, and even personal traits. Hence, here I ignore these secondary effects, and I discuss the benefit of access to the RI environment for ECRs in the next chapter.

Publication frequency and impact are specific to each scientific community or field; for instance, an RI related to life sciences can be expected to support more publications per year and per scientist than its counterpart in astrophysics. In any case, ultimately, from the perspective of a scientist, the direct or first-round economic effect of a publication is that it justifies the time spent and money that has been paid to produce it. Thus, the implicit price that society is willing to pay for a scientific publication is simply the cost of producing it.

This is not, however, the end of the story. The direct value of a publication is enhanced by its use in the scientific community and this can be measured, albeit imperfectly. This is a second-round effect, but it is an important one; ultimately, an RI produces a stream of knowledge outputs that has a value for the scientific community only if such outputs are read by peers and used to generate new outputs.

This argument can be summarized as follows: Given $\mathbb{E}(Y_t)$ as the expected social cost of producing knowledge outputs (not just publications, but any other scientific communications) at time t, s_t as the discount factor, and $\mathbb{E}(m) \geq 1$ as the expected multiplier of impact, the expected present value, PV, of this benefit is expressed as

$$\mathbb{E}(PV(Y_t)) = \sum_{t=0}^{T} (s_t \cdot \mathbb{E}(Y_t) \cdot \mathbb{E}(m)). \tag{3.1}$$

This is a generalization of equation (1.3) in chapter 1. Thus, the ex ante estimation of this benefit involves two operations. The first is the estimation of the social value of producing new knowledge as embodied in technical reports, proceedings, preprints or working papers, articles in scientific journals, research monographs, and so on, directly related to the RI. The second operation is the estimation of the multiplier of impact (influence)—that is, a synthetic multiplicative factor capturing the social value, attributed to the degree of influence of that piece of knowledge on the scientific community.

3.3   The Empirical Estimation of the Value of Knowledge Output

As mentioned in section 2.1 in chapter 2, in CBA, the marginal social value of a good can be estimated by considering the users' marginal willingness to pay (WTP), the marginal production cost (MPC), or a combination of these two measures.[7]

The first empirical approach, WTP, is not very promising to estimate the value of knowledge output, either by revealed or stated preferences. The observable demand to publish and access scientific publications does not provide a set of market prices that can be used to estimate the marginal WTP of scientific insiders. For example, the subscription price of scholarly journals is usually paid by libraries for their users; the open access fee for some journals may be paid by research grants; and many papers are available free of charge (e.g., more than 1 million preprints are available in the ArXiv repository for physics and other fields). As for stated preferences, it seems difficult, if not impossible, to elicit the preferences of researchers to publish in a journal or read it, because any payment is usually in the budget of a third party. In the "Further Reading" section at the end of this chapter, I present some hints about WTP estimation in this area.

The usual alternative to WTP estimation is the marginal cost approach.[8] Using the marginal production cost to estimate empirically the socioeconomic benefit of a good is common practice in CBA for certain types of services, when market prices are irrelevant. Moreover, this approach is employed extensively in government accounting systems for gross domestic product (GDP), in accordance with the internationally accepted System of National Accounts procedures.[9] The marginal cost of publications is observable in terms of the value of scientists' working time devoted to preparing papers (as well as other expenditures incurred when communicating scientific results, including attending conferences and other actions). Thus, the social value of a knowledge output is basically a function of the salaries and working time of scientists at the RI. I obviously refer here to a generic statistical paper, as the effort to produce any individual papers can vary considerably (there may be other variable costs when the paper draws from unique experimental data, if such costs are not already accounted for in those of the RI, as discussed in the previous chapter).

As previously stated, a good has an economic value if somebody's welfare increases when its availability increases. What is special in science is that the demand for the knowledge output of an RI project is driven by scientists who are often, at the same time, both users and producers of such knowledge. This does not happen with most other infrastructure services. Passengers of high-speed rail demand the transport service, but they are not involved in its production (there are interesting exceptions, however, discussed later in this chapter).

The fact that scientists are the producers of knowledge as well as the consumers offers a different way to think of the value of this output. Most scientists are paid fixed salaries and are relatively independent in terms of the allocation of their time. Thus, when they spend some time on a research project, they have an opportunity cost, which equals the average scientist's hourly compensation multiplied by the hours spent on producing the paper. Hence, a reasonable proxy of the value of scientific output is its MPC. The social MPC is the first derivative of a production cost function, which has as its argument the time spent by scientists to carry out research and produce a paper, a preprint, or other knowledge outputs, valued at appropriate shadow wages.[10] To show this, I recall the RI's costs (see chapter 1, equation 1.2):

$$PV_{C_u} = \sum_{t=0}^{T} s_t \cdot (K_t + L_{st} + L_{ot} + OP_t + EXT_t), \tag{3.2}$$

which can be rewritten as

$$PV_{C_u} = \sum_{t=0}^{T} s_t \cdot (K_t + L_{ot} + OP_t + EXT_t) + \sum_{t=0}^{T} s_t (L_{st}), \tag{3.3}$$

where K_t is yearly capital costs, L_{ot} administrative/technical labor, OP_t other operating costs, EXT_t the value of negative externalities, and s_t the discount factor. These expenditures are needed for the RI to work, even if nothing is published. The first term, $\sum_{t=0}^{T} s_t \cdot (K_t + L_{ot} + OP_t + EXT_t)$, is mainly a fixed production cost (as it does not change with the number of publications), while $\sum_{t=0}^{T} s_t (L_{st})$, which captures the labor of scientists, is the main variable cost. As the RI could be described as a publication factory, the MPC of a publication is independent of the first term (formally the first derivative of PV_{C_u} in terms of L_{st}). Hence, it seems reasonable to assume that the marginal cost of a knowledge output is simply a function of the scientific workforce associated with the RI, whose role is to observe the data, elaborate on them, and eventually publish some results.

To simplify, let us assume that scientists spend all their time on research leading to publications and that the relation between publications and research time is linear. Thus, if there are n_{st} scientists affiliated with an RI (including staff and those in the collaborations) in year t, each of them produces y_{st} papers in that year, so the total output per year is $pub_t = n_{st} \, y_{st}$; that is, if each paper requires on average h_{st} hours of working time, and level of earnings in terms of gross-of-tax salary per hour per scientist is w_{st} (in that year, including a correction for the shadow wage, if needed), then $MSV(pub_t) = w_{st} \, h_{st}$, and the direct cumulated PV of publications of the RI is simply

$$PV_{pub} = \sum_{t=0}^{T} [(n_{st} \, y_{st})(w_{st} \, h_{st})] / (1 + SDR)^t, \qquad (3.4)$$

where the discount factor s_t from equation (3.2) is written in its explicit form $1/(1+SDR)^t$ and SDR is the social discount rate. It is important to observe that the MPC is far below the total average production cost of a publication, which is (PV_{C_u} / pub). As the fixed part of the production cost is so important for large-scale RIs and publications are sold at a price that is far below the average production cost, it is apparent why market mechanisms cannot support the RI in any case (similar to the microeconomics textbook natural monopoly case, with marginal costs tending to zero and positive average costs). In other words, without a government subsidy, scientific publications would not be produced if they needed data produced by a capital-intensive RI.

However, equation (3.4) estimates only the value of the RI knowledge output (direct or first-round effect), as it does not capture the cumulative influence of publications on the scientific community and should be augmented by another term that captures such influence, as in equation (3.1).

3.4   Measuring the Influence of Publications

Bibliometric techniques and pattern analysis of the scientific literature generated over time around experiments at an RI (e.g., through keywords, citations, and other pointers) can be conveniently exploited to associate a measure of scientific output with the RI.[11] In practice, this consists of forecasting (or assessing ex post facto) the knowledge output generated by the scientists affiliated with the RI (taken as level 0), papers written by other scientists not affiliated with the RI and citing the insiders (level 1), other papers citing level 1-papers, and so on. From the ex post perspective, Carrazza, Ferrara, and Salini (2016) analyze the citation distribution of journal articles related to different high-energy physics infrastructures over a wide time span. Table 3.1 shows the number of papers from the four main LHC experiments from 2008 to 2012, and by comparison, from the four main LEP experiments (1989–2000, the predecessor of LHC).

For ex ante projections, one could adopt appropriate empirical curves (calibrated on ex post regularities) describing the dynamics over time of knowledge outputs. There are several empirical curves that may fit into the specific circumstances of an RI. An example of such an equation includes a logistic function, leading to a differential equation of the Bernoulli form:

$$\frac{dpub_t}{dt} = \theta \cdot pub_t \cdot \left[1 - \frac{pub_t}{\omega} \right],$$
(3.5)

where $\theta > 0$ is an instantaneous growth rate parameter and $\omega > 0$ the equilibrium limit size of knowledge output growth.[12] There are several versions of this simple, nonlinear differential epidemic equation, such as in the literature on innovation or mathematical biology. The growth process is initially exponential, and then slows down and asymptotically reaches a steady state (an S-shaped process).

Perhaps one could argue instead that the steady state will never be reached, and direct citations will continue forever. Alternatively, one could suppose that because knowledge is subject to obsolescence, after some time of stabilization, citations will decline. There are, however, several other statistical functions that can describe the dynamics of the impact, and only empirical calibration would allow for selecting the most appropriate one, case by case.

Curves describing the dynamic of knowledge diffusion over time can be proxied using citation patterns, which can vary significantly depending on the journal article. Some articles may never be cited, whereas others receive citations in the years immediately after publication before becoming obsolete. Additionally, some articles are rarely cited in the years following their publication but then start to become recognized (Andrés, 2009). A typical citation curve illustrates the history of an article that receives a few citations in the first years after publication and then peaks, but subsequently becomes gradually less cited (Sun, Mi, and Li 2016; Larivière, Archambault, and Gingras, 2008). In some cases, lognormal functions best fit typical citation curves (Egghe and Rao, 1992).

Some studies on obsolescence find that the use of the literature declines exponentially with age, and they parameterize this phenomenon with a single

Table 3.1
Cumulative number of papers to 2012 from LHC and LEP experiments, compared

Experiment	Experiment Papers (Including Preprints)	Published Experiment Papers	Experiment Papers Cited in the Literature	Literature Cited by Experiment Papers
LEP				
ALEPH	636	589	383	3,233
DELPHI	736	670	417	3,644
L3	605	549	381	3,563
OPAL	694	634	475	4,037
CDF	3,077	2,386	1,641	6,616
D0	2,383	1,769	1,176	4,744
LHC				
ALICE	1,579	945	382	2,963
ATLAS	2,529	1,921	1,195	4,862
CMS	2,580	1,603	1,030	4,640
LHCb	735	585	248	1,608

Source: Adapted from Carrazza, Ferrara, and Salini (2016). Data extracted in September 2013 from the INSPIRE website (http://inspirehep.net).

number, often called the "half-life" (Burton and Kebler, 1960). However, some argue that an exponential increase in citations is sometimes recognizable over a long period, thereby leading to an exponential function (Li and Ye 2014).

More recently, Galiani and Gàlvez (2017) have estimated with quantile regression data the trends over time of 5 million citations of 59,707 research articles (1985–2000), in selected journals, in twelve fields. They find a single peaked life-cycle pattern that varies across fields, and present plots of citation count up to fifteen years after publication for the median and average papers. For astronomy and astrophysics, biochemistry, biology, medicine, and physics, the peak value is reached before year 5, and then there is a sharp fall in the number of citations. In social sciences, the peak is reached later, and the decline is milder. Figure 3.1 shows the average numbers of citations by field, as recorded by the Web of Science,[13] with molecular biology papers getting about twice the cumulated citations of pharmacology or space science, and three times more than physics.

This brief discussion suggests that there is no generic empirical curve of the citation trajectories of RI-related literature, and one needs to study this issue, looking to the specific field as far as possible, but also that there is an empirical basis for forecasting or ex-post assessment.

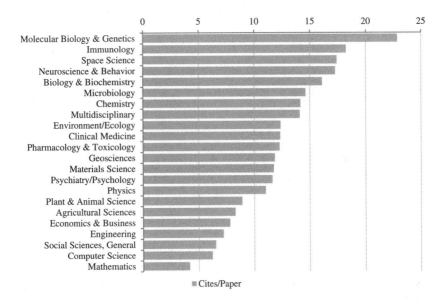

Figure 3.1

Average number of cites per paper in a number of fields (Essential Science Indicators). *Source*: Author's elaboration on Clarivate Analytics data.[14]

Note: Last updated May 2018.

3.5 The Value of Influence

To enter a CBA model, citations and other measures of influence should be valued in the same numeraire of any other benefit. One can extend this notion by considering that there is a chain of citations that extends itself in the literature in subsequent rounds. Moreover, influential papers are frequently cited by papers, which in turn are cited, and so on. This situation is similar to the assessment process of the value of patents, where the trajectory of an invention is studied by examining citations in other patents.[15]

In other words, the MPC of a paper [the first derivative of the production cost function in equation (3.2)] would capture only part of the total value of the knowledge output produced by the RI. The social value of a publication per se should be augmented by the social value attributed to the influence of that piece of knowledge on the scientific community. If the former is captured by the number of papers written and valued through the MPC, the latter is reflected by the number of people who would read the paper (e.g., by the number of downloads from an electronic repository), and eventually, the number of citations a paper gets.

Using citations as a measure of the significance of a scientific paper is an imperfect but widely accepted approach if we adopt the view that, on average, citations reflect the social recognition that the scientific community gives to the paper[16]. Therefore, it is reasonable to attribute a statistically higher MSV to a frequently cited paper in one field. However, this is unrelated to the evaluation of the quality of the paper by peers, which is not a statistical measure. Then a value (shadow price) of a citation is needed. By analogy with the MPC of papers, this could again be the opportunity cost of time employed by a scientist to download, read, and understand someone else's paper and decide whether to cite it or not. This time can vary from a few minutes to many hours or days, depending on the type of paper, its length, the topic, the experience of the citing scientist, and other variables.

In summary, after considering the time allocated and shadow wages, not only of scientists directly working in the RI but also of those who use and elaborate on it to produce new knowledge, in order to capture the cumulative process of knowledge output production, then the value of citations received should be added to the value of paper produced in the first, second, ... nth wave.

However, it would be an exaggeration to think that there is a one-to-one relation between knowledge units produced in the first round and those produced subsequently; thus, we need a decay function. For example, as a shortcut, we may just divide the value of papers produced by outside scientists by the number of references contained in the same papers, as if each contributed in the same way to the new knowledge output. This method is crude but simple. Abt and Garfield (2002) analyzed forty-one research journals and found that the number of references is generally between twenty and seventy in biochemistry and molecular biology, between twenty and fifty in physical sciences, and between five and sixty in medicine. Also, they noted that there is a linear relationship between the average number of references and the normalized paper lengths. There also may be more sophisticated estimates, not discussed here.

Bibliometric techniques analyzing the patterns of the scientific literature, generated over time around a similar RI or its collaborations, can be conveniently exploited to associate a measure of scientific output with the RI. What is needed is to fit the publication function parameters.[17] However, although the use of bibliometric techniques is a well-established approach to provide quantitative characterization of scientific activity, relying on publication records available in online repositories may lead to bias when the dominant mode of production in a specific scientific domain is not the journal article. The limited coverage of particular scientific fields by reference databases is a well-known issue in certain disciplines, such as the social sciences and humanities,[18] law, and computer science, in which peer-reviewed conferences are a major arena of communication. This issue must be carefully taken into account when evaluating the considered benefit, and if necessary, the bibliometric analysis needs to be complemented by a detailed analysis of unpublished scientific outputs. Also, attention should be paid to refraining from double-counting articles, as the same article is often produced in different preprint versions.

3.6 A Step-by-Step Tutorial

The previous discussion boils down to the following practical steps. Suppose that you need to forecast the benefits to scientists (not to society) of a specific RI. The first step consists of forecasting the knowledge production potential of the infrastructure: estimating the number of scientists working with the RI (permanent scientific staff and collaborations) over the time horizon and their yearly average productivity, such as the average number of knowledge outputs per author, per year. Clearly, predicting the number of knowledge outputs produced within the RI can be influenced by the standards of the personnel expected to be recruited, as well as by the scientific field. Hence, the scientists' yearly average productivity and the average number of authors per paper may largely differ from one discipline to another due to the various practices in use. In cases of multiple authorship, an issue to be solved when estimating the average productivity[19] is the definition of the individual contribution to an article (e.g., life sciences have different styles in listing the sequence of authors in comparison to other fields).

Attention should be paid to two particular situations. Experiment collaboration articles are typically signed by a long list of authors, including many scientists who did not directly contribute to writing the text. Conversely, in other cases, the article is signed only by those who wrote it, although many more contributed to the experimental work in certain roles. In both cases, the authors' productivity estimate can be distorted if these aspects are not appropriately reflected in the calculation. Practical solutions should be found case by case.

The second step consists of forecasting the median number of citations of the abovementioned scientific output. The median value is considered a more accurate indicator instead of the average number of citations. Also, an analysis of the median individual *h-index* of scientists involved in the infrastructure, which captures the *n* number of articles that have received at least *n* citations, could be useful.[20] The *h-index* is considered a conservative measure that avoids overestimating benefits. In some cases, however, citation data may overstate the extent to which the scientific literature has been consulted. In fact, an

Table 3.2
Metrics comparisons across disciplines

Scientific Field	Average *H-index*	Average Number of Authors	Individual *H-index*
Cell biology	24	3.90	15
Computer science	34	2.57	22
Mathematics	15	2.95	8
Pharmacology	39	3.08	23
Physics	30	2.66	18

Source: Adapted from analyses presented in Harzing (2010).

author may cite from the abstract of an article or simply copy a reference from another paper.[21]

A number of studies correct the *h-index* for the number of coauthors,[22] the scientific field,[23] and how recent the article is.[24] In particular, if the *h-index* is corrected for the number of coauthors, the resulting metric is called the *individual h-index*. According to the literature, the two indexes show wide differences between disciplines. As an example, table 3.2 presents the analysis provided in Harzing (2010).

As in figure 3.1, the citation statistics show high variability in different scientific domains.[25] This variability results from a number of factors. For instance, the chance of being cited is related to the number of publications (and the number of scientists) in the field;[26] thus, papers in specialty fields attract fewer citations than those in more general fields.[27] Therefore, bibliometric comparisons should be conducted only within a field unless a normalizing factor is applied.

The third step consists of estimating the average value of a statistical scientific output in monetary terms. The unit production cost per knowledge output may be estimated using the ratio of the gross salary of the author to the number of scientific outputs produced per year. Clearly, only the salary amount related to the time dedicated to research within the infrastructure should be considered in the calculation. Data and issues on scientists' salaries according to different scientific fields are discussed in the next chapter. Table 3.3 provides illustrative benchmarks for various scientific fields and countries. Corrections for the shadow wage may be country-specific.

The fourth step consists of assigning a monetary value to citations by deriving the value from the time that scientists need to download and read someone else's paper, and then decide to cite it.[28]

There may be several empirical variants of this forecasting approach. Moreover, ex post, finding the data is obviously easier than forecasting them ex

Table 3.3
Scientists' annual salaries for selected countries and years

Country	Scientific Field and Experience Level	Median (EUR)	Ref. Year	Source
Austria	All fields: senior researcher	66,038	2010	Ates and Brechelmacher (2013)
Finland	All fields: senior researcher	48,387	2008	Ates and Brechelmacher (2013)
Germany	All fields: entry-level	49,810	2018	https://www.payscale.com /research/DE/Job=Research _Scientist/Salary
United Kingdom	All fields: scientist	34,509	2018	https://www.payscale.com /research/UK/Job=Research _Scientist/Salary
France	All fields: scientist	38,736	2018	https://www.payscale.com /research/FR/Job=Research _Scientist/Salary
Italy	All fields: scientist	30,492	2018	https://www.payscale.com /research/IT/Job=Research _Scientist/Salary
Poland	All fields: senior academic	32,078	2010	Ates and Brechelmacher (2013)
United States	Biotechnology	64,932	2015	https://www.payscale.com/
United States	Material science	74,744	2015	https://www.payscale.com/
United States	Clinical research	59,504	2015	https://www.payscale.com/

Source: Based on cited sources.

ante, but the important point is the logic behind the computation of the benefits arising from publications and other knowledge outputs.

3.7 On Scientists' Salaries

Salaries of scientists enter into the model of chapter 1 in several ways; hence, it seems important to ask the question of whether they reflect the social opportunity cost of this type of labor. A *shadow wage* can be seen as a correction of the actual wage to account for distortions in the labor market.

Before discussing this point, it is useful to restate why this parameter of the model is potentially important. First, on the cost side, the financial expenditure for scientific personnel by an RI would change in economic terms if there is a wedge between the observed market price and the shadow wage. Second, in this chapter, the social value of publications is proportional to the wages of scientists. Third, in the next chapter, the human capital effect of the RI is estimated in terms of potential effects on future compensation of students and postdoc researchers trained in such environments along their career.

While on the cost side, I have adopted the cancellation hypothesis between the value of RI publications and the cost of producing them, there may be circumstances where such effects are only partial. In any case, as mentioned, the second-round effects of publications are strictly dependent upon the wages of scientists. Eventually, the social benefits for ECRs are entirely estimated through a wage premium, and it will be shown later in chapter 4 that in some cases, this benefit is the most important one for RIs.

In a precise sense, scientists are on average undercompensated for their work, which would imply that that the shadow wage is greater than the market wage. To see this, one may consider the seminal paper "Endogenous Techno-logical Change," by Paul M. Romer, a US economist to whom the Nobel Prize in Economics was awarded in 2018. Romer (1990) departs from the Solow (1956) framework by introducing the idea that research creates knowledge embodied in "designs." These are nonrival goods, but their economic utiliza-tion is partly excludable (in other words, they are not pure public goods of the Samuelson type).

Designs are produced by R&D in private, profit-maximizing firms, but a design is nonrival, as it can be used by everybody to create other knowledge and goods. Research needs human capital only: knowledge attached to edu-cated researchers (adding capital and non-R&D labor costs would not change the reasoning).

Every researcher has free access—at any time in the past and in the present—to the knowledge produced by other researchers. Firms may appro-priate the profits by temporary monopolies, such as in the form of patents, but cannot avoid the fact that knowledge spills over from one researcher to another without cost (or at negligible cost, such as the publishing cost of an article, in the context of the discussion in this chapter). In other words, firms or inventors cannot entirely appropriate the social value of a patent. Hence, a researcher working today will benefit from the knowledge cumulated in the past or elsewhere and is more productive the larger the stock of knowledge is. Owners of patents have exclusive economic rights to the use of designs for some time, but they cannot avoid that others invest their own resources in reading patent files and elaborating on them to invent new things.

One implications of this setting is undercompensation of researchers. As Romer (1990, S96) says:

There are two reasons to expect that too little human capital is devoted to research. The most obvious reason is that research has positive external effects. An additional design raises the productivity of all future individuals who do research, but because this benefit is nonexcludable, it is not reflected at all in the market price for designs. The second and an equally important reason why too little human capital is devoted to research is that research produces an input that is purchased by a sector that engages in monopoly pricing. The markup of price over marginal cost forces a wedge between the marginal social product of an input used in this sector and its market compensatio. Both of these effects cause human capital to be undercompensated... the marginal product in the research sector is higher than the wage because the price of the patent captures only part of the social value of the patent. ... As a result, in equilibrium, the marginal value of an additional unit of human capital is higher than the market wage.

From this perspective, what RIs (and universities joining the experimental collaborations) do is to increase the stock of human capital, in excess of the quantity that private firms would employ. RIs hence indirectly act as complements of upstream-to-downstream profitable applications by firms' R&D. While not profit-maximizers, RIs are cost-minimizers (i.e., their budget is constrained), and they tend to look at market wages of researchers in the private sector as a benchmark. Usually, they do not pay their scientists more than what they would earn elsewhere, which is Romer's equilibrium level with undercompensation of researchers within firms. In most cases, in fact, RIs and universities pay scientists even less then firms do because they can leverage on the intrinsic motivation of scientists, who accept a job in science mainly because of its nonmonetary rewards.

As a consequence, the shadow wage of scientists in an RI is possibly higher than the market wage. Interestingly, while in the economic analysis of a generic infrastructure project, such a wedge would decrease its social profits and desirability, the opposite happens in the special case of the RIs in the model described in this chapter. The effect is usually immaterial on the cost side because the shadow wage would enter in scientific personnel costs, but also in the value of the publication they produce, and because of this cancellation of the effects, nothing happens. However, the downstream effects in the literature—and, most important, in the social value of human capital—would increase the net present value (NPV) of the RI project.

Estimating the level of the shadow wage based on future applications is very difficult because they are usually unpredictable (see chapter 9). Then, using the actual wage for the value of publications and of human capital effects in the next chapter is certainly on the lower bound of the possible estimates of the RI benefits. Whenever possible, downstream economic effects on firms' innovations should be directly included in the analysis (chapter 7), but usually

it seems prudent not to try to estimate a shadow wage of scientists in excess
of their actual wage.

3.8   Case Study: The LHC Influence on the Literature

Drawing from Florio, Forte, and Sirtori (2016) as an illustration of the foregoing approach, a statistical publication produced by LHC scientists employed
or collaborating with European Organization for Nuclear Research (CERN)
has a marginal social value that is, on average, equal to (or at least not less
than) the average marginal cost of producing such a publication. These costs
are mainly related to the scientific personnel because other investments and
operating expenditures for the accelerator and detectors are not influenced by
the generated number of publications. Assuming linearity in the publication
production with respect to time, which is well documented by the stability of
the average coefficient of the number of publications per researcher, per year
in each field (see Carrazza, Ferrara, and Salini, 2016, and section 3.6), the cost
of the LHC and Collaborations' scientific personnel is approximately balanced
by the benefit of LHC-related publications. Hence, after excluding from the
LHC benefits the first round of publications produced by LHC insiders, only
the additional benefits arising from papers authored by non-LHC scientists
and citing papers authored by LHC scientists are considered.

The direct benefit of further waves of papers is in turn considered negligible;
however, the first round of citations received by non-LHC research papers
is included. The MSV of such papers is observable using the gross salary
received by an average scientist for the time spent on undertaking research and
writing a paper (a shadow wage is not used here) as a proxy. The benefits of
the first round of LHC research publications—that is, the publications signed
by the LHC users (around 10,000 particle physicists at any time, including
experimental Collaborations)—are excluded, as they balance with personnel
costs. The forecast of outputs is based on an estimate of publication trajectories
obtained through a statistical model over a period of $N = 50$ years, starting in
2006. The results are summarized in figure 3.2.

The past (1993–2012) number of LHC-related scientific publications was
extracted from the INSPIRE database[29] by Carrazza, Ferrara, and Salini (2016).
Data include both published articles and preprints. Citations up to 2012 have
been retrieved from the same source. To forecast the number of publications
(2013–2025), a double exponential model (Bacchiocchi and Montobbio, 2009;
Carrazza, Ferrara, and Salini (2014) was applied. This model is based on a
calibration of the publishing trajectory of the LHC predecessor at CERN, the
LEP accelerator. It takes the following form:

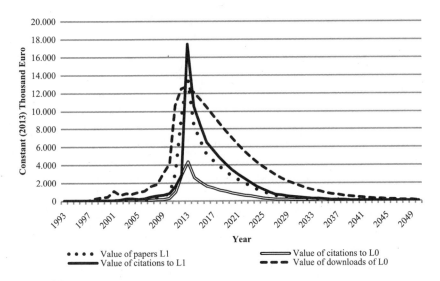

Figure 3.2

Economic value per year of LHC literature: constant thousand EUR 2013. *Source*: Florio, Forte, and Sirtori (2016).

Legend: Citations to LCH papers (L0); citations of the first wave of external papers (L1); value of the first wave of external papers (L1); value of downloads of LHC papers (L0).

$$\mathbb{E}(pub_t) = \alpha_1 \alpha_2 \, \exp\left[-\beta_1(T - t)\right]\left[1 - \exp\left[-\beta_2(T - t)\right]\right], \tag{3.6}$$

where:

$\alpha_1 = 65,000$, the expected total number of authors of publications during the entire time span considered;

$\alpha_2 = 2$, a proxy of their yearly productivity;

$\beta_1 = 0.18$ and $\beta_2 = 0.008$, two parameters determining the shape of the curve, based on the observed pattern of publications related to the LEP;

$T = 50$, the total number of years considered;

$t = (0, \dots, 50)$, the number of each of the remaining years from 2006, the start year of estimations, to the end of the simulation period (2056).

All these parameters are estimated from the data except β_1 and β_2.

The model actually underestimates the steady flow of LHC-related publications that can be observed in the years 2013–2018 in the INSPIRE database; hence, it should be considered rather conservative, as a peak has not yet been reached.

The forecast of the number of pub_1 publications, those citing publications pub_0 over the years 2013–2050, has been based on the observed pattern of the average number of citations per paper (INSPIRE data), without assuming any new spike after the one related to the discovery of the Higgs boson. This is again a (very) conservative assumption because it amounts to saying that nothing of importance will be discovered by the LHC until 2025. This is unrealistic because even rumors of discovery generate waves of publications. For example, in 2016, ATLAS and CMS experimental data suggested the possible existence of a massive particle not predicted by the Standard Model. The result was not confirmed by further data analysis, but in the meantime, hundreds of theoretical papers appeared in arXiv to explain how the discovery would have changed the landscape of physics. These and other spikes in publications are not considered in the following discussion.

The citations to pub_1 papers by pub_2 papers have been estimated as follows. Again, the number of pub_2 papers until 2012 is based on INSPIRE, while to forecast 2013–2050, four citations per paper are assumed, in line with previous years. To these figures, the total arXiv downloads for the field of High Energy Physics have been added, which are used for 1994–2013, while to forecast future citations until 2050, the same average of the past period (sixty-four downloads per paper) is assumed. This average number of downloads has been applied to pub_0 papers.

To sum up, the direct benefits for scientists are the value of pub_1 papers; the value of pub_1 citations and downloads to pub_0 papers; and the value of pub_2 citations to pub_1 papers. As previously mentioned, the value of pub_0 papers cancels out their production cost, and so this is not included. The value of pub_2 papers and beyond, and citations to them, are considered negligible. All values are discounted at the 3 percent social discount rate, as per the European Commission (2014).

After the baseline estimations, risk analysis was performed on the total PV of the publications. To perform a Monte Carlo simulation, a probability density function for the following variables was assumed: the number of citations to pub_0 papers in papers pub_1; the percentage of time that non-LHC scientists devoted to research papers produced per year, per average salary; time per download, time per citation. The resulting total PV of the publications has a mean of EUR 277 million (in 2013 constant price terms). The value of publications (net of pub_0) per se reimburses only a tiny fraction of around 2 percent of the total LHC cost (net of scientific personnel costs), or something more if the post-2013 steady flow of public cations is considered.

A very different issue is how publication performance is compared among different laboratories. In principle, one can use INSPIRE and compare several bibliometric indicators for high-energy physics organizations: *h-index,* the average number of citations per paper normalized by the number of years of data taking at experiments, and the number of presentations at international accelerator conferences by affiliations. In principle, one can also consider the ratio of the total costs of a laboratory, for each such indicator or similar ones and obtain cost-effectiveness (CE) indicators that can be considered as a crude efficiency measure for benchmarking an RI scientific performance (see the "Further Reading" section).

The main problem with CE is that a project such as the LHC generates various types of social benefits, and relating all its costs to the production of publications alone would severely bias the evaluation (and could also have perverse effects in terms of an incentive to multiply the number of publications to show higher CE).

3.9 Further Reading

This section selectively presents some background material on specific points: knowledge production in general, the role of universities, the relation between scientific literature and patents, obsolescence of knowledge, scientometric data and their use as an evaluation indicator, and publishing costs of scientific articles.

3.9.1 Knowledge Production

Foray (2004) is a good starting point for a discussion of knowledge production in general, including the classic examples that he gives of pure basic research ("Bohr-type"), use-inspired basic research ("Pasteur-type"), and applied research ("Edison-type"). The author also distinguishes between knowledge created "off line," in the laboratory, and "on-line," through learning by doing. Foray (2004) also discusses the mechanisms for the provision of knowledge as a public good: government subsidies to private organizations (Patronage), in exchange for full disclosure of discoveries; government production in mission-oriented laboratories (Procurement); and protection of intellectual property rights (Private business). Six institutional arrangements are identified in a matrix that has, on one side, the three former "Ps," and on the other side, public versus private access to knowledge. Universities and not-for-profit bodies are institutions combining patronage with public access, while corporate R&D units, combine private business objectives and private access. Intermediate combinations include government civilian research labs, university-industry research centers, and government defense labs. For a survey about the conceptualization of the knowledge economy, see Powell and Snellman (2004).

Concerning the topic of why RIs are not mainly promoted or owned by universities, one may consider what universities do. Trow (2007, 1) updates the classic discussion of Kerr (1963) about different models of universities as producers of knowledge. He depicts three ideal types: "(1) elite-shaping the mind and character of a ruling class; preparation for elite roles; (2) mass-transmission of skills and preparation for a broader range of technical and economic elite roles; and (3) universal-adaptation of the 'whole population' to rapid social and technological change." It is apparent that the RI paradigm is not compatible with any of these models.

3.9.2 The Role of Universities

On the measurement of the role of universities in promoting knowledge, there is a wide range of empirical studies. For example, Bacchiocchi and Montobbio (2009) estimate the diffusion of knowledge from patents filed by universities, public laboratories, and corporate patents in six countries. They find that university and public research patents are cited more relative to companies' patents, but mainly in the chemical and drugs and medical fields and at US universities, while in Europe and Japan, they do not find evidence of a better performance of university and public laboratory patents relative to firms. The relationship between scientific articles and patents is studied, for example, by Narin and Olivastro (1992). They use the fact that granted patent documents in the US cite previous scientific literature on which some aspects of the invention is based. This creates a link between academic literature and patents (see also chapter 7).

3.9.3 Patents

The need to study decay functions of knowledge, to be seen as an intangible asset (Pakes and Shankerman 1979), is well developed in the literature, mostly concerning patents. For example, Bacchiocchi and Montobbio (2009) study the pattern over time of citations in the United States and the European Union and propose certain functions to capture the process. Hall (2007) estimates that for the manufacturing industry in the United States, the depreciation rate of R&D can vary from zero to 40 percent per year, according to the estimation approach. Li W. (2012), in a study of the depreciation rate of ten R&D intensive US industries, finds that such a rate should be in excess of the usually reported 15 percent or less in earlier studies. For example, Nadiri and Prucha (1993) estimated a rate of between 6 and 12 percent. In this chapter, some ideas from this literature have been applied to the empirical study of citations of scientific publications and to the obsolescence of the value of the knowledge they embody.

3.9.4 Scientometrics

The book by de Solla Price (1965) is credited as the starting point of the measurement of scientific productivity. In the last several decades, an interdisciplinary field has bloomed, with its own specialized journals (such as *Scientometrics*); see Hicks et al. (2015). Ferrara and Salini (2012) discuss ten challenges (and pitfalls) of using bibliometric data. In general, the evaluation of the scientific performance of an individual by bibliometric data is misleading. However, statistical analysis of large samples of scientists can be more meaningful, if it is properly done and interpreted. Some findings in the scientometric literature, in fact, are useful for the CBA of RIs. For example, Abt (2007) finds that in physics, astronomy, geophysics, mathematics, and chemistry, the growth of the literature is only a function of the numbers of scientists and that in the past few decades, the yearly number of published papers per scientist is constant. He focuses only on US scientists and US journals, and concludes: "Papers may be becoming much better overall, but the numbers of papers depend only on the numbers of scientists. We learn from this that if organizations wish to produce better papers, they should employ better equipment and techniques, but if they wish to publish more papers, they should employ more scientists" (Abt, 2007, 287).

Abt and Garfield (2002, 1106) report that in around forty journals in the physical, life, and social sciences, "there is a linear relationship between the average number of references and the normalized paper lengths ... because papers of average lengths in various sciences have the same number of references, we conclude that the citation counts to them can be inter-compared within that accuracy."

Normalized paper lengths consider pages weighted in terms of standardized indicators of citable content. This finding is a clear advantage in terms of the way the impact of publication is measured in this section, when these relations are combined with a third one: the constancy, field by field, of the average number of citations.

Citation rates are chiefly determined by the average number of references made in articles in a field, rather than the size of the field or how popular it may be. The rank order of the fields has been constant for many years. Molecular biology and genetics, immunology, and neuroscience are the areas with the highest average citation rates, whereas computer sciences and mathematics exhibit the lowest averages.[30]

Some authors have tried to measure the performance of specific RIs in terms of their track record in publications. This is only one step of the procedure presented in this chapter, but it is an important one. Irvine and Martin (1984), in the second of their three papers on CERN accelerators, assess the scientific results of the CERN Proton Synchrotron and the Super Proton Synchrotron (SPS) compared with a similar machine at the Brookhaven National Laboratory and at the Fermi Lab (and other accelerators), respectively. They present comparative tables with the number of experimental papers, citations, citations per paper, and number of highly cited papers. They also present the evaluations of scientists interviewed about their appreciation of the strong and weak points of the compared facilities and organizations. The analysis turned out to be rather unfavorable to CERN compared with the United States. Subsequently, Martin and Irvine (1984b) discuss the expected scientific performance of the LEP (the collider that preceded the LHC in the same 27-km underground tunnel) and explicitly state that they want to develop a predictive approach. Their conclusions were pessimistic about LEP, but all this was before the demise of the Superconducting Super Collider (SSC) in the United States.

3.9.5   Publishing Costs

Finally, there is some empirical analysis of the cost of publishing scientific articles, which is a topic obviously related to but different from the discussion in this chapter. SQW (2004), in a report commissioned by the Wellcome Trust, after explaining that the market for scientific publications is imperfect, estimated that the average total publication cost per article by a high-quality, subscription-based journal was $2,750, plus a contribution to overhead and profit, while the author-pay fees would generally be lower ($1,950). On journal publication costs and revenues, Van Noorden (2013, 427) reports: "Data from the consulting firm Outsell in Burlingame, California, suggest that the science-publishing industry generated $9.4 billion in revenue in 2011 and published around 1.8 million English-language articles—an average revenue per article of roughly $5,000. Analysts estimate profit margins at 20–30 percent for the industry, so the average cost to the publisher of producing an article is likely to be around $3,500–4,000." On the other hand, Porter (2012) discusses the selling prices of individual articles in learned journals (typically around $30).

On the benefit side, according to Hargreaves (2011), an article in a peer-reviewed journal is valued at EUR 3,420 in some assessment exercises. See also Swerdlow, Teichmann, and Young (2016), who write on the "publication point" system, adopted by some governments for budgetary reasons. For instance, the funding of higher education in Norway is performance-based: "The value of a publication point for the budget year 2017: NOK 25,550 (this value is based on the sector results reported for the publication year 2015)."[31] An article in a higher-prestige journal is rated at two points, and this would bring around EUR 5,200 to the department.

For a survey of the literature on academic productivity, see IVA (2012) and Rørstad and Aksnes (2015).

4   Students and Postdoctoral Researchers: The Effects of Research Infrastructures on Human Capital

4.1   Overview

Contemporary research infrastructures (RIs) are magnets for graduate and doctoral students, postdoctoral fellows, trainees, and young scientists in general. For them, being involved in the vibrant environment of a foremost laboratory means having direct access to firsthand scientific evidence, having the opportunity to deal with interesting and challenging problems that arise in the context of large-scale experiments, and being included in a network of brilliant minds. Such hands-on experience influences the skills acquired during the early career stage, contributing to a signaling effect in the job market, with an ultimate effect on their professional path and future salary.

This experience in an RI is often a welcome complement to the university education, particularly for universities in small countries. For example, the University of Eastern Finland (15,000 students and 2,800 staff) has adopted its own Research Infrastructure Programme 2015–2020 that combines internal centers and laboratories with access to more than thirty RIs, including some in the ESFRI Roadmap and others in the National Finnish Roadmap. The university directly participates in twelve of these RIs, including XFEL (material sciences), ELIXIR Finland (biodata), BBMRI Finland (biobank of samples), and others in the health sciences, environmental sciences, and humanities.[1]

After reviewing survey data on the skills required of physicists when they enter the labor market, two American professors of physics (McNeil and Heron 2017, 40) point out the inadequacy, in certain aspects, of the standard academic curriculum:

We concluded that physics graduates are generally already prepared to pursue many careers and are sought for their flexibility, problem-solving skills, and exposure to a range of technologies. But most would benefit from a wider and deeper knowledge of computational-analysis tools, particularly industry-standard packages; a broader set of experiences, such as internships and applied research projects, that engage them with industrial work; and a closer connection among physics content, applications, and innovation.

In fact, in most cases, these skills are acquired through experiential learning in an RI environment. According to Anderson et al. (2013a, 16), there were nearly 1,500 US scientists who had been working at the Large Hadron Collider (LHC) detectors and involved in various activities, from design and upgrade to data collection and analysis. One-half would later accept a job outside particle physics: "So the bulk of the people with our unusual technical training take their experience and skills and apply them in some other areas of the U.S. economy." They report on two typical examples from the United States, and I am summarizing the stories as presented by them.

The first case is a student from the Southern Methodist University (SMU), who developed his PhD research at ATLAS collaboration (2010–2012) at LHC. He contributed to designing the monitoring panel for the trigger (a data-filtering system of the detector)[2] and developed a software interface of the control room. Then he adapted the control panel to subsequent software releases and became a shift leader for the liquid argon system, one component of the detector. After having been involved in the search for the Higgs boson decay to four electrons (one of the possible events predicted by theory), he was involved in activities related to the signal selection criteria used in the paper announcing the discovery of the Higgs boson.[3] Finally, after having received his PhD at SMU, he accepted a post at Lockheed Martin Space Systems in Houston, to work with the International Space Station (ISS) on controls and data flow for communication.

A second example is a student who completed his PhD from SUNY Stony Brook, in 2009, and was initially involved in data analysis research at the Tevatron collider at Fermilab; he then moved as a postdoc with the SLAC National Accelerator Laboratory and became a member of the ATLAS collaboration. After developing statistical techniques, he eventually contributed to the search for the Higgs boson decaying to two b-quarks, in association with a Z boson decaying to a pair of leptons (another event predicted by theory). After completion of his PhD, he worked at SLAC and European Organization for Nuclear Research (CERN) for three more years, becoming a trigger operations manager. He also developed some techniques for computer vision at SLAC. Then he left particle physics and became a senior data scientist in AT&T's Big Data division.

These two examples are typical of how an experiential learning process in RIs is instrumental in creating skills needed outside the basic research field. A survey[4] was conducted by the CERN Alumni initiative[5] among current and past members of LHC experiments, mostly doctoral students (85 percent) and covering a range of academic backgrounds, such as experimental physics (69 percent), theoretical physics (11 percent), and others (20 percent).[6]

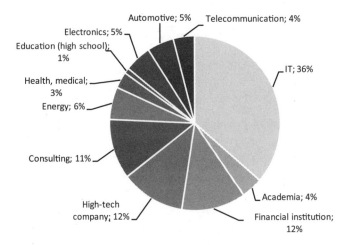

Figure 4.1

Respondents of the CERN Alumni survey (2018), by current employment in the private sector. *Source*: Authors based on Giacomelli (2019).

Figure 4.1 suggests that out of the 2,859 respondents, many currently work in the private sector, specifically in engineering, consulting, information technology (IT), and other domains. The respondents declared that they had acquired a variety of skills at CERN that they consider to be important in their current jobs, specifically related to programming, working in international groups, data analysis, logical thinking, and communication. The more time they spent at CERN, the higher their satisfaction was with their current position and the larger and more diverse skills they acquired.

Another example of a major hub of human capital in its domain is the European Molecular Biology Laboratory (EMBL), a unique intergovernmental organization for life sciences in Europe. While the numbers of PhD students cannot be compared with CERN, at each time, EMBL hosts 240 doctoral students from over forty countries.[7] The attraction is high: in 2016, over 1,800 applications were received, and only 56 were selected.[8] Joint degrees are given in partnership with twenty-four universities in seventeen countries, an interesting example of integration between RIs and traditional higher education institutions.

According to the EMBL Alumni (2017), a survey of the 3,684 members (which include former staff), with a 33 percent response rate, revealed that around one-third of the respondents moved from science to industry (mostly in pharmaceuticals and biotechnology companies). Moreover, of all the respondents (including the majority, who still are employed in academia or research institutes), 28 percent have been cited as an inventor in a patent application, 9 percent were involved in research that led to a start-up company, and 6 percent have created their own start-ups, in fields such as diagnostic services, software for biodata, proteomics, electron microscopy, drug development, and deoxyribonucleic acid (DNA) engineering. Around 39 percent have used the EMBL facilities under an arrangement that allows continued access to such resources to former researchers.

This chapter shows how the benefits of training for young scientists in advanced experimental research contexts should be valued. From the perspective of the economics of education, RIs build human capital and improve scientists' abilities, which in turn will be (on average) rewarded in the labor market because of increased labor productivity (albeit with some market failures that I discuss later in this chapter). Empirically, the core effect is in terms of the expected incremental lifelong salary, earned over the entire career, by an early career researcher (ECR) who spent part of his or her formation in an RI. This is often the most important socioeconomic effect of the project, beyond science.

The structure of the chapter is as follows. Section 4.2 presents evidence of large RIs as hubs for ECRs and offers information on the mechanisms that attract young talent; section 4.3 discusses how an RI contributes to ECRs' experiential learning and placement after spending a period of time in the laboratory. Section 4.4 explores the literature on social returns to higher education, finding some underpinnings for the RI context and suggesting empirical approaches to measure RI's effects on human capital. Section 4.5 illustrates an empirical approach with a case study on current and former students at CERN. After the "Further Reading" section, an appendix shows some empirical estimates supporting the case study.

4.2   Hub of Talents

An RI typically employs four broad groups of people under a variety of contractual arrangements and proportions: (1) scientific personnel, (2) technical personnel, (3) administrative and support staff, and (4) PhD students, postdoctoral researchers, young visiting academics, technical trainees, and other short-term junior staff. According to the online handbook of Realizing and Managing International Research Infrastructures (RAMIRI),[9] the personnel needed during the design, construction, operation, and finally the decommissioning or upgrading/reorientation of an RI involve a wide range of different skills, attitudes, and ages.[10] However, the proportions can be very heterogeneous across fields and the size of the RI (see the section "Case Study: The LHC Premium" for an example of a large-scale case).

While some human capital effects may accrue to the scientific, technical, and management/support staff, the focus here is on ECRs only. Such personnel are always critical to the day-by-day operations of RIs because they often perform most of the data collection work, preparation of experiments, software updating, statistical work, and the drafting of knowledge output. Moreover, the direct expenditure for hiring such talent is often modest for RIs' managing bodies because many fellowships are paid by third parties.

At any given moment in time, tens of thousands of ECRs are involved worldwide in hundreds of major RIs. In contributing to the hands-on training of young scientists worldwide, most RI projects are, in fact, similar to laboratories in good research universities. Students do not pay a fee for their training at the RI, as they are often supported by a fellowship funded by third parties. This fact creates an externality because a valuable learning experience is offered for free. What is the marginal social value of such on-the-job training? From this perspective, insights from the economics of education can be utilized to gauge the contribution of RIs to the increase of human capital available to society. Both theoretical and empirical analyses[11] suggest that higher education and training positively contribute toward economic growth by increasing the productivity of the labor force.

How big could this effect be? No systematic data are available for Europe, where probably the largest number of RIs of any size are located, but the relationship between RIs and academia in the United States can be gauged by the following figures, provided by a recent report of the U.S. Department of Energy (DoE)[12] on the support of the National Laboratories to education in science, technology, engineering, and mathematics (STEM): Overall, every year, more than 250,000 STEM students (including high school) have access to the facilities, together with 22,000 educators; there are 2,950 undergraduate and 2,010 graduate interns and 2,300 postdoctoral researchers. These educational activities at the National Laboratories involve more than 450 academic institutions in the United States and Canada. Moreover, in the National Laboratories, there are 57,600 full-time equivalent employees, 1,285 joint faculties, 33,000 users, and 10,600 visiting scientists. These figures make the US National Laboratories a good example of the scale of human capital formation in the environment of RIs and of the interplay with universities.

4.3 Experiential Learning

The main reason why a large-scale RI often attracts PhD students and junior scientists from different countries is that it offers something that, in many universities, is more difficult to achieve from within: learning by doing, a mechanism that I shall also discuss in the next chapter, in relation to firms entering in a procurement relation with the RI. For students, postdoctoral researchers, and visiting young scientists who enjoy the possibility of spending time in a major RI, the main expected benefit is the opportunity to acquire new skills and competences, even when part of their motivation may also lie in the willingness to be "part of the show, to be a player associated with one of the world's biggest scientific experiments" (Bressan and Boisot, 2011, 206).

The concept of experiential learning in itself goes back to research in social psychology, notably by David Kolb,[13] who in turn drew on earlier intuitions by famous experts of education such as Jean Piaget and John Dewey. The basic idea is that somebody engaged in a personal experience has an opportunity to observe, reflect, and ultimately create concepts to be tested, in a virtuous circle of four steps (experience–reflection–abstraction–testing) that can be repeated many times, with added value at each step.

A famous quotation attributed (perhaps wrongly) to Confucius says: "I hear and I forget. I see and I remember. I do and I understand." The Kolb theory goes beyond that—it suggests that doing something is more than understanding it, because learning has a creative potential that may lead to new testable ideas, and hence to new knowledge. Kolb (1984) suggests that experiential learning is, in fact, not just an approach to education, but a more general theory about the interrelations among formal education, work, and development of the personality of an individual.[14]

To what extent is this theory relevant to scientific education and the job market of young scientists? Science has a long tradition of teaching students how to experiment on what they are learning, whether in the laboratory or in fieldwork. In fact, scientific departments of many good universities tend to emphasize experiential learning as an essential part of their curricula. As just one example, the website of the Department of Biology at Northwestern University in Boston[15] claims that students can join cooperative learning as a cornerstone of the curriculum. This is based on spending six-month periods, two or three times over five years, in one hospital in the Boston area, working in a university laboratory, or in a company abroad.

From this perspective, RIs are ideal partners of universities, particularly for graduate and doctoral students, as they offer exactly the international learning-by-doing environment at the frontier of science that is often difficult to create in the classroom, even in the best academic institutions. It is both a matter of scale and of difference in missions between RIs and academia, as discussed in the previous chapter. Contemporary universities balance careful teaching and research, but after all, they cannot entirely focus on capital-intensive frontier research, as they need to supply classes on a day-to-day basis—a labor-intensive activity. Students are enrolled in universities with the objective to acquire certain formal qualifications that are associated with a curriculum. Universities in this role compete among them and, from this perspective, their teaching activities are a private good (rival and exclusive). In contrast, RIs focus entirely on their research mission and offer a public good, and their ethos is such that hosting a variety of cooperating teams is of the essence.

Training at international collaborations may improve technical and problem-solving capacities as well as teamwork capabilities and management and communications skills. As previously mentioned, the latter have often been found poor in science graduates without such advanced experimental training.[16] Skills, acquired hands-on at the RI, could then find a practical application in careers outside scientific research. Schopper (2009) states that around 40 percent of students working at CERN eventually go into industry. By analyzing the careers of more than 600 diploma, master's, and PhD students involved in one of the experiments (DELPHI) at the CERN electron-positron collider [the Large Electron-Positron Collider (LEP)], from 1982 to 1999, Camporesi (2001) found that while 57 percent of them continued doing research and teaching in an academic context, 43 percent found their first occupation in the private sector, especially in the fields of technology and computing.

The CERN case is part of a broader picture. For students in physics,[17] the laboratory is the standard experiential learning environment.[18] Camporesi (2001) suggested that the skills that students acquire at CERN are of interest for business, and more recently, OECD (2014b) suggests that "the intellectual environment at high-energy physics (HEP) laboratories can be compared to that of the most advanced high-technology firms." The skills of students are improved by a range of activities, including problem-solving challenges during experiments, participating in meetings and other events, and experiencing a plurality of cultures in an international setting. Similar mechanisms can be observed in RIs in other domains, such as genomics or material sciences.

Salary expectations arising in such contexts are influenced by the information available to communities of students and their supervisors about future job opportunities.[19] According to Van Maanen, Eastin and Schein (1977), the career of any individual is a sequence of experiences and changes, which in turn influence the decisions about the next steps. As a result, any precareer information or further information acquired during each step of the career determines subsequent salary paths.[20] Hence, being part of the RI learning environment shapes skills, expectations, and ultimately long-term salaries.

4.4 Signaling and Human Capital Effects

Ultimately, a signaling effect of the RI would lead to better job opportunities and a possible premium on future salaries for ECRs involved in large laboratories. The premium results from the acquisition of new capacities, skills, and experience, but it also may be associated with social capital because of the ECRs' wider network of contacts in the scientific community and reputational effects. Similarly, capacities and skills can be acquired by scientists, engineers, and technical staff working at the RIs. Once they leave the facility, the increase in earnings they receive compared with what they would have received without that experience is a premium similar to that enjoyed by young researchers. The evaluation approach presented for young scientists holds for former employees as well.

This context is relevant to the literature on the human capital effects of higher education. The estimation of the return to human capital, for the individual (private return) and society as a whole, has been the focus of considerable debate in the economic literature. In particular, a *private return* is defined as the extra salary earned due to an increase in human capital, with years of schooling typically acting as a proxy. The benchmark model for an empirical estimation of the returns of education is the relationship derived by Mincer (1974), which includes on-the-job training and experience beyond schooling as a driver of earnings over a life cycle. Mincer modeled the natural logarithm of earnings as a function of years of schooling and years of labor market experience. The most broadly used version of Mincer's human capital earnings function takes the form

$$\log w = \log w_0 + \gamma_1 NH + \gamma_2 EXP + \gamma_3 EXP^2 + e, \tag{4.1}$$

where w is earnings, w_0 is the level of earnings of an individual with no education or experience (the intercept of the function), NH is years of schooling, and EXP is the years of potential labor market experience, γ_1. γ_2, γ_3 are the return parameters to be estimated, and e is an error term. The equation is nonlinear in experience, with a negative sign of the parameter γ_3 possibly reflecting diminishing returns after some years (i.e., a parabolic shape).

However, the literature on the impact of education on earnings reveals a broad range of empirical approaches that have been adopted to estimate the returns and an equally broad range of estimates.[21] Table 4.1 presents the results of three empirical studies that have attempted to yield comparable results, by using cross-country data sources on the returns to tertiary education.[22] See the "Further Reading" section for an update on this approach.

Table 4.1
Average return on tertiary education (%)

Country	Harmon, Oosterbeek, and Walker (2003), 1995	Blöndal, Field, and Girouard (2002), 1999–2000	Boarini and Strauss (2007), 2001
Austria	6.8	–	6.4
Denmark	5.6	11.3	9.1
Finland	8.7	–	7.8
France	7.8	14.8	9.0
Germany	8.8	8.7	6.3
Ireland	11.3	–	13.1
Italy	6.9	6.5	5.1
The Netherlands	5.7	12.3	6.2
Portugal	9.7	–	12.2
Spain	7.8	–	5.7
United Kingdom	10.4	16.1	12.0

Source: Florio et al. (2016), adapted from cited sources.

The aforementioned premium in the RI context is the *incremental lifelong salary* earned by ECRs over their entire careers, compared with the without-the-project scenario. Conceptually, two slightly different effects contribute to the formation of a salary premium. First, the premium reflects the marginal salary increase, gained by a former student who has spent time in an RI, relative to the salary that would have been earned without such experience. Second, after their experience working in RIs, ECRs tend to increase their chances of being hired in labor markets that offer higher average wages, such as the financial sector.

The expected present value of human capital accumulation benefits, $\mathbb{E}(HC)$, can be defined as the sum of the expected increasing earnings, $\mathbb{E}(Dw_{zt})$, gained by the RI students and young scientists and commonly indexed by z, from the moment (at time φ) they leave the RI. Thus:

$$\mathbb{E}(HC) = \sum_{z=1}^{Z} \sum_{t=\varphi}^{T} s_t \cdot \mathbb{E}(Dw_{zt}), \qquad (4.2)$$

where s_t is the discount factor at year t.

In principle, assessing the effect produced by RIs on ECRs requires a quasi-experimental perspective. Such perspective would imply tracking the careers of cohorts of students in the long run, matching data for young people with experience in the RI with those who lack such experience. This may be difficult because of the lack of suitable comparators. In the absence of quasi-experimental evidence, the standard approach would be to set up an econometric model to estimate the marginal effect of human capital formation on the earnings gained over the entire lifetime. As previously mentioned, Mincer's (1974) human capital earning function and its variants disaggregate individual earnings into a function with an education term (given, for example, by the number of years of education, or the degree) and experience (measured by the number of years of work since completion of schooling), a constant parameter and an error term.

Instrumental variables (IV)[23] are usually used to reduce the correlation between the explanatory independent variables and the error term. These could relate to the students' country of origin, gender, race, parents' level of education, quality of the education, and so on. A review of the literature, carried out by Card (1999), shows that IV estimates of the return to education are in the range of 2.4–11 percent. A European survey by Psacharopoulos (2009) reports a minimum private return of higher education of 2.1 percent in Croatia, compared to more than 20 percent in the Czech Republic, Poland, and Portugal (2004 data), with an average of 10.2 percent in thirty-one European countries.[24]

Private return on education is defined as the increased earning (after tax) of an individual that has achieved a tertiary education net of what he or she has paid to attend the education institute, relative to the control group of people with a secondary level of education. A study of UK universities (O'Leary and Sloane 2005) indicates high returns associated with mathematics and computing (21.1 percent), education (19.4 percent), medical-related (17.4 percent), and engineering (15.8 percent) degrees; returns on education in sciences, business and economics, and social sciences are around 12 percent; the lowest return is associated with arts (4.1 percent).

To sum up, the following operational steps should be applied to estimate ex ante the human capital benefits: First, forecasting the number of incoming young researchers by category (e.g., master's students, PhD students, fellows, postdoctoral researchers); second, estimating possible professional sectors in which students leaving the infrastructure are expected to find a job, such as in other research centers, academia, or different industry sectors; third, assuming the probability distribution of different categories of students who find a job in the previously identified professional sectors; fourth, estimating the median gross annual salary for each of the identified professional sectors at different career levels (entry level, midcareer, experienced, late career) and the distribution around the median value, and then using an appropriate empirical function to estimate the continuous salary curve progression (from starting level to end of career) for each relevant professional sector; and finally, estimating the premium salary associated with having spent a training period at the considered RI (i.e., the incremental lifetime earnings compared with the average salary curve previously described).

In practice, from the perspective of an ex ante project appraisal, the estimation of a future premium on salary may be based, with due caution, on benefit transfer approaches (e.g., inferences from benchmarks in other contexts), interviews with students, former students, their supervisors, and expert opinions from specialists, such as recruiters in the relevant labor market, professional bodies, and other groups.

4.5 Case Study: The LHC Premium

According to Florio, Forte, and Sirtori (2016), from whom part of this section is drawn, the intake of ECRs for LHC experiments over the period 1993–2025 amounts to approximately 36,800, comprising approximately 19,400 students and 17,400 postdoctoral researchers (and not including participants in summer schools or on short courses).[25]

Consistent with the literature on marginal returns to education,[26] the benefit arising from the learning experience at the LHC is valued as the present value of the incremental salary, earned over an entire working career, compared with peers who had not been at the LHC. The survey reported in Florio, Forte, and Sirtori (2016) (see appendix 4.A.1 for details) covers two samples: current and former students involved with LHC experiments. The survey elicits the expectations of current students at the LHC and evaluations from past students, subsequently employed in various jobs, including posts outside academia. Information from approximately 400 interviewees from more than fifty countries has been collected.

Each respondent answered questions on a number of individual characteristics, his or her perception of the skills acquired at the LHC, and finally on an ex ante (students) or ex post (former students) basis, perceived LHC premium on their salaries. The research design assumes that former ECRs have acquired firsthand information on job market opportunities and can compare their expectations with those of their peers. The two samples' averages for the premium effect are strikingly similar. The average premium declared by the respondents who have already found a job is approximately 9 percent. In particular, salaries are classified by experience level (entry, midcareer, experienced, and late career) for different jobs in the United States[27] and grouped in four broad sectors: industry, research centers, academia, and others (the latter including, for instance, finance, computing, and civil service). The distribution of CERN students across these broad sectors has been retrieved based on earlier work by Camporesi (2001) and other sources.[28] An additional small premium of 2–3 percent has been applied because of the composition effect of job opportunities across occupations. The resulting combined premium has been attributed to an average student over a career spanning forty years, with the implication, for example, that the cohort of 2025 students would enjoy the benefit of the premium up to 2065.

The estimated premium per capita by Florio, Forte, and Sirtori (2016) is well within the range of the returns related to higher education, which are reported in the literature. Discounting future earnings because of the long time span approximately halves the cumulative benefit compared to its undiscounted value. The resulting expected value of the corresponding human capital benefits repays approximately 40 percent of the total LHC social cost and is the largest component on the benefit side (see chapter 10).

These results have been used by Bastianin and Florio (2018) in the cost-benefit analysis (CBA) of the High-Luminosity LHC, a major upgrade of the accelerator worth an investment of around CHF 950 million. This investment, expected to offer some additional years (2025–2038) of scientific life to the LHC, will also increase the intake of students and postdocs with long-term human capital effects. The analysis considers only technical students, doctoral students, and postdoctoral researchers younger than 30 years old who are enrolled in a CERN education program. Also included are those CERN-registered users who are between 30 and 35 years old. A maximum career length of 42 years is assumed.

Given this assumption, the 2038 cohort of ECRs will enjoy a salary premium due to their experience at the High-Luminosity LHC until 2080. That represents the horizon for benefits related to human capital formation. The career length is calibrated to the years of contributions required to get a pension.[29] In the case of the High-Luminosity LHC, the value of training (integrated to the initial LHC year) represents 33 percent of total benefits, or a discounted figure of about CHF 8.4 billion.

A more detailed statistical analysis of the abovementioned survey data is provided in Camporesi et al. (2017), who show that experiential learning is the key determinant of incremental salary expectations. Camporesi et al. (2017) confirm the results of Florio, Forte, and Sirtori (2016). The premium is estimated in the range of 5–12 percent over the entire career, compared to peers without the opportunity to be involved in the LHC experiments, hence with a multivariate analysis. Some details are provided in appendix 4.A.1.

More recently (April 2018), a survey of CERN team leaders asked them to confirm the abovementioned findings.[30] Team leaders are senior scientists, usually at a university or other research institution, who are supervising the work of doctoral students and hence may be able to compare the career outcomes of different students in their group, including those who had not enjoyed practical training at CERN or with other RIs. Overall, 332 team leaders responded, a response rate of about 29 percent of the potential target. Out of the supervised doctoral students, 38 percent spent one year or more at CERN, 26 percent instead stayed at their university or went elsewhere for a research period, while the remainder spent a shorter period of time at CERN (financial constraints are cited by some team leaders as the main reason of limitation in the length of stay).

For the majority of the respondents, the attraction for students of a period at CERN was rated "very important" because of "Working in an international environment"; "Possibility to work with world-class scientists and engineers"; or "Deepening the knowledge and competences in the domain of interest." In comparison with students not going to CERN, the most important effect in terms of skills acquired was "Developing, maintaining, and using networks of collaborations" and "Technical and communication skills." Team leaders were finally asked their agreement with this statement: "In a recent survey (Camporesi et al., 2017), current and former students at LHC and experiments (some of them now employed outside HEP) put a price tag on their learning experience: a 'salary premium' ranging from 5 percent to 12 percent compared with what they would have expected for their career without such experience at CERN."

One-half of the team leaders (54 percent) stated, "The range sounds reasonable to me"; one-third (31 percent) stated, "I would have expected a greater impact," whereas only 3 percent would have expected a lower impact; and the remainder responded, "Do not know." Hence, overall, 85 percent of the respondents confirm that the findings reported in Camporesi et al. (2017) on the salary premium for ECRs having spent a research period at CERN are either in a reasonable range or may have been even more optimistic. Overall, there is a clear convergence of opinions about the premium from current students, former students, and their supervisors who know different cohorts of students and also students who did not go to CERN. Further research would be needed, possibly with control groups of peers, to measure the premium more precisely, but it seems difficult to deny the social benefit of this human capital effect of CERN.

4.6   Further Reading

This section briefly suggests some additional material to read about the returns on education, expected wages of scientists, and PhD and skills premiums.

4.6.1   Returns on Education

Dickson and Harmon (2011), in introducing a special issue of the *Economics of Education Review*, briefly explores the empirical literature on the returns on education, going beyond the seminal Mincer (1958, 1974) approach. He considers multiple rates of return, uncertainty, option values, and some econometric problems in this context. Other useful references include Lemieux (2006); Heckman, Lochner, and Todd (2006); Rosen (1992); and Polachek (2007). For a review of more than fifty years of empirical research, see Montenegro and Patrinos (2014), who find high returns for tertiary education.

Private returns on education may be lower than social returns if there are externalities of human capital. For example, Blomquist et al. (2014) study the willingness to pay (WTP) for the Kentucky Community and Technical College System through a contingent valuation survey, and then compare the elicited WTP with the estimation of increased earnings. The difference is an education externality. They find that the social value of expanding higher technical education exceeds the private value by approximately 50 percent. It would obviously be interesting to study whether, from the perspective discussed in this chapter, there is a similar effect for the combined university education and RI training. This would increase the social welfare effect of human capital formation in the RI environment.

4.6.2   Expected Career and Wages of Scientists

Research on salary expectations in general include Frick and Maihaus (2016), Schweitzer et al. (2014), and Maihaus (2014). On the job market for graduates in science (mainly physics), see Islam et al. (2015); Nielsen (2014); Institute of Physics (IOP, 2012); Jusoh, Simun, and Choy Chong (2011); Hazari et al. (2010); and Sharma et al. (2007). The British Office of National Statistics (ONS) provides detailed data for average annual wages for hundreds of professions, drawing on tax records. The average for 400 professions in the United Kingdom in 2014 was about 27,000 pounds.[31] Physical scientists would have earned, on average, 49,336 pounds, slightly more than an IT manager or a biological scientist, while the average for sciences was 38,370 pounds.

Mulvey and Pold (2017) report the results of a large yearly survey of jobs and salaries of physics and astronomy bachelors (2013–2014), managed by the American Institute of Physics. Out of around 7,300–7,500 bachelors in each of the two years, the survey collected information for 34 percent of students (in one-fourth of the cases from students themselves, and in other cases from their advisors). The segment pursuing graduate studies abroad amounted to 4 percent and were not included. The main findings are that 54 percent of the bachelors went ahead with some graduate studies. The majority, around two-thirds of them, were in the private sector (mostly in engineering, computer, and information systems), while under 20 percent were employed in the education sector. Team work and problem-solving were mentioned as the core skills. Satisfaction about the position was clearly lower in students who accepted a job outside the STEM sectors.[32]

The Payscale website reports salary information by college and major,[33] for those with a bachelor's degree, and at the beginning and at the end of their careers. For example, in 2013–2014, a major in physics is reported to lead to an entry-level salary of $53,000 and midcareer salary of $101,000; this amounted to being among the best ten fields, along with some engineering degrees. Biotechnology would lead to midcareer salaries of $84,000, molecular biology $76,000, and social sciences $54,800. These are median values for over 1,000 colleges in the United States.

4.6.3 The "PhD Plus Skills" Premium

According to Mulvey and Pold (2015, 2016, 2017), who report the results of a survey managed by the American Institute of Physics, starting salaries for PhD in physics (classes 2013–2014) have a median between 25 and 75 percent of $71,000 per year ($99,000 permanent full-time jobs in the private sector—31 percent of respondents, $66,000 for those working in government positions—14 percent of respondents, and 48,000 for those working in academia—52 percent of respondents). The equivalent median salary for those holding a master's degree after one year was $53,000 (2012–2014; with 65,000 in the private sector—53 percent of respondents, and 41.000 in the academia—19 percent of respondents). Although these samples are small, the reported data would imply a PhD premium of around $18.000 at the starting level in the US job market compared to the master's level (around 34 percent higher salary).

Lower returns to doctoral studies are found by referring to the National Association of Colleges and Employers (NACE) Salary Survey in other scientific sectors. However, these figures are not based on the 25 and 75 percentiles and refer to base salaries only; hence, they do not include bonuses, commissions, fringe benefits, or overtime. Data are obtained by surveying NACE employer members from August–November 2016, for a total of 243 surveys (25.3 percent response rate). Results show a total PhD premium of 24 percent in math and science, 28 percent in engineering, and 37 percent in computer science.

In continental Europe and for natural sciences in general, the PhD premium seems to be lower (data are uncertain), perhaps in the region of 20 percent:[34] the yearly salary for a bachelor's degree in the natural sciences is reported to be on average EUR 42,000 a year, a master's degree EUR 47,200 a year, and researchers and computer scientists EUR 55,266 a year.

Another data source on the RI premium would be in terms of the differential value of the skills acquired. According to Payscale survey data,[35] a physicist with data analysis skills (a typical feature of the particle physics environment) has an average salary of $93,140, while the average salary for a job in physics is $89,000 or $75,000 for a physicist with generic skills in the field. The total premium for a physicist with specific skills is in the range of 5.5 percent to 26.7 percent, compared to a physicist with generic skills. Anderson et al. (2013b) had a survey of young scientists in high-energy physics. The objective of the survey[36] was to record attitudes and interests, including career outlook, for those employed in nonacademic fields after training in high-energy physics.

A total of 1,112 responses were collected for the 2013 survey, around 750 from graduate students or postdocs, with 57 percent of the respondents being US citizens, and the remainder from mainly European countries. Around 30 percent of the respondents were at CERN, 40 percent in universities, and the remainder in other laboratories. Two-thirds wanted to pursue an academic career despite a similar share that feared that funding for their field would decline. If a major experiment was not taking place in the United States, in most cases young US physicists would not leave the country to pursue it. According to the authors, the United States is still attractive, but will not be so forever: "However, all this can shift if the U.S. misses an opportunity to build the next major physics experiment most relevant to the various frontiers in HEP, helping support the idea that the most compelling science will attract the best and brightest minds" (Anderson et al., 2013b, p. 31). For those in a nonacademic career, "many of the skills learned in HEP are seen as valuable skills" (p. 31).

The Institute of Physics (2011) presents a report on the careers of young physicists in Ireland. There were 810 respondents (with 227 holding a PhD), with a median annual salary between EUR 50,000 and 60,000 (EUR 35,000 is the average for the country for all economic sectors, excluding agriculture). While over 41 percent were working in the education sector, 14 percent were employed in IT/electronics, 6.5 percent in health, 5.6 percent in finance, and the remainder in a number of other sectors. One-fourth of the respondents worked abroad. Again, there is evidence of the value of acquired skills.

4.A.1 Appendix: Experiential Learning in High-Energy Physics

The content of this appendix draws heavily on Catalano et al. (2016) and Camporesi et al. (2017).[37] The survey involved both current and former students at LHC. The questionnaire is available in Camporesi et al. (2017).

Let $expw_z$ be our dependent variable, measuring the range of salary expectations and taking integer values from 1 to F. Suppose, also, that the underlying process to be analyzed is:

$$expw_z^* = X_z \delta + \varepsilon_z, \tag{4.A.1}$$

where $expw_z^*$ is the exact but unobserved (i.e., latent) dependent variable (i.e., the exact level of agreement with the statement proposed by the interviewer); X_z is a vector of independent variables (as discussed next), aimed at explaining the range of salary expectations; δ is the vector of regression coefficients we wish to estimate; and ε_z is the random disturbance term that follows a logistic distribution (Balakrishnan, 1992). The variable $expw_z$ relates to the latent variable ($expw_z^*$) according to the following rule:

$$expw_z = 1 \text{ if } expw_z^* \leq \tau_1$$
$$expw_z = f \text{ if } \tau_{j-1} < expw_z^* \leq \tau_j \text{ } f = 2,\ldots, F-1,$$
$$expw_z = F \text{ if } \tau_{J-1} < expw_z^* < \infty$$

where $\tau_1 \leq \tau_2 \leq \ldots \leq \tau_{f-1}$ are unknown thresholds (cut points) to be estimated. The conditional distribution of $expw_z$, given X_z, is expressed by

$$\Pr(expw_z = f | X_z) = \Lambda(\tau_f - X_z \beta) - \Lambda(\tau_{f-1} - X_z \delta), \tag{4.A.2}$$

where Λ (.) denotes the logistic cumulative distribution function. Equation (A.4.2) tells us what the probability is that the respondent selects one of the proposed range of salary expectations, given the value of the independent variables. The beta coefficients are estimated by maximum likelihood.

Figure 4A.1 shows the variables in the conceptual model.

Starting salary or end-career salary expectations as dependent variables (Schweitzer et al., 2014) in the sample are positively correlated (coef=0.62, $p<0.05$). Therefore, only end-career salary expectations are used as a dependent variable. The results are shown in table 4.A1. Columns 1 and 2 of the table reveal that experience at LHC positively and significantly correlates with salary expectations when it is proxied both by the acquired competences and by *Length of Stay*. These variables also keep up their statistical significance when they are plugged simultaneously into the same model (column 3). Column 4 adds the interaction term between *Technical Skills* and *Length of Stay*. The

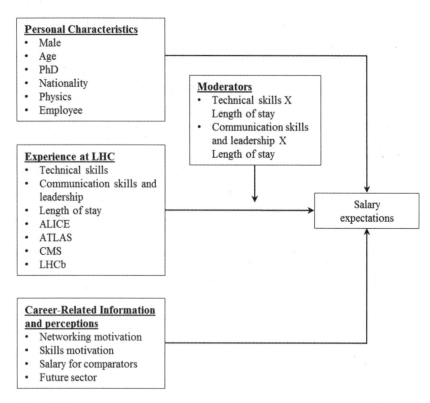

Figure 4.A.1
Conceptual model of salary expectations. *Source*: Camporesi et al. (2017).

positive and statistical coefficient indicates that the skills acquired at LHC increase as the time spent on the experiments increases, which in turn generates higher reward expectations.

Further, the estimated association between salary expectations and experiential learning at LHC remains robust after adding respondents' personal characteristics, as well as their career-related information and perceptions (column 5). The coefficient on *Male* is positive and statistically significant,[38] particularly among physicists.[39] The positive coefficient of variable *PhD* confirms that salary expectations increase with educational attainment.[40] There are no significant differences in end-career salary expectations between employees and students.

Column 5 also shows that the variables *Salary for Comparators* and *Future Sector* enter the model with a significant and positive coefficient.[41] Higher salaries are expected in sectors such as industry and finance.

Table 4.A.1
Ordered logistic estimates of end-career salary expectation

Variables	(1) coef	se	(2) coef	se	(3) coef	se	(4) coef	se	(5) coef	se
Experience at LHC										
Technical Skills	0.103*	(0.062)			0.110*	(0.061)	0.004	(0.145)	0.135	(0.134)
Length of Stay			0.009*	(0.005)	0.009*	(0.005)	0.011**	(0.005)	0.017**	(0.007)
Technical Skills X Length of Stay							0.004**	(0.002)	0.004**	(0.002)
Personal Characteristics										
Employee	0.814***	(0.282)	0.455	(0.346)	0.493	(0.352)	0.500	(0.354)	0.444	(0.409)
Male									0.946***	(0.349)
Age									−0.035	(0.043)
PhD									2.653***	(0.924)
Physics									−0.294	(0.449)
Career-Related Information										
Networking Motivation									−0.098	(0.157)
Skill Motivation									0.272	(0.239)
Salary for Comparators									0.342***	(0.130)
Future Sector									0.495***	(0.155)
Nationality-specific Effects	Yes		Yes		Yes		Yes		Yes	

(continued)

Table 4.A.1 (continued)

Variables	(1) coef	(1) se	(2) coef	(2) se	(3) coef	(3) se	(4) coef	(4) se	(5) coef	(5) se
Experiment-Specific Effects	Yes		Yes		Yes		Yes		Yes	
Interview-Specific Effects	Yes		Yes		Yes		Yes		Yes	
Observations	318		318		318		318		318	
McFadden's R2	0.036		0.035		0.043		0.050		0.159	
Log Likelihood	−254.3		−240.8		−237.4		−235.9		−172.8	
Likelihood Ratio Test	16.87		17.99		19.17		22.75		52.20	
Proportional Odds hp Test (p-value)	0.291		0.276		0.227		0.205		0.182	

Source: Camporesi et al. (2017).

This table shows the determinants of the probability of falling in one of the expected salary categories. Robust standard errors in parentheses. ***, **, * denote significance at the 1%, 5%, and 1% levels, respectively.

Marginal effects according to Camporesi et al. (2017) are such that one additional month of training spent at LHC increases the probability of declaring an expected salary in the two highest categories (50,000–60,000 EUR and > 60,000 EUR).

For an average individual who declared an expected salary between EUR 50,000 and 60,000, the experiential learning at LHC is worth about 5 percent excess salary (3 percent for a student and 7 percent for an employee). For those respondents whose expected salary is in the category > 60,000, the stay at LHC is worth, on average, about 12 percent (6 percent for a student and 16 percent for an employee). This 5–12 percent is the range of our final estimation of the expected "LHC premium."

These results have been recently confirmed by a survey with the same questionnaire to the CERN Alumni Community from March–May 2018.[42] Considering the average *Length of Stay* (fifty-nine months for employees and twenty-four for students), the CERN LHC premium effect for researchers belonging to the highest category of the end-career salary expectations is, on average, 13 percent. The students' supervisors (team leaders) confirmed the results as well (see section 1.5 in chapter 1).

5 The Direct Effect on Firms: Knowledge Spillovers and Learning

5.1 Overview

The construction and operation of a research infrastructure (RI) partially pays back its social costs thanks to knowledge spillovers to its suppliers and other firms and organizations. A positive externality arises from technology transfer, which has been defined (Zuniga and Correa, 2013, 1) as "the movement of know-how, skills, technical knowledge, or technology from one organizational setting to another. Technology transfer from science occurs both formally and informally, as the technology, skills, procedures, methods, and expertise from research institutions and universities can be transferred to firms or governmental institutions, generating economic value and industry development."

In fact, the transfer of knowledge may go beyond technology. It could be related to research and development (R&D) management, organization practice, project accounting, commercialization techniques, etc. Several channels have been identified in the potential transfer of knowledge from public research organizations, including universities, to the industry. These channels include, inter alia, joint research projects between the industry and scientific institutes, contractual relations such as consultancy and certification services, technology licensing to established firms, start-ups, spin-offs and academic entrepreneurship, scientific publications, conferences, seminars, and other forms of intercourse between science and business.[1]

An important mechanism of knowledge spillover is public procurement for innovation (PPI), a mission-oriented policy tool.[2] Large-scale RIs are important drivers of innovative procurement relations because they often require entirely new technologies to be built and operated. In general, economic externalities in this context arise because of a divergence of objectives between the two sides of a procurement relation. Scientific objectives intrinsically motivate engineers and scientists working at the RIs. Instead, firms' managers want to create value for shareholders through profitable operations. Scientists enter business relationships with profit-motivated firms, either directly (e.g., with suppliers of technology) or indirectly (e.g., with second-tier suppliers or under specific joint projects). These firms, in turn, exploit any acquired knowledge in their transactions with third parties. This chapter does not cover all the range of knowledge spillover mechanisms mentioned here; instead, it mainly focuses on technological procurement, corporate spin-offs, and other formal knowledge transfer relations.[3]

The initial core impulse is a learning event (e.g., the exposure of the firm to new technological problems in the context of a procurement contract, or to other forms of knowledge spillovers from the RI). This event may produce a chain of effects: a firm's R&D increases, leading to knowledge production, and then possibly (but not always) the filing of new patents. R&D would facilitate the creation of innovative products and stimulate productivity and, ultimately, profitability for the firm, according to a well-known sequence, established in the empirical literature of innovation economics.[4]

How is it possible to study this economic impact of the RIs empirically? The idea of tracking innovation linked to the development of scientific projects, through names of inventors associated with either patents, firms, or keywords, was suggested many years ago.[5] Patents could provide a useful, but only partial, indication of the total knowledge creation effect. As a matter of fact, not all innovation generated both by the RI scientists and technical staff and by firms in their supply chain is patentable or actually might be protected by a patent.

More generally, the increase of business profits ascribable to the RI (against a realistic counterfactual scenario) should provide a comprehensive measure of technological and other learning spillovers, accounting for the benefits related to the production of new marketable products, increase of productivity, and also increase of visibility and corporate image.[6]

The approach suggested here to value learning externalities (technological, commercial, organizational, etc.) cannot be confused with a simple estimate of sales growth, increased efficiency, and performance immediately generated by procurement contracts. Some earlier studies defined the economic benefits of technology transfer as the sum of the increase of turnover and saving in production costs, generated by but independent of the procurement contracts. In the context of European Organization for Nuclear Research (CERN), for example, Schmied (1977, 1982) and Bianchi-Streit et al. (1984) analyzed its supply chain in the periods 1955–1978 and 1973–1982, respectively. The former study suggests that the "economic utility" ratio (i.e., the ratio between sales to the RI as a customer and further sales to third parties or cost savings) was in the range of 1.4–4.2, with an average of 3. This figure would indicate that for every euro spent by CERN in a high-tech contract, a company receives around EUR 3 in the form of increased turnover or cost savings. As stated by Schopper (2009, 150): "[T]his implies very crudely that in a laboratory such as CERN, about one-quarter of the budget is spent on high-tech products and consequently around three-quarters of the overall public spending is eventually returned to industry."

From the cost-benefit analysis (CBA) perspective, however, it is not the change of sales per se that needs to be considered, but the change of net output (i.e., gross profit or producer surplus) at shadow prices. The general CBA principle in all the abovementioned cases (procurement, patents, spin-offs, etc.) would be first, to identify externalities, and second, to value them in terms of incremental shadow profits accruing to firms and other economic agents. *Shadow profits* are the difference between a firm's discounted revenues and costs, evaluated at the shadow prices of input and output, gross of interests and taxation, as mentioned in chapter 1. If shadow prices are simply estimated as equal to market prices, this would be the net present value (NPV) of the additional profit gross of depreciation, taxes, and interests.

Because of its specific aspects, innovation in information technology (IT) is presented in chapter 6. Benefits to firms and other organizations deriving from the use of some RI services (e.g., synchrotron light sources in pharmaceutical research or hadrontherapy treatment of cancer) and the final benefits to consumers are discussed in chapter 7. Innovation that currently remains commercially unexploited and does not yet produce an actual increase in firms' profits or consumer surpluses cannot be valued ex ante as a use benefit of the project (see chapter 1 for the definition of use and nonuse benefits). Such unexploited innovation is part of the quasi-option value of the RI, a nonuse benefit effect that will be discussed in chapter 9. Thus, it is important to consider that this chapter considers only the direct effects of RI activities on certain firms, while other indirect or potential effects on firms are discussed later. From this perspective, the analysis here is prudent and partial.

Having set these boundaries to what I will present here, the structure of the chapter is as follows: After this overview, section 5.2 defines knowledge spillovers; section 5.3 addresses the economic effects of technological procurement; section 5.4 considers knowledge spillovers leading to patents filed by third parties; section 5.5 describes the valuation of innovations leading to technology transfer initiatives and spin-offs; and, finally, section 5.6 presents a case study on the economic effects of CERN procurement. A "Further Reading" section at the end of the chapter mentions some literature on the empirics of R&D impact and patents.

5.2   Knowledge Spillovers

The concept of knowledge spillovers is an old one. It goes back to Alfred Marshall's *Principles of Economics*, in which, in a discussion of industrial districts ("the concentration of specialized industries in particular localities" such as the textile industry in Lancashire), he wrote that in such places: "The mysteries of the trade become no mysteries; but are as it were in the air, and children learn many of them unconsciously" (Marshall, 1890, 198).[7]

Marshall pointed to learning mechanisms within a community, beyond market relations. Daily, we learn something from each other, and proximity between individuals and organizations enhances mutual learning. Most of such transmission of knowledge is given freely, as it is donated among individuals in many ways. An *economic externality* arises when an agent can acquire some utility from knowledge received as a gift from another agent, a fact that may be intentional or unintentional. The public good feature of knowledge stems from intrinsic nonrivalry in the consumption of ideas and information and only partial or costly excludability of third parties.[8]

Yet ideas have an origin. They may be "in the air" only after they exist. To cite Arrow (1962, 155): "Learning is the product of experience. Learning can only take place through the attempt to solve a problem and therefore only takes place during activity ... [L]earning associated with repetition of essentially the same problem is subject to sharply diminishing returns." I already discussed one aspect of this process in the previous chapter: A new problem to be solved creates learning; the second time that one has to solve a similar problem, one learns less, and after many repetitions, one learns nothing.

In our context, the key concept is an "attempt to solve a problem," as RIs often pose unprecedented technological problems to the firms in charge of constructing them. The formal model initially proposed by Arrow, which posited increasing returns to cumulative gross investment,[9] was instrumental in creating the endogenous theory of growth in macroeconomics (see section 1.1 in chapter 1), although the learning-by-doing effect ultimately depends on microeconomic, firm-level mechanisms. In this regard, Thompson (2010) reviewed some of the empirical literature on learning mechanism drivers and concluded that passive learning is not so important, and that accordingly, other factors should be considered. Irwin and Klenow (1994), in a study of the semiconductor industry, also concluded that under normal market circumstances of cumulative knowledge accumulation, learning spillovers might be limited.[10] Something else was needed to create the learning effect. Hence, Solow (1997) expanded Arrow's original model to comprise discontinuous innovation, arising from experience related to new investment. For a firm, solving new problems requires more R&D investment and innovation. Product and process innovation then spurs productivity and may generate changes in output prices and profit margins through the Schumpeterian mechanism of the temporary monopoly. Several theoretical models of innovation, based on this mechanism, have been proposed.[11]

Innovation was defined by Dosi et al. (1988, 222) as "the search for, and the discovery, experimentation, development, imitation, and adoption of new products, new production processes, and new organizational set-ups." They see innovation as a cumulative process, where prior knowledge influences the chances of exploiting new opportunities. This is a complex process that takes time and is influenced by many factors.[12] Such complexity is likely to induce firms to interact with other organizations for knowledge exchange and technological learning,[13] making interactive learning a fundamental driver of innovation.[14] From this perspective, Chesbrough (2003) defines *open innovation* as an intentional exchange of inflows and outflows of knowledge between firms and external parties to accelerate firms' internal innovation, while Von Hippel (1986, 2005) observed that users and suppliers of innovation-related components sometimes were more important, as functional sources of innovation, than the product manufactures themselves. As a result, and to emphasize the necessity of communication with the users of innovations, he introduced the term *lead-users*, defined as "users whose present strong needs will become general in a market place months or years in the future" (Von Hippel, 1986, 796). Of course, the absorptive capacity of the firm, being the ability to recognize the value of new information or skills and to assimilate them and apply t to increase its profits, is a critical factor.[15]

From this perspective, the RI is similar to any scientific project that is not motivated by profitability. Vast literature analyzing the relationship between academic research and industrial innovation activity exists. For example, an econometric analysis by Jaffe (1989) found a significant positive impact of university R&D on industrial patenting in twenty-nine US states.[16] Along the same line of thinking, Cowan and Zinovyeva (2013) analyzed the effects produced by the opening of new universities in Italy, during the period 1985 and 2000, on regional innovation, in terms of the number of patents filed, and confirmed the existence of a positive relation. Other studies show that university research also positively affects firms' product and process innovation.[17]

5.3 Public Procurement and Innovation

In this context, one spillover mechanism is public procurement of innovative solutions. This "happens when the public sector uses its purchasing power to act as early adopter of innovative solutions which are not yet available on large scale commercial basis."[18] Public procurement for innovation (PPI),[19] defined as a policy tool, implies that, in general, the development and diffusion of innovation depend on user-producer interactions occurring in the procurement process.[20] Moreover, major science organizations, such as the National Aeronautics and Space Administration (NASA) or CERN, can be seen as "lead-users," acting as learning environments for supplier companies, which often strive to meet the stringent technological specifications, demanded by the projects, to be carried out.[21] In this sense, Autio, Hameri, and Vuola (2004) argue that communication and interaction in the big science industry come mainly in the form of technological learning from big science to the industry: firms would benefit by obtaining pioneering information or technology or by being stimulated in solving new problems.

Indeed, building a new large and complex RI or carrying out an experiment at the scientific and technological frontier can be an important source of innovation per se. For example, a wide range of new materials and tools stem from space technologies needed for the NASA projects, such as "memory foam," able to deform and absorb pressure and return to its original shape; invented to improve the safety of aircraft cushions, it is used nowadays for helmets, mattresses, and wheelchair seats.[22] In this vein, Giudice (2010, 109) reported that companies involved in Large Hadron Collider (LHC) procurement use the skills they acquired in other markets, such as superconducting materials for magnetic resonance imaging in medicine or manufacturing automobile parts. There are several examples of this, such as the Daresbury Synchrotron Radiation Source (see STFC, 2010b). Also see the discussion of NASA spin-off products later in this chapter.

Technological spillovers might also occur among the firms and laboratories along the RI's supply chain. When a procurement contract for an RI is signed, an intense collaboration process between the RI staff and the suppliers begins, aimed at effectively designing, testing, and manufacturing the required product or service. These efforts give firms the opportunity of learning something new. In particular, high-tech suppliers involved in the design, construction, and operation of infrastructures at the forefront of science or technology can enjoy spillovers from working with or for an RI's procurement process. Indeed, several other firms involved in the supply chain of an RI typically face the challenge of providing customized solutions to a number of complex technological questions.

This situation gives firms the opportunity to cooperate with the scientific and technical staff of the RI and acquire new knowledge and technological skills from them. In addition, suppliers are incentivized to expand beyond their current states of knowledge. There may also be learning-by-doing benefits for suppliers because some sizable orders require them to reorganize the production on a larger scale. These circumstances can yield different types of innovations, ranging from improvements to already-existing equipment and manufacturing processes to the invention of new tools that may find uses in other areas of science, services, and industry. Further, there may be organizational and market learning in this context. Reputation is an additional bonus, but it can be seen as intimately connected with the assumption that a firm working with a large-scale scientific laboratory is itself knowledgeable and trustworthy.

In the 1970s, the first attempts to measure the benefits from NASA R&D programs were also undertaken (e.g., Midwest Research Institute, Chase Econometric Associates, and Rockwell International[23]). However, these studies usually estimate the productivity changes in the national economy and are not based on firm-level production function or cost-benefit approaches (see chapter 10). Some of these studies typically focus on quantitatively estimating output multipliers or the average economic utility to supplier firms, implicitly assuming that the value of learning should be measured by increased sales and decreased costs. As previously mentioned, *economic utility* is defined as the sum of the increased turnover and cost savings arising directly from the contract, but excluding the value of the contract itself. Thus, a sales multiplier is estimated to assess the procurement, which is likely to generate learning benefits in terms of increased turnover (or decreased costs). Table 5.1 presents the results of earlier studies that aim to estimate the economic utility ratio for some research bodies.

In a CBA framework, however, the net social benefit should be considered; hence, firms' revenues must be taken into account after production costs (i.e., economic profits before tax and interest) and from an incremental perspective. In principle, an increase in gross economic profits should be assessed against a counterfactual group of firms operating in the same sector and sharing similar characteristics with the firms that actually worked for the infrastructure.

In some cases, this comparison is impossible because of the uniqueness of the technology; however, a practical way to value ex ante the incremental increase in profits consists of using a benefit transfer approach,[24] exploiting the results of an ex post survey of firms within and outside the supply chains of similar infrastructures. The term *benefit transfer approach* refers to the process of extrapolating the results of existing primary studies (i.e., surveys or other ad hoc analyses) and transferring them to different populations and contexts. The section "Case Study: How CERN Procurement Creates Value," later in this chapter, offers an example of this concept, as data from other accelerators

Table 5.1
Economic utility ratios and output multipliers in the literature

Organization	Method	Average Values	Source
CERN	Survey of firms	3	Schmied (1977)
CERN	Survey of firms	1.2	Schmied (1982)
CERN	Survey of firms	3	Bianchi-Streit et al. (1984)
European Space Agency	Survey of firms	3	Brendle et al. (1980)
European Space Agency	Survey of firms	1.5–1.6	Schmied (1982)
European Space Agency	Survey of firms	4.5	Danish Agency for Science (2008)
NASA Space Programs	Input-output model	2.1	Bezdek and Wendling (1992)
National Institute of Nuclear Physics	Input-output model	2–2.7	Salina (2006)
John Innes Centre, United Kingdom	Input-output model	3.03	DTZ (2009)

Source: Florio et al. (2016), based on cited sources.

at CERN were extrapolated to evaluate externalities from procurement in the LHC construction, and then they were double-checked using other approaches and more recent data.

The analytical issues involved in estimating the technological impact of RIs include two aspects: (1) how to identify and measure spillover effects, and (2) how to value them. If the R&D cost is fully internalized by the firm, and it is then repaid by the procurement contract, there is no identifiable first-round externality. However, this does not bar second-round effects from occurring. Innovation spilling over the scope of the initial procurement contract can be attributed, at least to some extent, to the knowledge acquired on the job.

The empirical literature focusing specifically on technological spillovers of RIs is less developed. Several studies usually rely on a qualitative methodology of analysis and case studies, developed through desk research, in-depth interviews, and surveys. Autio, Bianchi-Streit, and Hameri (2003) investigated the learning benefits gained by European firms that had participated in CERN's procurement activity between 1997 and 2001. Other data refer to the ATLAS experiment.[25] I report on more recent evidence from about 650 suppliers in the case study later in this chapter.

Ideally, a CBA of an RI should look at the social profits generated by the spillovers. A possible approach is to look at the company's return on sales (corrected with shadow prices of inputs and outputs, as needed). With j being the number of companies benefiting from technological spillovers over time $\mathcal{T}$, Π_{jt} is their *incremental* shadow profits (i.e., profits at shadow prices) directly imputable to the spillover effect, and given the discount factor, the present value of technological learning externalities is expressed as

$$TE = \sum_{j=1}^{J} \sum_{t=0}^{T} \frac{1}{(1+SDR)^t} \cdot \Pi_{jt} = \sum_{j=1}^{J} \sum_{t=0}^{T} \frac{1}{(1+SDR)^t} \cdot (\Delta r_{jt} - \Delta c_{jt}), \quad (5.1)$$

where the last term is the difference between incremental revenues Δr and costs Δc for firm j over years $1, \dots t, \dots \mathcal{T}$. If costs decrease thanks to innovation, then profits increase.

The average firms' return on sale (ROS), possibly using income gross of taxes, interests, and depreciation to be closer to a cash-flow frame, reported in balance sheets can be taken as a proxy of social profit in competitive markets. In distorted markets, where observed prices do not reflect the real opportunity cost of resources, the profit has to be derived as the difference between the firms' total income or cash inflow and operating costs, all valued at shadow prices (such as the shadow wages, discussed in chapter 2). This approach is broadly consistent with the empirical literature, where R&D spillovers and externalities are captured through variations in the private profit margins,[26] and it can also be adopted in a CBA framework, subject to the important proviso that only variations in profits, ascribable to the activities carried out by the RI's supplier, are considered.

As previously mentioned, the increase of profit, in principle, should be assessed against a counterfactual group of companies, operating in the same sector and sharing similar characteristics with the companies that actually worked for the RI, to control for selection bias. The set of techniques typically used for implementing a counterfactual impact evaluation,[27] which is well established, especially in the evaluation of the effects of government subsidies on private R&D in the European Union,[28] can also be relevant in the RI context. If this is not possible, a "before-after" empirical approach to the learning event can be tried (see appendix 5.A.1).

5.4 The Social Value of (Not) Patenting Inventions

Another source of externalities for firms, which may or may not directly arise from procurement relationships, originates from the fact that RIs cannot or are not always willing to patent the knowledge that they generate. In some cases, legal constraints or formal agreements among the participants prevent RIs from doing so, as in the case of the Human Genome Project (HGP; an example discussed in the next chapter), while in other cases, such limitations constitute a condition for government funding. This situation creates patenting opportunities for third parties that build on the discoveries at the RIs. Thus, the value of nonpatented invention might be estimated if somebody else files a patent on a derived concept.

The social benefit is created in $(1 + n)$ rounds. In the first round, the value of the potential patent is donated to somebody else. In the subsequent n rounds, there is the incremental impact of the patent itself. In fact, when a patent is registered, it produces a private return to the inventor, as well as potential externality spillovers to society. A public document is issued, which contains information on various aspects of the invention, including citations to existing patents and previous literature. This document grants the inventor exclusive rights to the commercial use of the invention for a predetermined period and serves to delimit the scope of the property rights granted to the patent owner. According to the Organisation of Economic Co-operation and Development (OECD, 2009, 107):

There are basically two kinds of citations. Patent references are citations to previous relevant technology protected by or described in other patents filed anywhere in the world, at any time, in any language. References categorised as non-patent literature (NPL) are scientific publications, conference proceedings, books, database guides, technical manuals, standards descriptions, etc. .... Backward citations—citations to previous patent documents—can help to track knowledge spillovers in technology. They make it possible to estimate the curve of obsolescence of technologies, the diffusion of knowledge emanating from specific inventions to institutions, areas, regions, etc. .... Forward citations—the citations subsequently received by a patent—can be used to assess the technological impact of inventions, e.g., their cross-technology and/or geographical impact. The technological impact of inventions can indicate the economic importance of patents. The value of a patent and the number and quality of its forward citations have repeatedly been found to be correlated. Citation-weighted indicators (e.g., patent stocks of companies) have been seen to have a close relationship to economic indicators (market value of companies).

To understand the relationship between *backward citation* (i.e., patents cited by a new patent) and *forward citation* (i.e., the number of times a patent has subsequently been cited), consider the following example: If Patent A in 2005 is cited by Patent B in 2015, then Patent A is a backward citation of Patent B, whereas Patent B is a forward citation of Patent A. Some databases allow both backward and forward citation searching. Some examples of citations in patent documents and issues of interpretation of the data are discussed in OECD (2009, 117).

The OECD manual also provides a discussion of the role of science in the NPL citations, which include scientific articles, conference proceedings, deoxyribonucleic acid (DNA) databases and gene sequences, chemical structures, and other material. Caution is needed in interpreting the data, which are more abundant in patents filed in the US Patent and Trademark Office (USPTO) than elsewhere, and in some industries such as biotechnology, pharmaceuticals, organic chemistry, and IT. The basic idea, in earlier studies and subsequent papers, was that the higher the number of NPL citations, the closer the inventions are to basic science.[29] Thus, both cited patents and NPL represent the previously existing knowledge on which the new patent builds its idea.[30] Therefore, patent citations have become a broadly used proxy for estimating the social value of patented technologies.[31] The intuition is that a filed patent that receives many citations is worth more than one that does not.

The rest of this section briefly deals with how to take advantage of the extensive literature on the social value of patents from the specific perspective of CBA of RIs. In a CBA framework, the private returns and knowledge spillovers caused by patents granted by an RI represent a benefit that should be considered. Thus, the *marginal social value* of the patent generated by an RI should be forecast; this is the shadow price of a new patent in the perspective of society as a whole, which may be different from the financial value to the inventor.

The fact that patent citations reveal prior art that an inventor has previously learned makes them potential measures of the knowledge spillovers from past to current inventions. In other words, citations of a patent by many subsequent patents suggest that the patent generated significant spillovers because numerous developments build on the knowledge that it embodies. Given p as the set of patents over time $t = 0 .... T$, $MSV_{(PRIV_{pt}, EXT_{pt})}$ as the patent marginal social value, and s_t as the discount factor, the expected present value of this benefit is expressed as

$$\mathbb{E}(\text{PV}Patent) = \sum_{p=1}^{P} \sum_{t=0}^{T} s_t \cdot \mathbb{E}(MSV_{(PRIV_{pt}, EXT_{pt})}), \qquad (5.2)$$

where the marginal social value (MSV) includes (linearly or nonlinearly) both the private value of patents ($PRIV_{pt}$) and the externality (EXT_{pt}) (i.e., the knowledge spillover brought about by patents granted by an RI). It is worth observing that the private and social values of a patent can also diverge because a patent allows the owner to earn a temporary monopolistic rent in excess of normal returns on investment. Such rent is a transfer from consumers to producers and should not be counted as a benefit.

Only a case-by-case analysis can ascertain the extent that the shadow profits diverge from the financial profits once the rent component is subtracted. For example, a research-intensive industry where marginal costs and prices (and hence social and private values of patents) widely diverge is the pharmaceutical industry, which often and controversially works around the monopolization of rents arising from patented drugs, allegedly to protect research investments.[32]

In practice, a possible empirical strategy for the estimation ex ante (or ex post) of this benefit would involve the following steps:

1. Estimating the number of patents that are registered over time by third parties correlated with the RI research (e.g., patents citing the scientific literature stream originated at the RI, as covered in chapter 3). This step involves either using previous data on the track record on patenting or referring to observational data related to similar RIs, if available. Alternatively, considering shortcuts may be useful, including the correlation between the existing statistics on the number of patents granted and the number of R&D personnel in a given area, industry, or domain.

2. Forecasting the average rate of usage of granted patents (*use*). This activity involves, again, either using previous data on patenting or referring to observational data related to citations of patents issued in the same scientific field or similar infrastructures, if available. The average rate of patent usage, proxied by the median number of lifetime forward citations per patent, is important to understand the actual rate of exploitation and, in turn, the knowledge spillovers resulting from patents related to an RI.

3. Forecasting the average number of references (Ref_p)—that is, backward citations to existing patents, which are typically included in patents issued in the relevant technological field.

4. Estimating the marginal *private* value of patents, carefully avoiding double-counting, given the change in expected profits from the sale of innovations. In fact, depending on the estimating method used, the value of a patent may or may not already include the market value of the patented invention. In principle, the patent value should be based on the discounted sum of the yearly profits that the patent holder expects to earn because of the patent, net of the equivalent discount stream of profits without the patent.[33]

5. Converting financial profits into economic profits in the ways discussed in chapters 1 and 2. In practice, often a return gross of tax, interest, and depreciation would be an initial benchmark value, and then further corrections for input and output value distortions may be considered (particularly shadow wages and monopolistic power).

6. Estimating the externality of patents in monetary terms. As has been mentioned, patent citations mirror the technological importance of a patent for the development of subsequent technologies.[34] In other words, a citation is a measure of the knowledge spillovers from past inventions to the current invention.

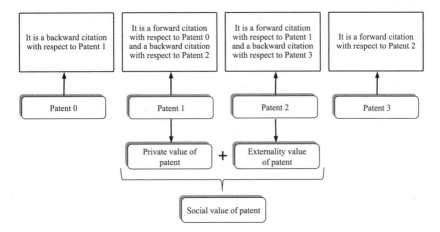

Figure 5.1
The social value of patents. *Source*: Florio et al. (2016).

However, simply counting citations does not provide information on a patent's value in monetary terms. To attach a monetary value to the stream of citations produced by a patent, the following formula can be used (but only as a statistical shortcut):

$$\mathbb{E}(EXT_{pt}) = \sum_{p=1}^{P} \sum_{t=0}^{T} use \cdot \frac{\mathbb{E}(PRIV_{pt})}{Ref_p}, \tag{5.3}$$

where EXT_p is an externality from patents, *use* is the average rate of usage of granted patents, Ref_p is the average number of references included in patents issued in the relevant technological field, and $PRIV_{pt}$ is the private value of patents granted in the relevant technological field.

A range of statistics (including patents and citation counts) useful for forecasting the number of patents can be retrieved from several repositories. However, importantly, only patents granted by patent offices generate both a private value and a knowledge spillover. Instead, patent applications or invention disclosures that are not granted may have a private value of zero, while their externalities could be positive. When patent-granted statistics are not available, some assumptions should be made about the number of patents that are eventually registered.

Concerning the estimation of the private value of patents, various empirical approaches have been suggested in the literature. The concept of private value takes into account only the economic value of the patent for its holder. Thus, the private value can be defined as for any economic asset, in terms of the discounted sum of the expected cash flows accruing from the patent to the owning entity. The following main lines of work have been followed by researchers to estimate or infer the private economic value of patents. Estimates based on a patent holder's behavior include methods that analyze either the decisions to renew patents (or not) and pay the related fees[35] or the economic terms of actual patent transactions.[36] Estimates based on inventors' surveys involve directly asking the inventor to provide an estimate of the value of his or her patents on the basis of the price at which he or she would be willing to sell the patent.[37] Estimates based on external investors' valuations include methods either based on stock market valuations of patent portfolios of publicly listed companies[38] or valuations made by venture capital firms of intellectual property-based start-up companies.[39]

At the European level, a reference study was published by the European Commission, in 2006.[40] See table 5.2. The analysis relies on a questionnaire survey of almost 10,000 inventors in eight European countries.[41] Patents belonging to a number of technology classes were considered. To obtain a measure of patent value, inventors were asked to provide their best estimate of the value of their patents on the basis of the price at which they would be willing to sell them.

Because patent values are acknowledged to vary significantly across sectors, technological fields, and geographic areas, in general, it is useful to consider country/region, sector, and technology-specific statistics when available. A final remark on this benefit: The previous discussion focused on nonpatenting policies by publicly funded RIs; however, there may be different situations. If an RI managing body decides to patent an invention, no externality is created in the first round. In a CBA framework, the revenues from holding the patent then will be entered as any other revenues of the RI, in the way discussed in chapter 2, and will contribute to decreasing the financial and the social cost of the RI, respectively. However, there may still be externalities in later rounds.

5.5   Technology Transfer Contracts, Start-ups, and Spin-offs

Beyond procurement for innovation and nonpatenting policies, the third set of benefits arising from learning and knowledge spillover mechanisms is related to formal policies of technology transfer, which may include certain R&D contractual arrangements between the RI and firms, and the promotion of start-ups and spin-offs by an RI. Sometimes these words have different meanings. For example, NASA has been tracking products that are considered its spin-offs since 1976 and now has a database including 2,000 spin-off case studies,[42] the majority of which are associated with the Langley Research Center, the

Table 5.2
Average patent values by country and technological area

	Average Patent Value (EUR thousands)	Median Patent Value (EUR thousands)
Technological area		
Pharmaceuticals, cosmetics	5,260	605
Macromolecular chemistry, polymers	3,980	449
Space technology weapons	3,854	414
Environmental technology	3,250	354
Biotechnology	3,134	336
Semiconductors	2,555	284
Telecommunications	2,331	247
Electrical devices, engineering, energy	1,938	211
Country		
Denmark	2,947	300
France	2,922	293
Germany	2,958	305
Hungary	3,647	408
Italy	3,007	297
The Netherlands	2,788	285
Spain	3,029	307
United Kingdom	3,355	332

Source: Adapted from European Commission (2006).

Johnson Space Center, and the Marshall Space Flight Center. In general, an average of forty-eight spin-offs is generated every year. NASA (2018b, 6) reports the most recent cases in such sectors as health and medicine, transportation, public safety, consumer goods, energy and environment, IT, and industrial productivity. These spin-offs are defined as follows:

A commercialized product incorporating NASA technology or expertise that benefits the public. These include products or processes that:

• were designed for NASA use, to NASA specifications, and then commercialized;

• are developed as a result of a NASA-funded agreement or know-how gained during collaboration with NASA;

• are developed through Small Business Innovation Research or Small Business Technology Transfer contracts with NASA;

• incorporate NASA technology in their manufacturing process;

• receive significant contributions in design or testing from NASA laboratory personnel or facilities;

• are successful entrepreneurial endeavors by ex-NASA employees whose technical expertise was developed while employed by the Agency;

• are commercialized as a result of a NASA patent license or waiver;

• are developed using data or software made available by NASA.

While the NASA definition is clear, there is a certain ambiguity of use of the word *spin-off* in the literature and policy documents. It could be similar to the abovementioned concept—that is, "A product that develops from another more important product"—for example, "The research has had spin-offs in the development of medical equipment."[43] In such a case, there may be no legal transaction between the RI and the innovative firm. Or a spin-off may be a *corporate spin-off*, which is defined as "the separation of a subsidiary, creation of an independent company through the sale and/or division of a corporation from its parent company by issuing shares in a new corporate entity."[44] The former case is discussed in chapter 7, while here I refer to knowledge transfer (KT) as legal arrangements between an RI and an existing or new firm.

From this perspective, KT is related to formal agreements through which the RI management involves third parties in the economic development of innovations. If an RI contributes to such agreements or the establishment of start-ups and spin-offs, the economic value of the social benefit of such activities is determined as the *expected shadow* profit gained by the new business during its overall expected lifetime, as compared with the without-the-project scenario.

Although the return on equity to investors and the operating revenues deriving from the sale of consultancy services, leading to the establishment of, a corporate spin-off in the form of a start-up, for example, are considered inflows in the financial analysis of an RI, these factors are not included in the economic analysis in order to avoid double counting. When an RI, through specific KT arrangements, contributes to increasing the survival rate of start-ups (see figure 5.2 for survival rates in general), the benefit is valued as the incremental expected shadow profit attained by businesses that survive longer than they would have in the without-the-project scenario. This topic has been studied for university spin-offs, but the approach is similar to what could be done in the case of RI spin-offs. Some relevant empirical studies include Oskarsson and Schläpfer (2008), for the survival rates of university spin-offs in the Netherlands, France, Sweden, and Northern Ireland, as well as at Oxford University and ETH Zurich (see table 5.3). The ex ante estimation of this benefit involves:

1. Forecasting the *number* of KT activities, start-ups, and/or spin-offs expected to be created by the RI during the entire reference period

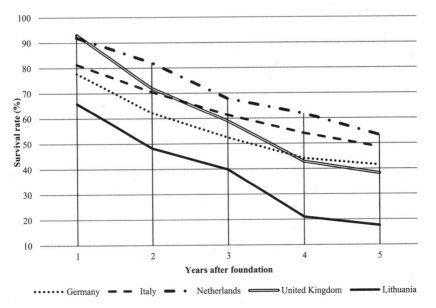

Figure 5.2

Average firm survival rates in different countries. *Source*: Author's elaboration on Eurostat business demography statistics (2012), from Florio (2016).

Note: All sectors of industry, construction, and services, except insurance activities of holding companies.

Table 5.3

Survival rates of university spin-offs in various countries and two universities

Country	Survival Rate (%)	Years	Source
ETH—Zurich	88	10	Oskarsson and Schläpfer (2008)
France	84	4	Mustar (1997)
The Netherlands	83	9	Shane (2004)
Northern Ireland	94	12	Shane (2004)
Sweden	87	34	Shane (2004)
UK—Oxford	81	9	Lawton Smith and Ho (2006)

Source: Adapted from Oskarsson and Schläpfer (2008) and Florio et al. (2016).

2. Establishing the *expected lifetime and survival rate* of start-ups and spin-offs (when the RI contributes to increasing the life expectancy of start-ups, the expected increase in their survival rate must be estimated)

3. Estimating the expected *shadow profit* generated by start-ups and spin-offs created by the RI

The median number of start-ups and/or spin-offs created by RIs in specific countries and sectors, as well as their expected lifetimes and survival rates, can be proxied by looking at similar infrastructures in other contexts or retrieved from an official statistical database or the literature.[45] Other more indirect mechanisms of knowledge spillovers from RIs to firms are discussed in chapter 7, as these spillovers ultimately benefit consumers if markets are reasonably competitive.

5.6   Case Study: How CERN Procurement Creates Value

As has been mentioned, previous research on the economic effects of CERN procurement has drawn mainly on surveys of suppliers. This literature has stressed the importance of industrial knowledge spillovers[46] and considered the important role of CERN as a risk-taker in the realization of complex scientific projects.[47] Florio, Forte, and Sirtori (2016) estimate the incremental profits based on LHC-related procurement orders (categorized according to activity and technological intensity codes) forecast up to 2025, and then use these values to determine the incremental turnover for the suppliers, through estimates of economic utility/sales ratios and earnings before interest, taxes, depreciation and amortization (EBITDA) margins (gross profits/sales) for companies in related sectors; data are extracted from the Orbis database (Bureau van Dijk) containing companies' balance sheets.[48]

The total value of CERN procurement, by year and by activity code, has been recovered from the CERN Procurement and Industrial Services Companies. A sample of 300 orders, exceeding CHF 10,000 in nominal value, has been extracted from a data set provided by the aforementioned CERN office. Each sampled order has been classified (with the help of expert CERN staff), according to a five-point scale: (1) very likely to be off-the-shelf products with low technological intensity; (2) off-the-shelf products with an average technological intensity; (3) mostly off-the-shelf products, usually high-tech and requiring some careful specifications; (4) high-tech products with a moderate-to-high specification activity intensity to customize the products for LHC; and (5) products at the frontier of technology, with intensive customization work and codesign involving CERN staff. An average technological intensity score has been attributed to each CERN activity code; high-tech codes are those with an average technological intensity class equal to or greater than 3. This led to the identification of twenty-three high-tech activity codes. Procurement value was then computed only for orders related to these codes, which turned out to be 35 percent of the total of procurement expenditures. This would be only 17 percent if we exclude orders below CHF 50,000, and 58 percent if we include orders below this threshold and for other activity codes. A share of 84 percent of yearly total expenditure of collaborations is attributed to external procurement, using the same share as CERN. This share has also been used for the future forecasts of both CERN and the collaborations up to 2025, based on the previous forecast of cost trends.

Then, 1,480 benchmark firms were identified from the Orbis database in the year 2013 and in six countries (Italy, France, Germany, Switzerland, the United Kingdom, and the United States). These countries were selected because they received 78 percent of the total CERN procurement expenditure between 1995 and 2013.[49] In selecting this sample, companies whose primary activity matches with the corresponding CERN suppliers were selected. The EBITDA sample average (13.1 percent) and standard deviation, weighted by country, were used to define a normal distribution of the EBITDA. The incremental turnover over five years was estimated by the Large Electron–Positron Collider (LEP) average utility/sales ratio to be equal to 3, based on the results of Bianchi-Streit et al. (1984) and Autio, Bianchi-Streit, and Hameri (2003), which, in turn, are within the range of other studies, as reported in table 5.1. This ratio has been applied to the high-tech procurement of both CERN and collaborations. Finally, the additional sales times EBITDA margin was computed, thus estimating the incremental profits of firms in the LHC supply chain in other markets. It was found that the incremental shadow profits of suppliers are estimated at around EUR 2 billion and pay back around 15 percent of the LHC total costs.

While the abovementioned estimation was based on simple extrapolation from previous research at CERN, including on the LEP (a form of benefit transfer method), other more recent studies by the University of Milan team use different approaches.

In one paper, Castelnovo et al. (2018) study LHC suppliers from thirty-five countries for orders above CHF 10,000. Changes in intangible fixed assets, patents, sales per employee, operating revenues, profits, and profit margins of each of around 350 firms are analyzed for a twenty-four-year period (1991–2014), before and after the year of the first LHC-related order.

Figure 5.3 traces the distribution over time of the orders assigned in the restricted sample, as well as the number of new suppliers and of initial procurement events, which is the variable of interest, as it marks the starting point of firms' collaboration with CERN for the LHC project. These first-time orders, initial events, mark the start of a potential learning process, a reputation effect, or both. The variable of interest is modeled as a simple binary code,

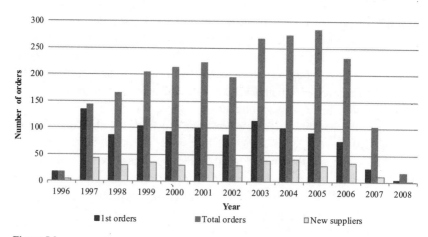

Figure 5.3

Yearly distribution of LHC procurement orders. *Source*: Own elaboration based on CERN data, from Castelnovo et al. (2018).

taking value 0 before and 1 after the year of the first order. The subsequent time profile is informative: the average number of years in which a company received at least one order is 2.2 (standard deviation 1.9, with a minimum of 1 and a maximum of 11 years). This implies that the direct impact of orders on company profitability could not last more than 2 or 3 years, on average. However, there are economic effects beyond these initial years. The first-year events in the study are evenly spread between 1996 and 2007 (unlike total orders, which peak in 2003–2005). Around ninety new suppliers were involved in the LHC in any of the dozen years considered.

The empirical analysis exploits the fact that unlike most event studies, there is not just a single "before-after" event, but a sequence of them. For example, a company entering the CERN procurement system in 2000 can potentially be observed for ten years before the event (including 2000), and fourteen years after (as we have financial data from 1991–2014). For a company entering in one of the last groups (e.g., 2007), we have potentially seven years of observations "after" and eighteen years "before." Using the Amadeus and Orbis databases, Castelnovo et al. (2018) searched for company performance data between 1991 and 2014, and six outcome variables in particular: intangible assets, number of patents filed, sales per employee (proxying for productivity), and three profitability measures, Earnings before interest and taxes (EBIT), operating revenue, and the EBIT margin (EBIT over operating revenue). The macroeconomic control variables analyzed are yearly gross domestic product (GDP) growth and inflation in the supplier's country.[50]

On the demand side, GDP growth may affect the outcome variables as faster growth presumably increases the demand for all goods, and hence, firms' sales and profits. Inflation (measured by the consumer price index) serves to control for price changes that may affect the real value of our performance indicators (which are in nominal terms), given the long time span covered, all the Orbis and Amadeus data being in current euros and at current exchange rates.

The results show a positive and statistically significant correlation over time between CERN procurement events and each of the outcome variables we consider, controlling for observable firm characteristics and macroeconomic conditions, as well as for unobserved time, country, industry, and firm-level fixed effects. After becoming suppliers, on average, companies experienced a rise in intangible assets (our proxy for R&D effort) and in annual patent filings (our proxy for innovation). Labor productivity, proxied by sales per employee, also increase, as revenue, EBIT, and EBIT margin. A simultaneous equation estimation model confirms the results. See appendix 5.A for details.

In another paper (Florio et al. 2018), information from more than 650 firms that received at least one order from CERN between 1995 and 2015 was collected through an online survey, conducted in 2017. Firms provided information on the types of service delivered to CERN, the relationship established with CERN, the technological intensity of the order, and the variety of effects produced by procurement on their performance as perceived by the firms themselves. Responses were processed by means of Bayesian network analysis, which enables the identification of the causal mechanisms through which CERN procurement generates learning and economic benefits for suppliers and, specifically, the identification of the role that different interaction modes between CERN and suppliers play in producing knowledge spillovers. The main finding is a network of statistical effects, which range from the intensity of the relationship with CERN staff during procurement to firms' innovations and, ultimately, their performance in other markets. In more detail, the survey resulted in 669 questionnaires collected from suppliers located in thirty-three countries. These firms processed 8,247 orders in the period 1995–2015 for a total amount of EUR 0.75 billion. Suppliers were asked about the innovation level of the products and services supplied to CERN. The governance structure was investigated by asking suppliers about the type of relationship established with CERN during the procurement activities. See table 5.4.

As a result of their relationship with CERN, supplier firms were asked about its impact on their company: 55 percent of firms agree that thanks to their relationship with CERN, they increased their technical know-how, followed by improvement of products and services (48 percent). As a result of new knowledge acquired and these improvements, the interviewed firms stated

Table 5.4

Innovation level of products and the governance structures

Question	N	%
Innovation level of products and services (a)		
Products and services with significant customization or requiring technological development	331	–
Mostly off-the-shelf products and services with some customization	239	–
Advanced commercial off-the-shelf or advanced standard products and services	172	–
Cutting-edge products/services, requiring R&D or codesign involving the CERN staff	158	–
Commercial off-the-shelf and standard products and/or services	110	–
Governance structure		
During the relationship with CERN, we carried out the project(s) on the basis of:		
The provided specifications, but with additional inputs (clarifications, cooperation on some activities) from CERN staff	349	52
The provided specifications with full autonomy and little interaction with CERN staff	181	28
Frequent and intense interactions with CERN staff	133	20

Source: Florio, Giffoni, and Giunta (2018).

Note: (a): multiple responses are possible. N: number of respondents.

that they had developed the following innovation outcomes (suppliers could choose more than one option): "new products" (282 suppliers); "new services" (203 suppliers); "new technologies" (138 suppliers); "new patents, copyrights, or other intellectual property rights (IPR)" (22 suppliers). Sixty-five respondents denied having experienced any technological outcome, and therefore, the option "none of the suggested options" was chosen. About 60 percent of firms used CERN as an important marketing reference and declared that they had improved their credibility as suppliers. Concerning the acquisition of new customers, on average, and regardless of the type of customer, 20 percent of the interviewed firms stated that they had acquired new customers.

Many studies show that technological spillovers caused by Big Science laboratories through their procurement activity are likely to spread out along the entire supply chain.[51] First, firms were asked whether they mobilized any subcontractors to carry out the CERN projects and, in that case, they were asked to list the number and the country of origin of the subcontractors, as well as the way that subcontractors were selected: 256 firms (38 percent) mobilized about 500 subcontractors (on average, 2 subcontractors per firm), located in twenty-six countries. Good reputation, trust developed during previous projects, and geographic proximity were the most quoted ways through which interviewed firms selected their partners. This study finds that knowledge spillovers are likely to be enjoyed by subcontractors as well.

The bottom line of the abovementioned studies is converging evidence (with varying quantitative methods and sampling procedures) of the positive long-term impact on firms' economic performance of entering a technological procurement relationship with CERN.

5.7   Further Reading

Selected material on the following topics is discussed: (1) returns on R&D in general; (2) spillovers to firms from academia; (3) the role of procurement; (4) KT activities; and (5) patent data. This section draws from Florio, Forte, and Sirtori (2016); Del Bo (2016); Florio et al. (2018); and Castelnovo et al. (2018).

5.7.1   Returns on R&D

For a survey of the wide empirical literature on the returns on R&D in general, see Hall, Mairesse, and Mohnen (2009). They find that the private returns to R&D are positive and higher than for other investments, and the social returns are higher, but not well measured. The core problem is how to measure the externalities of R&D. Two main approaches have been employed to estimate the magnitude of R&D spillovers empirically. The *technological flow* approach positions firms or industries within a matrix of technological linkages, using either input-output or technology matrices based on patent data. The spillovers from the R&D activities of one firm or industry on the others are examined.[52] The *cost function* method is an econometric approach that estimates the impact of spillovers on the cost and production structure of the receiving firms or industries.[53] For an extensive review of the studies using these two alternative procedures, see Nadiri (1993).

5.7.2   Spillovers from Universities

In the case of private firms' relations with nonmarket organizations, such as universities and research centers, the learning spillovers are enhanced (Bernstein and Nadiri, 1991). Jaffe's (1989) seminal paper empirically demonstrates spillovers from university research to business innovation, measured by corporate patents. The beneficial side effects of academic research are also found in business start-ups (Bania, Eberts, and Fogarty, 1993) and high-technology innovations (Anselin, Varga, and Acs, 2000). Mansfield (1991, 1998) demonstrated that a significant portion of firms' product and process innovation in the United States could not have come in the absence of university research, save with a substantial delay. The closer firms are to major academic research centers, the greater the benefits (Mansfield and Lee, 1996). More recently, Helmers and Overman (2017) have discussed the correlation between proximity to the Synchrotron Diamond Light Source in the United Kingdom and the productivity of academic research (see also Hamaguchi et al., 2017).

Publicly funded basic research at universities does not substitute for private R&D, but rather stimulates and enhances it (Rosenberg and Nelson, 1994).

5.7.3   The Role of Procurement

Procurement is a possible source of learning from discontinuous innovation in the Arrow-Solow frame, as information is never symmetrical between the contracting parties ex ante, which requires a delicate balancing of risks and incentives (Bajari and Tadelis, 2001). Accordingly, procurement has been described as a learning process itself (Newcombe, 1999), and public procurement in particular has been studied from various standpoints as a driver of innovation (Edquist et al., 2015; Edquist and Zabala-Iturriagagoitia, 2012). Edquist and Hommen (2000, 5) define public procurement for innovation as a situation in which "a public agency places an order for a product or system which does not exist at the time, but which could (probably) be developed within a reasonable period. Additional or new technological work is required to fulfill the demand of the buyer." Public procurement is also considered as an important demand-side innovation policy (Aschhoff and Sofka, 2009; Martin and Tang, 2007), particularly when the development of sophisticated products is required (Salter and Martin, 2001). For this reason, the role of PPI in promoting radical innovations (including general-purpose technologies) is relevant in economic fields characterized by the high risk that cannot be borne entirely by the private sector (Mazzucato, 2016).

A concrete example concerns Swedish military jets (Eliasson, 2010). The development of a new aircraft involved a complex network of suppliers and close public-private cooperation, leading the aircraft industry to become similar to a technical university where continuous learning occurs. PPI has a positive effect on firms' R&D investment, with a demand-pull effect that is larger than that of other private contracts (Litchtenberg, 1988) and is a possible complement of, or even an alternative to, supply-side policies (Edler and Georghiou, 2007). Guerzoni and Raiteri (2015) show that PPI has a major impact on firms' expenditure in innovative activities, stronger than that of R&D subsidies and tax credits.

The context of procurement contracts for large-scale RIs is considered by Autio, Bianchi-Streit, and Hameri (2003). The first paper is based on a survey of 154 major suppliers (1997–2001), representing around half of CERN's total procurement budget in those years. The main findings are as follows: 38 percent of the suppliers designed new products, 13 percent created a new R&D team, 14 percent created a new business unit, 17 percent entered new markets, 42 percent reported increased international operations, 44 percent reported technological learning, and 36 percent said that there was market learning. Earlier research at CERN (Bianchi-Streit et al., 1984), related to the construction of the Super Proton Synchrotron (SPS) accelerator (1963–1987), is also based on interviews with suppliers (160 out of 519 providing high-tech components). Åberg and Bengtson (2015) and Bengtson and Åberg (2016), in two recent papers, also critically examine the CERN procurement process as an innovation driver, based on around 100 interviews, and find that, in certain cases, CERN does not provide the right innovation incentives to suppliers because of the need to ensure a fair return to members' states.[54]

Important insights are also provided by Salter and Martin (2001); for a more general discussion, see Stephan (1996). For qualitative case studies, see, inter alia, Fahlander (2016), on the impact on local firms of two accelerators (HIE-ISOLDE and ESS); Pero (2013), on three accelerators for material science (Elettra Synchrotron in Trieste, ESRF synchrotron in Grenoble, and XFEL in Hamburg); and OECD (2014b), with a specific focus on LHC magnets, a core feature of the accelerator, as discussed in some detail by Rossi and Todesco (2009), and hadrontherapy (Battistoni et al. 2016). Autio (2014) reviews the literature on the impact of Big Science and discusses seven case studies in the United Kingdom.[55]

5.7.4   KT Activities

To improve the relationship between industry and RI, an ESFRI (2016c) working-group report has suggested several instruments: to create and support the Industrial Liaison Officers and Industry Advisory Boards; to raise aware-ness of RI access and services for industry; to develop a transparent data man-agement policy; to anticipate the foresight of purchase of large equipment in European RIs; to support the predevelopment of highly innovative compo-nents; to define road maps and strategic agendas; to develop new collaborative frameworks for coinnovation between RIs and industrial companies; and to promote the development of local or regional ecosystems integrating RIs, T-infrastructures, technology and service providers, and industrial users. On knowledge transfer activities at CERN, including licensing intellectual property, releasing software and hardware under open licensing, and engaging in internal collaborations among firms, see Nilsen and Anelli (2016). More on this topic will be covered in section 6.8 of chapter 6.

5.7.5   Patent Data

Sources of patent data include the European Patent Office (EPO) Worldwide Patent Statistical Database (also known as EPO PATSTAT), the Eurostat statistics available under the "Science & Technology" category; the Trilateral cooperation website, which provides statistics from the EPO, Japan Patent Office (JPO), and USPTO, dating back to 1996 in the annual Trilateral Statistical Reports; the reports provide an overview of worldwide patenting activities. The World Intellectual Property Organization (WIPO) website provides patent and Patent Cooperation Treaty statistics. The OECD's Patent Indicators reflect trends in innovative activity across a broad range of OECD and non-OECD countries, with six main sections: EPO, USPTO, and JPO patent families; patenting at the national, regional, and international levels; patenting in selected technology areas; patents by institutional sector; international co-operation in patenting; and European and international patent citations. The OECD data shows how skewed the distribution of licensing revenue per license is. (OECD, 2009, 2013).The European Investment Bank (2013) reports an average yearly revenue per license in Germany of EUR 55,000, with a higher average (EUR 200,000) for licenses of the Max Planck Institute. This issue is also discussed by the CBA guide (European Commission, 2014). See OECD (2009) for further references.

5.A.1   Appendix: An Empirical Model of Procurement

Castelnovo et al. (2018) estimated a simultaneous equation model to highlight the complex linkage between procurement and economic performance. Building on the standard Crépon, Duguet, and Mairesse (1998) model, they added an initial trigger provided by the CERN commissioned procurement. This is done by including in the R&D equation a dummy variable that takes value 0 before and 1 after the year of the first order. In addition to productivity, earnings and profitability are investigated in the fourth equation of the system.

Because there are endogenous variables both on the left and right sides of each equation, the system was estimated by using a three-stage least squares (3SLS) procedure (see Zellner and Theil, 1962). Moreover, the 3SLS estimator allows cross-correlations in the residuals of the equations in the system and is thus more efficient than two-stage least squares (2SLS) (see Cameron and Trivedi, 2005).

Table 5.A.1
3SLS Estimation of the impact of procurement

	(1)	(2)	(3)
	Full Sample	High-Tech	Non-high-Tech
ΔR&D			
CERN	0.668***	1.256***	−0.250
	(0.050)	(0.129)	(0.398)
ΔEmployees (mln)	−0.0787	−0.0370	−0.125
	(0.067)	(0.085)	(0.111)
ΔTangible Assets (bln)	−5.950***	−1.664***	11.23
	(0.482)	(0.312)	(12.37)
Macroeconomic controls	Yes	Yes	Yes
Country FE	Yes	Yes	Yes
Time FE	Yes	Yes	Yes
Sector FE	Yes	Yes	Yes
Cons	−1.540	−25.79***	−1.089
	(3.671)	(4.090)	(5.636)
ΔPatents			
ΔR&D	7.446***	3.956***	−0.102***
	(0.459)	(0.330)	(0.034)
ΔEmployees (mln)	1.171***	0.675***	−0.0152*
	(0.099)	(0.086)	(0.009)
ΔTotal Assets (bln)	−4.290***	−2.170***	0.506***
	(0.306)	(0.220)	(0.140)
Macroeconomics controls	Yes	Yes	Yes
Country FE	Yes	Yes	Yes
Time FE	Yes	Yes	Yes
Sector FE	Yes	Yes	Yes
Cons	9.142***	103.7***	−0.384
	(3.466)	(9.051)	(0.362)
ΔProductivity			
ΔPatents	2,848.6***	2,535.5***	2,815.7***
	(731.7)	(731.2)	(443.9)
ΔEmployees (mln)	−0.0491***	−0.0369***	−0.0510***
	(0.013)	(0.012)	(0.013)
ΔTotal Assets (bln)	−15.52	−19.94	−113.0
	(42.28)	(42.35)	(118.8)
Macroeconomic controls	Yes	Yes	Yes
Country FE	Yes	Yes	Yes
Time FE	Yes	Yes	Yes
Sector FE	Yes	Yes	Yes
Cons	57.26	−258.1	801.7**
	(638.3)	(803.4)	(371.3)

(continued)

Table 5.A.1 (continued)

	(1)	(2)	(3)
	Full Sample	High-Tech	Nonhigh-Tech
ΔRevenues			
ΔProductivity	209.6***	221.2***	211.7***
	(25.68)	(30.16)	(73.97)
ΔEmployees (mln)	8,739.2***	16,710.0***	5,434.3***
	(1,415.5)	(2,678.4)	(2,015.8)
ΔTotal Assets (mln)	21,310.4***	8,888.5	62,310.0***
	(4,663.1)	(6,481.9)	(11,839.7)
Macroeconomics controls	Yes	Yes	Yes
Country FE	Yes	Yes	Yes
Time FE	Yes	Yes	Yes
Sector FE	Yes	Yes	Yes
Cons	−1,949.1	−11,787.3	−9,095.6
	(61,762.9)	(96,571.7)	(23,364.0)
N	2,852	1,876	976

Standard errors in parentheses; * $p < 0.10$, ** $p < 0.05$, *** $p < 0.01$.

Specifically, the system of equations estimated by Castelnovo et al. (2018) is the following:

$$
\begin{cases}
\Delta R\&D_{jt} = \beta_1 CERN_{jt} + \Delta Size'_{jt}\gamma_1 + \theta_1\Delta GDP_{ct} + \omega_1\Delta CPI_{ct} + \sigma_s + \eta_t + \rho_c + u_{jt} \\
\Delta Patents_{jt} = \beta_2 R\&D_{it} + \Delta Size'_{jt}\gamma_2 + \theta_2\Delta GDP_{ct} + \sigma_s + \eta_t + \rho_c + e_{jt} \\
\Delta Productivity_{jt} = \beta_3\Delta patents_{jt} + \Delta Size'_{jt}\gamma_3 + \theta_3\Delta GDP_{ct} + \omega_3\Delta CPI_{ct} + \sigma_s + \eta_t \\
\qquad + \rho_c + \varepsilon_{jt} \\
\Delta Revenues_{jt} = \beta_4\Delta Productivity_{jt} + \Delta Size'_{jt}\gamma_4 + \theta_4\Delta GDP_{ct} + \omega_4\Delta CPI_{ct} \\
\qquad + \sigma_s + \eta_t + \rho_c + \epsilon_{jt}
\end{cases}
$$

$$(5.A.1)$$

where $\Delta R\&D_{jt}$ is proxied by the yearly change of intangible fixed assets per employee; $CERN_{jt}$ is a dummy variable that takes value 0 before the first order is received and 1 hereafter; $\Delta Size'_{jt}$ is a vector including information on the yearly change of assets and number of employees; $\Delta Productivity_{jt}$ is proxied by the early change of sales per employee; $\Delta Patents_{jt}$ is the number of patents filed by company j in year t; $\Delta Revenues_{jt}$ is the yearly change of the operating revenues for the company j; ΔGDP_{ct} is the yearly percentage change of GDP in the firm's country c; ΔCPI_{ct} is the yearly percentage change in that country's consumer price index (CPI); η_t denotes time-fixed effects; and σ_s and ρ_c are time-invariant unobservable industry- and country-specific fixed effects, respectively. Finally, u_{jt}, e_{jt}, ε_{jt}, and ϵ_{jt} are the random error terms of each equation.

The variable *Patent* is linearized using the transformation $ln = (1 + patents)$,[56] thus allowing the knowledge production equation to be estimated in the framework of a linear system.

As can be seen from table 5.A.1, the coefficients obtained from the estimation of the system highlight the direct effect of procurement on R&D investments, as well as its mediated impact on company innovation output, productivity, and economic performance.

For high-tech companies, the estimates clearly show that the impact of procurement on innovativeness comes via R&D, which in turn affects productivity, whose rise finally enhances the economic outcomes.

For nonhigh-tech companies, by contrast, there is no positive influence of procurement on R&D and innovation output, suggesting that the significant association of productivity with revenues and profits that shows up in the final estimation stage is not driven by technological spillovers that boost technical know-how. Other factors like market penetration and reputational gains are likely to be more relevant.

6   Benefits to Users of Information Technology in the Big Data Era

6.1   Overview

The first sequencing of an individual genome by the Human Genome Project (HGP) took thirteen years to accomplish (1990–2003). It required around $3 billion of funding from the US Department of Energy and US National Institutes of Health (NIH) and later from the Medical Research Council and Wellcome Trust in the United Kingdom. The labs from these organizations then joined with other collaborations across six countries to take on the massive task.

On April 25, 2018, while I was writing this chapter, the Broad Institute of the Massachusetts Institute of Technology (MIT) and Harvard announced that their genomics team had sequenced the whole human genome of their 100,000th individual. That precise moment could be watched live on Facebook.[1] Since 2009, the Broad Institute has accumulated 70 petabytes of genetic data, and the stock of information is currently doubling every eight months, producing a whole genome sequence every 12 minutes, at the unit cost of $1,000.[2] According to some estimations (Stephens et al., 2015), given the current growth rate, by 2025 the total amount of genomics data will be greater than the cumulated data of social media such as YouTube and Twitter.

In fact, a distinctive feature of the new research infrastructure (RI) paradigm is the generation and analysis of an unprecedented amount of information. This feature is common across scientific fields, including life sciences, which, some decades ago, were still close to a traditional little science environment, as mentioned in the Introduction. The typical biologist's laboratory bench was endowed with test tubes arranged in racks, beakers, graduated cylinders, flasks, funnels, a Bunsen burner, a good scale, a crucible or mortar, reagents and reaction plates, and finally, an optical microscope. This instrument was the very icon, along with the telescope, of the seventeenth-century revolution in the study of nature—a revolution associated with Galileo Galilei (who was able to work with both instruments). Creating evidence of discovery in this context was a slow, largely manual, handcrafted process, mostly managed by an individual with a very small group of assistants.

Electron microscopes today may cost several million dollars. In their latest version, these microscopes generate a large amount of digital data. In the words of an NIH report (NIH, 2018, 1):

One of the revolutionary advances in microscope, detectors, and algorithms, cryogenic electron microscopy (cryoEM) has become one of the areas of science (along with astronomy, collider data, and genomics) that have entered the Big Data arena, pushing hardware and software requirements to unprecedented levels...As is the case with astronomy, collider physics, and genomics, scientists using cryoEM generate several terabytes of data per day.

The interesting aspect of this statement is its acknowledgment of a common trend among seemingly unrelated fields, such as physics, astronomy, and biology. The last generation of telescopes, microscopes, particle accelerators, synchrotron light sources, satellites for Earth observation, outer space probes, gene-sequencing platforms, and biodata repositories are all extremely data-intensive RIs, and they share similar data science challenges. This creates new challenges for information (and communication) technology (IT)—for example, in terms of data acquisition, storage, retrieval, analysis, interpretation, transmission, distribution, long-term data curation, and security of the information.

The communication dimension of IT is crucial because, as I discuss in this chapter, currently, the scientific community mostly accessess data through the Internet. Solutions to these IT challenges, given the open nature of contemporary science in RIs, tend to create free knowledge for other players, both within and outside the research environment. This spillover of knowledge is a positive socioeconomic externality. This mechanism is similar to the learning processes discussed in the previous chapter, but with higher speeds and greater pervasiveness of transmission from the RIs to users and third parties. To a certain extent, these externalities can be measured, valued, and then entered in the RI's benefits evaluation. This chapter first presents some examples of how the Big Data environment at major RIs has changed the IT landscape, even well beyond the RI's scientific scope, and then discusses how to value the social benefits of IT spillovers to third parties in a cost-benefit analysis (CBA) framework.

The core paths for the creation of socioeconomic benefits in this context are, on one side, data availability and, on the other, innovations in software and performance of IT. The first path is related to the direct impact on users of the information made available to them for free, thanks to the IT investment of the RIs. The second path instead relates to knowledge spillovers toward other scientific and professional fields of IT innovations, implemented within the RI. To simplify a complex story, data and software released in the public domain (in different forms) are actually donated by the RI to any potential user. The research question is how to value this gift.

A notable example of the first type of spillover mechanism is the impact on the life sciences from the HGP open-access database. The open-access policy, codified at an early stage of the HGP by a specific agreement among the main participants and made concretely feasible because of the digitization of the information, has contributed dramatically to changing the way that research is carried out in life sciences. Researchers can effectively build on existing knowledge, which is freely available as a public good. Certain types of research designs in life sciences in the last two decades just would not have been conceivable without the opportunities offered by such comprehensive and efficient biodata repositories, of which the HGP was the first and most important, and without the related informatics tools made available to data users. This double externality, from data and software, in turn has additional effects, as other research and applications draw from them for other projects in subsequent rounds, with a large cumulative effect. This impact is explored later in the chapter through the European Bioinformatics Institute at the European Molecular Biology Laboratory (EMBL-EBI), one of the core institutes providing computational and analysis support to the HGP.

One of the most famous examples of the second type of spillover, related to the creation of new technological knowledge in digital information management, is the invention of the World Wide Web (WWW) at the European Organization for Nuclear Research (CERN) in 1989. The Hypertext Markup Language (HTML), Uniform Resource Identifier (URL), Hypertext Transfer Protocol (HTTP), and WWW browser were initially conceived as means to improve the sharing of information among scientists working on particle physics experiments with a variety of computers. CERN decided to release all the software and specifications for free, and the WWW itself helped with the dissemination.

In the words of Tim Berners-Lee (the main developer of the WWW, along with Robert Caillau)[3]: "In those days, there was different information on different computers, but you had to log on to different computers to get at it. Also, sometimes you had to learn a different program on each computer. Often it was just easier to go and ask people when they were having coffee."

We are now so accustomed to using the WWW that perhaps we do not realize how recently this innovation was introduced—just slightly more than twenty years ago, and how dramatically it has changed our lives in a myriad of applications. The social and economic impact is huge (perhaps so much that nobody tried to measure it). Currently, another project at CERN, the Worldwide LHC Computing Grid (WLCG), allows physicists to deal with a volume of information that exceeds the capacity of any computing facilities in a single site.[4] While I am not sure whether the WLCG impact will be as revolutionary as the introduction of the WWW, it is still a good example of IT innovation in an RI context, spreading its effects in other contexts. More generally, RIs have been instrumental in the creation of both grid computing (the distributed combination of computing capacity for a common goal) and the data grid (an architecture to access distributed data by different users). These distributed IT architectures are now widely used in fields such as climatology, astronomy, and biology.[5]

Interestingly, the demise of the Manhattan Project–style, traditional, hierarchical, and "closed" Big Science model as the main way to organize research on a large scale (see the Introduction) paved the way for a new era: open scientific Big Data as a public good, shared by multiple communities. Contemporary RIs are key actors in this wide change in the global rules about the generation and management of information.

In this chapter, I discuss the following questions: To what extent does or will the involvement of large-scale RIs in the global production of Big Data have socioeconomic consequences beyond the immediate scope of science? Which are the specific mechanisms that may contribute to creating knowledge spillovers from the RI to data science, as well as advancements in IT in general? How is it possible to value empirically the positive externalities of IT and Big Data for science in different fields?

In this chapter, I exclusively focus on the *direct* effects concerning the benefits deriving from the RI to other private or public organizations, or to individual users of IT and databases. I do not consider the benefits conferred on the final users of innovation, such as patients affected by genetic diseases underpinned by life science research, because these final social effects are mediated by further actions and cannot be attributed exclusively to the availability of biodata.

Thus, it seems better to be conservative and focus only on the direct effects of IT innovations here, while returning to discussing the final effects on firms and citizens in chapter 7.

The structure of the chapter is as follows. The direct effects are explored first by highlighting three fields where changes in IT have had a significant impact: astronomy (section 6.2), high-energy physics (section 6.3), and biodata (section 6.4). Next, section 6.5 explains how externalities can be generated in these contexts and who benefits from them; section 6.6 presents empirical approaches to valuing such externalities; and section 6.7 presents two detailed case studies of valuation with the tools of CBA: impact on third parties of software (ROOT and Geant4) developed at CERN, and the value of open access to biodata at the EMBL's European Bioinformatics Institute. Finally, section 6.8 suggests some additional sources to read.

6.2 Astronomy Data from Galileo to the Square Kilometre Array

The amount of data collected by Galileo with his first rudimentary optical telescope was very limited. However, it was enough to ignite a revolution in knowledge. Stars that were eight times more distant than those visible with his naked eye now could be discovered (Cottrell, 2016). The gathered information was processed and stored in Galileo's brain in the first place—the only computer available to him. Observations collected in November 1609 with an improved telescope with a twentyfold magnification capacity were published a few weeks later in the *Sidereus Nuncius*, a slim booklet printed in 550 copies, written in new Latin to be readable by scholars outside Italy. It contained seventy drawings of the moon and other celestial bodies, famously including the graphical representation of the motion of Jupiter's moons. In total, these ninety-six pages forever changed our understanding of the world. The amount of textual information contained in the *Sidereus Nuncius* is equivalent to perhaps 300 kilobytes plus some kilobytes needed for each image, including the black-and-white, carefully penciled, but obviously low-resolution drawings by Galileo.

In approximately 2025, as well as in subsequent years, the Square Kilometre Array (SKA) radio telescope (see box 6.1 and figure 6.1) will collect electromagnetic information, not visible to any human eye, from space on the scale of zettabytes (1 ZB $= 10^{21}$ bytes),[6] which is perhaps the size of the amount of the yearly data traffic on the WWW at the end of 2016.[7] To put these statistics in perspective, a zettabyte is 1 million petabytes—or if you prefer, it is the equivalent of 1 *trillion* (one million million) gigabytes, the latter being a more familiar measure (as a comparison, this book is written on a laptop with a 16-gigabyte memory).

Indeed, the scientific program of SKA is very different from what Galileo was looking for. It includes, inter alia, the search for radio evidence of amino acids, the building blocks of life in exoplanets, and signs of extraterrestrial intelligence (should this exist), and it is able to emit radio signals that are visible to us (Cottrell, 2016). SKA is an amazing project that will be the largest radio telescope in the world, and IT technologies lie at its core.

The COST Office (2010) compares the data rate per second of SKA with the entire WWW traffic (which is predicted by a team at the University of

Box 6.1
The SKA Project in a Nutshell

According to the official website, SKA radiotelescope will observe a very wide radiofrequency range (from 50 MHz to 20 GHz) through artificial "eyes" placed over huge collecting areas between South Africa and western Australia. In its first phase, the South African section of the radio telescope will include about 200 paraboloid dishes scattered in clusters, separated by hundreds of kilometers around a central core. The Australian sections of the SKA will comprise an array of aerials, controlled by centrally based software (phased array feeds). Further, to catch a different range of frequencies (50–350 MHz), there will initially be over 130,000 aerials interconnected by fiberoptic cables, all controlled by unique software (Cottrell 2016). The data-processing components of SKA will include many field stations, a central station hosting a facility for data storage, and a central computer accessible to the international community.

According to the COST Office (2010), the SKA website, and Chrysostomou (2017), the IT progress needed for SKA will be significant. Each field station (a radio receiver with different types of antennas) will include sensors, digitizers, and data-processing units. Analog-to-digital converters will generate in parallel some petabits per second (equal to a 1 followed by fifteen zeros). The treatment of the information involves exascale computing power to reduce the information stream to some hundreds of terabits per second, and this information is fed to a central, single processing system after filtering the noise. The transmission of signals to the central station will require thousands of kilometers of optical cables. An exaflop computer (able to perform 1 billion billion operations per second) will be needed for mapping the sky, based on the collected data, and a system of processors of similar capacity will be needed to analyze the data received from the field components.

Minnesota[8] to be 1 petabit/second after 2020), and suggests that SKA will innovate relatively to existing commercial solutions in various fields. These include wireless communication; algorithms for signal processing (which may have applications in financial and retail markets, security and military intelligence, and weather and traffic monitoring); industrial collaboration in research and development (R&D) on reliability and maintainability of components (as SKA facilities will be located in deserts exposed to a wide range of temperature variability, with failure risk of the electronic components and the need for monitoring and self-repair routines); and software, which is predicted to include highly sophisticated programs for monitoring, control, radio interference mitigation, basic calibration, calculation, and special detection tasks (such as the identification of new pulsars and sky mapping). Moreover, while not specifically IT but crucial for its functioning, remote power generation for the SKA will rely on new solar and wind technologies that need to supply

SKA1 MID, South Africa	SKA1 LOW, Australia
Frequency range: 350 MHz to 14 GHz	**Frequency range: 50 MHz to 350 MHz**

~ **200 dishes** (including 64 MeerKAT dishes)	~ **130,000** antennas spread between **500 stations**
Total raw data output: **2 terabytes** per second, **62 exabytes** per year	Total raw data output: **157 terabytes** per second, **4.9 zettabytes** per year
Compared to the JVLA, currently the best similar instrument in the world:	Compared to LOFAR Netherlands, currently the best similar instrument in the world:
4x the resolution **5x** more sensitive **60x** the survey speed	**25%** better resolution **8x** more sensitive **135x** the survey speed

Figure 6.1

The SKA radiotelescope will be located in South Africa, Australia (in desert areas), and in the United Kingdom. The name refers to the huge collecting area of the device. In its final version, SKA will include thousands of dishes and up to 1 million low-frequency antennas. The objective of the astronomers is to monitor the sky using SKA with a high-image resolution, as a complement to optical and infrared observations by other projects. The computing resources needed to handle and store the astronomical data are unprecedented. JVLA is the Karl G. Jansky Very Large Array, located in New Mexico. LOFAR is a radiotelescope completed in 2012 for the study of radio frequencies below 250 MHz. *Source*: Courtesy of ©SKA, author's adaptation. https://www.skatelescope.org/.

electricity in a compatible way with highly sensitive radio telescopes and the associated computers, in such a way to avoid interference with radio signals or disturbances from current instability.

In the words of COST Office (2010, 9):

The SKA is more than a science project. It is a laboratory for innovations in radio communications, data transport, intelligent analysis, storage, and worldwide dissemination of knowledge…The team of scientists working on the SKA will inevitably come up with new ideas that could change the way we do things. If these ideas can be developed and their organizations receive patents and industrial partners, then SKA innovations could spread to the marketplace with considerable profit potential.

This citation exactly represents the second spillover mechanism mentioned in the introductory remarks in this chapter. Probably the most important future change is related to new artificial intelligence approaches that will gradually replace Galileo's eyes and brain to interpret data. Thus, machine-learning approaches will identify patterns of signals automatically. Memory will no longer be a passive storage facility, but rather a source of learning mechanisms. In a similar way as the human brain, artificial intelligence will be able to elaborate on information and evolve. The fact that the SKA data will be open to anybody for research purposes will produce a virtual laboratory to experiment on new artificial intelligence approaches, providing the raw materials for various teams.

Astronomy, however, is just one of many examples of the coevolution of the modern RI and the digitalization of information. The former could not exist without the latter, and the latter has been greatly stimulated by the needs of science.

6.3   High-Energy Physics on the Zettabyte Scale

As in astronomy, future data-intensive RIs in physics will need to process zettabytes of information per year. Currently, the Large Hadron Collider (LHC) at CERN generates scientific data from about 1 billion particle collisions per second, as observed by the detectors of the four main experiments and by other smaller ones. The experiments manage online data processing farms directly, close to the detectors, and then there are various tiers of offline reconstruction facilities. To put the statistics in perspective, according to some estimates, in 2014 the *total capacity* of 300 petabytes of content was available in the Facebook data warehouse,[9] while the CERN Data Center (in Meyrin), with more than 10,000 servers and over 174,000 physical processor cores and around 350,000 logical cores (an average of two threads per core), manages 1 petabyte per *day*—equivalent to 210,000 DVDs. CERN's Worldwide LHC Computing Grid receives up to 10 gigabytes per second.[10] Currently, a selection of all the acquired information of LHC's data collection in past years is stored: 150 petabytes on disks in Meyrin, another 250 petabytes on tape, and another 100 petabytes in a new secondary data center in Budapest. The WLCG distributes 30–50 petabytes of data per year to the scientific community of particle physicists for analysis, through 170 computing centers in forty-two countries.

Taking advantage of this challenging environment, CERN has developed, and releases into the public domain, software that was initially tailored for the needs of high-energy physics. In many cases, it turns out to be very useful for other application domains as well. Two examples of such software are ROOT, a library of tools for data analysis and visualization, and Geant4, software that simulates the effects of particles passing through matter, used in medicine, for simulating radiation damage in deoxyribonucleic acid (DNA), and in other industrial applications.

The benefits of such externalities were estimated by Florio, Forte, and Sirtori (2016) using yearly download statistics of the software code[11] and interviews with experts. In 2013, ROOT had approximately 25,000 users also outside the physics field, including the financial sector. With regard to Geant4,[12] there were approximately fifty research centers, space agencies, and firms that routinely used it (not including the substantial number of hospitals that use Geant4 for medical applications). As the CERN contribution to the total cost of the software is estimated to be 50 percent, the avoided cost for each of these thirty-eight centers is based on the contribution that they actually provided (the full Geant4 cost is applied to the remaining centers). The cumulated social benefits of the externalities from ROOT and Geant4, taken at their expected value, according to Florio, Forte, and Sirtori (2016), would repay approximately 20 percent of the total present value of the LHC's cost. Moreover, there is other software provided for free by CERN, but that is not included in this estimation.

The future IT scenario is even more challenging. According to the CERN Openlab's white paper (Di Meglio et al., 2017),[13] the High-Luminosity LHC, an upgrade of the existing particle collider, will generate data from 2025–2038, whose analysis will require 50 to 100 times the current computing capacity (see figure 6.2). In the meanwhile, the IT industry is expected to increase by only ten times the computing and storage capacity for the same cost. Something radical has to be invented—otherwise, the experiments will not be able to store and analyze the data they observe, at least without considerable extra funding to increase storage and computing.

The CERN Openlab is a public-private partnership (PPP), launched in 2001, currently involving some of the big names in the IT industry such as Huawei, Intel, Oracle, and Siemens. The EMBL joined in 2015, signaling the shared interest of biosciences and particle physics to tackle the future challenges of Big Data. In a white paper (Di Meglio et al., 2017), key areas of collaboration between CERN's scientists and engineers on one side and companies on the other, over a three-year cycle (2018–2020), are identified. More than forty use cases (i.e., projects arising from user needs) have been identified around four R&D topics (computing performance and software, machine learning and data analysis, and applications in other disciplines).

Openlab's first R&D topic is on how to design new distributed data-center technologies and infrastructures, in line with the scale of the information generated. The challenges identified are networking, to serve a global scientific community that is widely scattered but still needs reliable and secure local campus infrastructure; data-center architecture, given the trade-offs between own capacity building and recurring to commercial providers of clouding capacity; data storage, which is perhaps one of the fields where the fewest advancements in IT have been achieved in recent years (they are still mainly based on spinning disks and tapes, while new solid-state devices are explored); database technologies, including issues related to the management of a very large number of sensors for the control of information about the LHC, an extremely sophisticated machine in itself; and finally, cloud infrastructures and the possibility that global platforms may be a collective solution for major data-intensive RIs.

A second R&D topic concerns the computing and software innovation needed to match the gap between what will possibly be offered by the industry in the next ten years and what is actually necessary. An interesting issue here is the topic of code modernization, as creating software for high-energy physics is a collective venture involving literally thousands of contributors, from doctoral students to senior scientists, as well as developers who change over time. This implies that an alignment is needed between software created for physics objectives and optimization opportunities arising from computer science. Other challenges in this area include the development of heterogeneous platforms, alternative architectures, and high-performance accelerators.

The third topic, and perhaps the most fascinating area of breakthrough potential coming from the R&D at CERN, is related to machine learning and data analytics. In high-energy physics, this area started more than twenty years ago because the analysis of the high number of events produced by accelerators could not rely on humans, so automated identification with certain strategies was needed. However, machine learning is an area where tremendous progress recently has been generated outside science by big names in the industry, and the software we use is increasingly designed to learn from what we do (e.g., every time we start a search process on Google, the process changes). Thus, challenges in data acquisition, data processing, Big Data applications to high-energy physics from social media and web applications, and data engineering have been identified as clusters of use cases. Some of these issues may lead to the convergence of problems to be solved in fields such as self-driving cars, face recognition for security reasons, and simulation of events in LHC detectors.

Finally, a fourth R&D cluster jointly identified by the CERN Openlab and private companies is related to applications to other disciplines, including platforms for open collaborations, applications for life sciences and medicine, environmental and traffic monitoring, and astrophysics data management. Not surprisingly, in 2017, SKA and CERN announced a collaboration agreement on *extreme computing* because they have a similar scale of data problems.[14]

Another interesting issue for data-intensive RIs is long-term data curation, a problem arising from the fast evolution of ways that data are archived. This is something that all of us have experienced if some of our data are recorded on floppy disks (a technology probably unknown to younger readers), and we no longer have a driver to read them.[15]

I would mention here three use cases in areas ranging from particle physics IT to life science (here, the EMBL participation may be key), as they reveal how spillover mechanisms potentially can have an impact outside physics. First, the simulation of data on the development of biological tissues (including interactions of cells and their mechanical, chemical, and electrical parameters) can take advantage of techniques designed for new versions of the Geant4 tool. This is already the case, but the situation may greatly be enhanced by dynamic self-scaling systems working on commercial platforms and new deep-learning algorithms. Second, the analysis of increasing volumes of genomic data could take advantage of tools originally developed for high-energy physics (such as ROOT). Third, machine learning and data-quality monitoring for the LHC may be applied to the booming availability of personal data on human health, including ways to assist the diagnostic capacity of physicians and improve prevention initiatives.

Nobody knows which of these collaborative research projects at CERN Openlab, if any, will make a difference in the future of the companies involved. After all, historians of the WWW suggest that its invention at CERN was not initially in the mainstream of IT research in the laboratory.[16]

6.4 Life Science and Bioinformatics

The HGP has dramatically changed public policies on knowledge creation and data access in the life sciences. The ways in which genomic and related data are generated, stored, and made accessible are based on the development of new sequencing techniques, and subsequently in the building of bioinformatics infrastructures, continuously updated by the hosting organizations.

In a nutshell, sequencing a whole genome[17] means reading the DNA in the nucleus of a cell. Four bases are the chemical alphabet of the DNA: G, A, T, and C (guanine, adenine, thymine, and cytosine), famously organized in two complementary strands: the double helix. All the information is ultimately determined by the ordering of such bases in the DNA. The number of bases in the genome of a species (around 3 billion pairs in humans) is its size. The information contained by its nuclear DNA is organized in pairs of chromosomes, one from each parent. With currently available technologies, reading over 100 million base pairs (a very long string of GATC in any possible positions and repetitions) from its beginning to its end is impossible, and sequencing means breaking down DNA into much smaller segments, analyzing them by certain reagents and methods, and finally, reassembling them by computational approaches like a jigsaw puzzle. The initial HGP reading was compiled as a reference whole genome sequence (based on a mixture of data from a sample of anonymous individuals), and subsequent genomes of individuals were compared to it.

The HGP pushed technological advancement in sequencing (ending up in a small oligopoly of firms producing the equipment), with an associated spectacular drop in unit costs. Figures 6.2a and 6.2b show the trend in two ways: cost per megabase of DNA sequence and cost per genome (based on the size of the human one), respectively. A conventional whole genome is in fact a reading with 30x coverage, which implies $30 \times 3,000$ megabases to ensure statistical robustness (where the 30 factor is arbitrary to a certain extent).

The production costs included are labor, administration, management, utilities, reagents, and consumables; sequencing instruments and other large equipment (amortized over three years); informatics activities directly related to sequence production (e.g., laboratory information management systems and initial data processing); submission of data to a public database; and some indirect costs (Wetterstrand, 2013). Figure 6.2a, for example, shows that as of March 2003 (when the first whole genome of a human was available), the cost of processing 1 million bases (a megabase) was $2,609, while in July 2017, it was just $0.013. Similarly, the cost for sequencing a whole human genome was around $47 million in 2003 and went down to around $1,150 in 2017 (figure 6.2b).

The computing resources needed to deal with the resulting exponential growth rate in data generation will probably be at the exascale level in the next decade, when millions of whole genomes will be available (and not only for humans). The computing exascale level is 10^{18} calculations per second, while the most powerful supercomputers in the world currently work at the petascale level (10^{15} calculations per second). According to NIH (2018), it will be possible to create virtual models of life processes, and perhaps also large-scale virtual clinical trials, by exascale supercomputers.

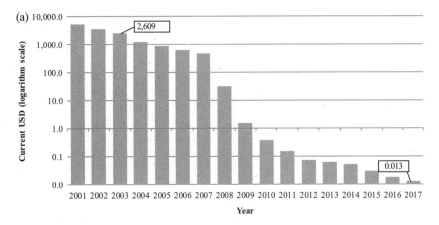

Figure 6.2a

DNA sequencing costs per raw megabase of a DNA sequence in dollars. *Source*: Adapted from Wetterstrand (2013).

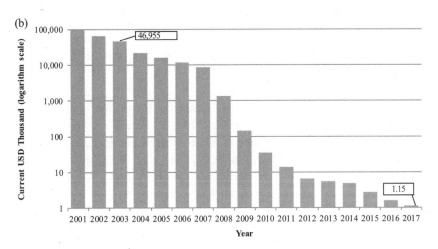

Figure 6.2b

DNA sequencing costs per genome in thousands of dollars. *Source*: Adapted from Wetterstrand (2013).

Beyond sequencing technological issues, the question of how to release the data was a crucial and contentious one from the start. The general principles for access to HGP data were established more than twenty years ago in a meeting in Bermuda, on February 25–28, 1996,[18] which dramatically changed the way that scientific results are made available as a public good. Cook-Degan, Ankeny, and Maxson Jones (2017) provide an interesting historical overview of the debate on genomics data management before and after that meeting, including the issue of the patenting policy. The summary of the Bermuda agreement is still worth reading:[19]

The following principles were endorsed by all participants... It was agreed that all human genomic sequence information, generated by centres funded for large-scale human sequencing, should be freely available and in the public domain in order to encourage research and development and to maximise its benefit to society. Sequence assemblies should be released as soon as possible; in some centres, assemblies greater than 1 Kb would be released automatically on a daily basis. Finished annotated sequence should be submitted immediately to the public databases. It was agreed that these principles should apply for all human genomic sequence generated by large-scale sequencing centres, funded for the public good, in order to prevent such centres establishing a privileged position in the exploitation and control of human sequence information.

Similar arrangements are now common in major publicly funded life-science infrastructures, but this was a dramatic change compared with the traditional procedures, where each laboratory tended to protect its data as far as possible. The current panorama in life sciences is, in fact, increasingly based on the opportunity to access biodata freely through IT. The NIH, the European Commission, and several other government agencies support such large-scale projects, with the understanding that public funding implies public data (with different specific implementation rules).

New infrastructures have been created to support this process. The European Molecular Biology Laboratory–European Bioinformatics Institute (EMBL-EBI), already mentioned in relation to the HGP, started its operations in 1992 and is one of the six sites of EMBL which, in turn, is the leading research body in the field in Europe, an intergovernmental organization (similar to CERN), comprising twenty-five member states (EMBL, 2016).[20] The EMBL-EBI core missions (see box 6.2) include collecting and distributing, for free, data and analytic tools to researchers in life sciences, offering training on how to use such resources, and developing their own data analysis and research.

In 2016, the EMBL-EBI websites received 27 million biodata requests per *day*, with 82 percent of data usage through web interfaces, and the total volume of data downloaded in the scale of 3.5 petabytes. For instance, nucleotide sequence data stored are 5.91 petabytes. The total data storage capacity at the institute is 120 petabytes. The latter data are particularly interesting if one considers that the CERN central data center stores 150 petabytes of data on disks (i.e., without considering tapes). This can be compared with the abovementioned 300 petabytes of Hive data from Facebook.[22]

Box 6.2
EMBL-EBI in a Nutshell

The EMBL-EBI, located at Hinxton, near Cambridge in the United Kingdom, offers data accessibility for twenty-four hours a day, seven days per week—a welcome feature to researchers who often work in their laboratories even during the night and during weekends, or work in different time zones (the user base is global). Its core databases are the result of collaborative research with the National Center for Biotechnology Information in the United States, the National Institute of Genetics in Japan, the SIB Swiss Institute of Bioinformatics, the Wellcome Trust Sanger Institute, and other organizations. EMBL-EBI is funded by contributions from the EMBL member-states. It also receives a large amount of competitive funding from the European Commission, the NIH, the Wellcome Trust, Research Councils UK, and some EMBL-EBI industry partners.[21] According to its annual report (EMBL-EBI, 2017), in 2016 it employed a staff of 574 full-time equivalents from sixty-four countries, of which 395 come from EMBL member-states and associate member-states, and the remaining from other countries. In addition, there were 141 scientific visitors from thirty-four countries.

Also related to the EMBL is ELIXIR,[23] the European Life-Science Infrastructure for Biological Information. It is a pan-European distributed RI established in 2014[24] with the mission to coordinate life science resources across Europe and improve data access, exchange of expertise, and best practices.[25] It is made up of twenty-one member-states and EMBL.[26] ELIXIR is based on a hub-and-nodes architecture, with the hub hosted alongside EMBL-EBI. ELIXIR nodes comprise around 200 research centers in the participating countries. Each node, mainly funded by national contributions, offers services as data storage, added-value databases (or knowledge bases), biocomputing facilities, services for integration of data, software, tools and resources, and training and standards. For instance, the ELIXIR-IT lead institute is the CNR Institute of Biomembrane and Bioenergetics, located in Bari, Italy.[27]

The services are currently structured according to five platforms and seven communities (see figure 6.3). The platforms[28] include *Compute* (access, storage, transfer, and analysis of large amounts of life science); *Data* (the identification of key data resources and linkages between data and literature; *Tools* (enabling researchers to find appropriate software to analyze the data; *Interoperability* (establishing Europe-wide standards for life science data recording); and *Training* (to increase data analysis skills of researchers). The communities[29] include specific research activities in certain areas, such as accessibility of

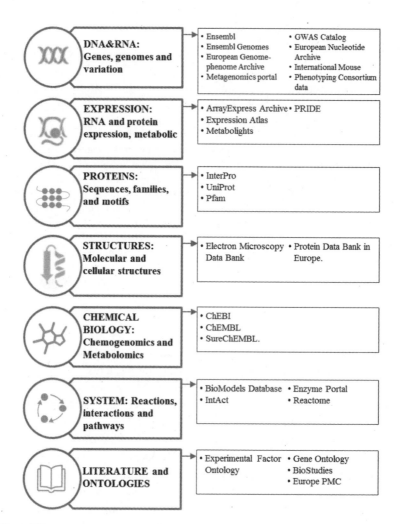

Figure 6.3

EMBL-EBI provides genomics data and bioinformatics services. The molecular data resources available are the world's largest range and services released for free. These include Ensembl (genome), UniProt (protein sequence), PDBe (three-dimensional structural data on biological macromolecules), Europe PMC (worldwide life sciences literature), Expression ATLAS (a database on the expression of genes and proteins), and ChEMBL (bioactivity data). *Source*: Courtesy of © EMBL- EBI, own adaptation. https://www.ebi.ac.uk/.

PLATFORMS   **COMMUNITIES**

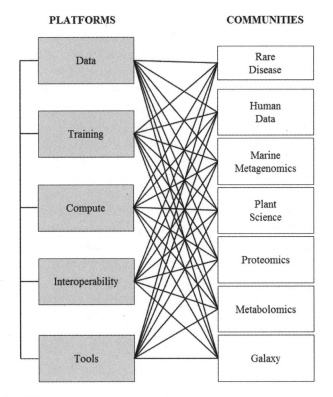

Figure 6.4

ELIXIR is a distributed digital research infrastructure with its hub located in Hinxton, United Kingdom. Its activities are structured around five platforms and a portfolio of communities. The platforms are built on the needs of established research communities. The ELIXIR communities drive the work of the ELIXIR platform by defining their bioinformatics needs and requirements. *Source*: Courtesy of ©ELIXIR, own adaptation.[30]

sensitive human data, rare diseases, marine metagenomics (standards, databases, tools, and training on the genetic material sampled from the sea), plant sciences (genetics of crops and trees), and others.

Within the Data platform, a set of core data resources have been selected by ELIXIR, according to such criteria as scientific focus and research quality of science, size of the community served by the resource, quality of service, governance, legal and funding arrangements, and potential impact. Among such core data resources, those owned and operated by EMBL-EBI include, inter alia, Ensembl (a genome browser on vertebrates), Ensembl Genomes (non-vertebrate species), and EGA[31] (personal genetic and phenotypic data from biomedical research projects).[32] In terms of the number of users, while tracking is difficult given the typically open-access policy where users access the data directly without the need for users to register, significant examples include UniProt (a curated protein database) and the European Nucleotide Archive (ENA), which probably have a cumulative 600,000 yearly users, an estimation based on accesses from unique Internet Protocol (IP) addresses. The website of the Human Protein Atlas (HPA),[33] the core data resource (managed by ELIXIR Sweden), received over 750,000 visits during 2013, of which approximately 60 percent came from outside Europe.

A report on the use of public bioinformatics data by Small Medium Enterprises in Europe,[34] based on a review of business models and the various types of companies that use open biological data, presents case studies of innovation-driven European SMEs. An earlier ELIXIR study on life science database citation patterns in industry patents[35] has found citations of bioinformatics databases in more than 8,000 patents from 2014.[36] For example, the proportions of patents using bioinformatics databases are 14 percent for industrial enzymes, 16 percent for biomarkers, and 16 percent for vaccines.

The abovementioned figures suggest that high-particle physics and astrophysics will not be alone in the next years in needing very large-scale data storage, accessibility, and analytics. However, biodata related to humans pose additional ethical, privacy, and cybersecurity concerns, as well as potentially sizable costs to protect the data.

6.5   Externalities from IT Innovation

As mentioned in the introductory remarks to this chapter, a specific mechanism that generates positive externalities in this context arises from the technological challenges that need to be solved by data-intensive RIs. I have already listed some of these challenges. A visionary computer scientist may identify others or suggest other ways to explore the RI future landscape from this angle. The pace of change is so fast in IT that forecasting is particularly challenging. Whatever the future scenario, there are some common features that suggest a simple empirical valuation strategy of the IT externalities from the perspective of CBA.

First, one may distinguish between the case of the release of free software, on the one hand, and learning mechanisms arising either from a collaboration between RI teams and external partners or other forms of knowledge transmission, on the other. By *free software*, I refer here to any kind of tool (with or without free access to source code) released at no charge in the public domain by the IT teams on behalf of the RI managing body. The value of this externality can be empirically measured from the perspective of the software producer and the software consumer. Free software and open-source software (OSS)[37] are different concepts. Both free software and OSS have economic impacts, and the latter may also be free. However, free software may not be OSS, and the latter may not be free.

On the producer side, if we use the marginal cost approach, there will certainly be an underestimation of the marginal social value of the good. What the RI managing body is actually giving for free is often an unlimited right to use, modify, redistribute, and even sell a good that has perhaps a very low marginal cost in the first place. In fact, with one mouse click, you can obtain the knowledge embodied in the software and all its potential value. Your additional click costs close to nothing to the inventor (and to you), but there is an opportunity cost for the inventor, or for the RI body for which he or she is an employee. This is the potential income arising from any possible use of the good. Thus, it seems more credible to turn to the perspective of the user for a valuation strategy.

On the demand side, one may ask the question: What is the maximum price that an average user would find acceptable to pay to obtain software for which she or he is currently charged nothing? This shifts the issue to the potential value of the software to the user. The answer may obviously depend on some variables that would determine a notional demand curve: the price of substitute goods (e.g., commercial software, if it exists, available with similar purposes); the professional position of the user (e.g., a student versus a professional developer); and economic opportunities arising from modifications. Thus, from a CBA perspective, one may focus on the information given by revealed preferences, such as the market price for similar goods, or other objective information, such as the *value of time saved by users* who would have to invest their own effort to get the knowledge embodied in the free software. The latter is often a sensible question to ask to professional software developers or other expert users (see section 6.7).

If this information is difficult to find, one may turn to a stated preference approach, and ask to reveal directly either the willingness to pay (WTP) for using the software (acquiring it is currently not the typical arrangement), or the compensation required to be denied the free access (willingness to accept, or WTA). Obviously, given the variability of user characteristics, the WTP or the WTA may be quite different, for mainly psychological reasons, an issue well known in the CBA literature (see, e.g., Boardman et al., 2018). The empirical analysis requires appropriate statistical data, for example, based on contingent valuation experiments or other stated preferences. These methodological issues will be discussed elsewhere in this book in greater detail, particularly in chapter 9, as they are also related to the empirical analysis of the social value of knowledge as a public good (a different type of social benefit).

Free software, in fact, is not strictly a Samuelsonian public good: There is no rivalry in consumption (your use of software does not constrain my own use), but exclusion is feasible, as shown by a large market of commercial software and even larger market of services based on IT. Hence it is a "club good" (see Cornes and Sandler, 1996). The difference is often immaterial from an empirical perspective in this context, as the empirical approaches to valuation are similar for club goods and public goods.

This perspective may change, however, if the innovation in software (or in certain cases in hardware or in other IT aspects) is the result of an explicit form of mutual learning between the RI staff and external teams of users. The reasoning here is similar to what was discussed in the previous chapter. The externality arises, ultimately, because of a divergence of objectives between the RI, typically a nonprofit organization and private partners, which are profit maximizers. Thus, if there is no formal contract regulating future revenues originated by IT innovation arising by a collaborative project, the RI is donating the time and efforts of its staff, and any potential profit, to the private partners.

The same applies when there are learning spillovers from IT procurement, similar to the situation discussed in the previous chapter. Hence, the R&D cost avoided by the private partner is the most conservative estimate of the benefit, and a share of any future profitability is the opportunity cost for the RI and the actual maximum externality for the company. The more conservative approach is particularly relevant when the partner is, in turn, another RI or public-sector body, with the avoided cost as the actual benefit of the externality. An example is IT innovation in high-energy physics, which is relevant for radio astronomy or vice versa, as the abovementioned agreement between CERN and SKA suggests.

6.6   Externalities from Open Data

The other mechanism that creates an externality is the adoption by the RI of an open-data model. The RI's scientific community has often turned to the open-data model either as a legal obligation related to government funding or, in other cases, prompted by nothing more than generosity, embodied in the mission of scientific projects to create and spread knowledge. In other cases, it is just a matter of efficiency: The amount of information generated by some RIs is so large that storing it somewhere for the usage of one specific subset of the scientific community is inefficient. In some scientific fields, no individual scientific team would be large enough to analyze the data on their own without the collaboration of a large share of all the scientists active in the field. This is a general feature of professional community networking through open-science arrangements and a crucial feature of the RI paradigm: A cosmopolitan, decentralized community arises because a more proprietary and hierarchical structure would be inefficient, or just impossible.

The open-data model ideally should be "F.A.I.R." This approach is described as follows by NIH (2018) for biodata (but, in fact, the definition is applicable to several other domains): Biomedical research data should adhere to F.A.I.R. principles, meaning that it should be Findable, Accessible, Interoperable, and Reusable. In the NIH (2018) definition, to be *Findable*, biodata must have unique identifiers, effectively labeling it within searchable resources. To be *Accessible*, biodata must be easily retrievable via open systems and effective and secure authentication, and authorization procedures. To be *Interoperable*, biodata should "use and speak the same language" via the use of standardized vocabularies. To be *Reusable*, biodata must be adequately described to a new user, have clear information about data-usage licenses, and have a traceable "owner's manual," or provenance. For more information on the F.A.I.R. concept, see Wilkinson et al. (2016).

Having said this, what is open data worth to outsiders? If a biologist some years ago wanted to collect all the existing evidence on a gene involved in a disease, she or he had to spend months or years of research. Today, the genomic information available in bioinformatics databases may be obtained in a matter of hours through just a few clicks.

This example suggests that the most important direct social benefit of the open-data revolution in science, supported by the RI, is simply a function of the *time saved* by scientists. Thus, a straightforward approach to evaluate the costs and benefits of this dimension of the new RI paradigm is to estimate the investment and operational costs per unit of information released, against the social benefit that can be gauged in terms of the hourly salaries of the scientists, multiplied by the research time saved.

While the mechanism creating the externality is different from what has been discussed in the previous section, the empirical approach to its valuation is similar. A survey of data users may be structured directly, as a contingent valuation experiment, in order to gauge the time saved and, implicitly or explicitly, the WTP to access the data or the willingness of compensation to accept to renounce such access. Alternatively, one may assume that there may be some factual, observable information, revealing the value of time devoted by external users to search, access, and download the data, and the time they would have needed without the information.

Some aspects of this approach will be further developed in chapter 8 on cultural goods, as the value of visiting a website, either just for personal entertainment or for acquiring information for professional reasons, can be investigated in terms of the opportunity cost of time, even if the objectives of the search differ. Again, the producer perspective here is less relevant, as the marginal costs (opposed to the total cost) of producing and disseminating the data may be very low compared with their marginal social value on the demand side.

It is important here to discuss a possible confusion that can arise if one tries to go beyond the *direct* effect and wants to attribute to the free data the further advancement in research that was made possible thanks to their accessibility. Let us consider again the case of the HGP. Building on it, more than 1,800 disease genes have been discovered, more than 2,000 genetic tests invented, and around 350 biotech-based drugs tested in clinical trials.[38] Would it be reasonable to attribute to the free data availability of a specific defective gene the chain of events that may have ultimately led to finding a diagnostic and therapeutic approach to a genetic disorder? Clearly, if this was the case, the value of the statistical human lives saved, or additional quality-adjusted life-years, would have entered into the evaluation of the benefits of the RI project and would have increased it highly.

While one may be tempted to do so, there is a high risk of overestimating the benefit in this way. This issue of the value of innovations is discussed in chapter 7. It is also related to the discussion in chapter 3 on the value of publication of information. To anticipate the discussion here, it is important not to confuse knowledge and the way that it is communicated. A specific gene mutation discovery can be obtained in different ways, and not necessarily within a major RI project such as the HGP. Even when a specific telescope is strictly necessary to see a distant galaxy, this incremental astronomical knowledge is not the same as the way that the information is communicated to third parties. The information can be written in a scientific paper or in a preprint, announced in a conference with some slides, or released with a data set.

What should be valued is the difference that one specific communication arrangement makes to third parties as opposed to another one. From this perspective, thus, the value of the open-data model is the difference (in welfare terms) to users of either having direct access to the data or alternatively, having more limited access (e.g., one where the data supporting the experimental results published in a journal are available from the authors only upon request). It is not the difference between the open-data arrangement and the absolute exclusion forever from the knowledge created with them, but between concrete models of data availability to users, that is significant.

6.7   Case Study: The Value of Open Biodata

The exact number of individual researchers who take advantage of the EMBL-EBI databases is not known, but 3.2 million unique IP address accesses to EMBL-EBI websites are recorded *every month* for over 60 million requests per day (figure 6.5). These users have potential access, inter alia, to 42,529 genomes of various species and strains: to 2.2 million gene expression assays, to 71 million protein sequences, and to 125,463 macromolecular structures. These numbers grow every day (those reported here are to January 2019). The activity is also supported by extensive training. In 2016, the EMBL-EBI TrainOnline, a web-based resource for training in bioinformatics[39] offering twenty-seven webinars and twelve new courses, was accessed by 358,905 unique IP addresses (a 60 percent increase as compared to 2015).

What is the cost? In terms of yearly funding, in 2016, it was EUR 77.1 million, of which EUR 40.3 million came from EMBL member-states (59 percent), EUR 5.3 million from UK Research Councils, EUR 4.9 million from the European Commission; EUR 7 million from NIH, EUR 5.8 million from the Wellcome Trust, EUR 4.2 million from other sources (e.g., EMBL-EBI Industry Programme), and an additional EUR 9.6 million funding from the UK government's Large Facilities Capital Fund. The total cost breakdown is 69 percent for the staff, 27 percent operating costs, and 4 percent capital expenditure and depreciation.

What is the benefit to society of the bioinformatics services provided for free? EMBL-EBI periodically carries out a survey of users to get feedback, both on the extent to which they help their research work and their attitudes toward the services. In November 2016, 1,191 respondents answered 28 questions about EMBL-EBI. The main findings were that 72 percent of respondents stated "It is essential to his/her research"; 72 percent, "It is useful for daily work"; 84 percent, "It is informative, helpful, and logical"; 90 percent, "It reduces the time required to find relevant/required data"; 90 percent, "It has improved efficiency in exploiting publicly available data"; 75 percent, "It enabled undertaking a greater quantity of research activities"; and 79 percent, "It enabled research to go ahead that might otherwise not have."

While this is a qualitative assessment, Beagrie and Houghton (2016) have quantitatively estimated the socioeconomic benefits related to open access to the EMBL-EBI data. They define two broad categories of benefits: value to the users' community and wider societal impact. As for the former, they apply two methods: a direct estimate of the cost of producing or obtaining/using the service. As for the latter, they recur to contingent valuation (WTA to forgo

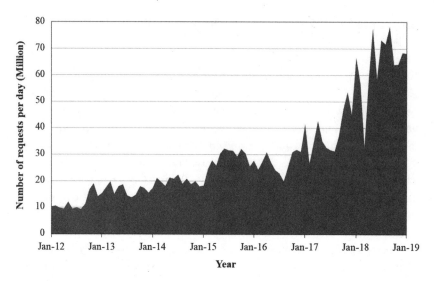

Figure 6.5
Usage of EMBL-EBI websites. *Source*: Courtesy of EMBL-EBI, ©EMBL.

the service or WTP to access it if it is not free). The wider impact on society is estimated by efficiency impacts (value of efficiency gains because of the service, including time savings) and return on investment in R&D using the service. Here, I do not report their findings on the wider societal impacts, as this chapter focuses exclusively on the direct impact on users of data and tools, not on the ultimate effects of their research (e.g., I do not consider whether the use of EMBL-EBI data helped to discover a new drug).

Beagrie and Houghton (2016) report the following estimations (all values are per year). The annual operation expenditure to produce the service (average 2012–2014) was 47 million pounds. Around 2015, there was an estimate of 198,000 unique users for thirty-nine online services (out of several million unique hosts in one recent year, according to EMBL-EBI log data reports which, for several reasons, cannot be considered actual unique users). The population of users is estimated as about 13 percent of potential users (1.7 million full-time equivalent life-science researchers worldwide). A user survey sample (4,509 responses, 17 percent response rate over the available email addresses) suggests 445 data accesses per user per year (88 million access and use events). Then average access time was estimated from the survey at 51 minutes (with large variance). Using the product of the median time of access times the mean hourly cost times the frequency of access, the authors estimate 270 million pounds per year as the benefit. The authors estimate that the hourly labor cost is 37 pounds per researcher (based on data from the Times Higher Education Salary Survey), and an adjustment for nonwage labor costs.

The authors then also report a much larger benefit, which they define as *use value*, which is the following product: (mean number of hours with EMBL-EBI data per week) times (mean hourly cost times weeks per year) times (number of estimated users) = 3 billion pounds. This concept of use value is apparently much less conservative. Interestingly, the contingent valuation approach, based on the survey itself, where there were questions trying to elicit the respondents' WTP for the service, led to a mean WTP of 1,628 pounds per user, equivalent, for 198,000 users, to 322 million pounds per year. However, the authors report that the median is much lower (126 pounds) and that the frequency of access events in the sample was higher (631 events) than in the population (an estimated 513 events per year). In any case, the benefit/cost ratio would apparently look very favorable.

In general, this case study suggests that the value of time of access is the most conservative estimate, the avoided cost methods probably being a mid-range value, while it seems advisable to test the robustness of WTP survey data carefully, as it is well known that their variance may be high—an issue that will be discussed later in chapter 9. Indirect effects on innovations need different empirical approaches, discussed in the next chapter.

6.8   Further Reading

To illustrate the convergence of IT challenges in particle physics, astronomy, and life sciences further, this section briefly mentions some additional aspects of open software (and hardware) policy of CERN, as well as some other examples of Big Data generation in astronomy and in life sciences. Finally, some references are provided on the economic value of free and open-source software in general.

6.8.1   CERN Policy on Software

Nilsen and Anelli (2016) report that CERN has officially adopted an open-source approach for its software, with limited exceptions related to excessive effort needed to develop the software for public release; inadequate quality; specific collaborative agreements with third parties, preventing open-source licensing; and when proprietary licensing in special cases is more effective than open source for dissemination. Otherwise, in the normal case, widely used open-source public licenses are adopted. An unfortunate consequence of the way that the policy is implemented is that CERN is unable to track the actual dissemination of its software precisely because users do not need to register when downloading the codes.

Perhaps less well known is the CERN Open Hardware policy, which authorizes anybody to study all the details of a specific design, such as in 2009 some electronics components, developed by the Beams Department and placed in the Open Hardware Repository. Subsequently, a CERN Open Hardware Licence (CERN OHL) was created to grant anybody the legal entitlement to use, copy, modify, and distribute hardware design documentation and related products. The CERN OHL also is now used outside particle physics, while the Open Hardware Repository hosts more than 100 projects, with seventeen companies involved. According to Nilsen and Anelli (2016), until 2014, more than 1,200 units were produced for almost 100 users.[40]

6.8.2 Big Data in Astronomy

The case of SKA in astronomy is far from unique in terms of data intensity. The Large Synoptic Survey Telescope (LSST) in Cherro Pacho, Chile, which is also expected to deliver sky observations in the mid-2020s, will collect and store about 200 petabytes of data overall. Its website[41] claims that:

This telescope will produce the deepest, widest image of the Universe:

27-ft (8.4-m) mirror, the width of a singles tennis court

3200-megapixel camera

Each image the size of 40 full moons

37 billion stars and galaxies

10-year survey of the sky

10 million alerts, 1000 pairs of exposures, 15 Terabytes of data … every night!

Elsewhere, the LSST website mentions that a challenge to be solved is the software needed to process and store more than 30 terabytes of data each night, "producing the largest non-proprietary data set in the world."

The James Webb Space Telescope,[42] which will be a successor of the Hubble Space Telescope from 2019, at a budgeted life-cycle cost of around $9 billion (much increased from the initial forecast), is developed jointly by the National Aeronautics and Space Administration (NASA), the European Space Agency (ESA), and the Canadian Space Agency. It will combine eighteen hexagonal mirror segments in a primary mirror with a 6.5-meter diameter. It is intended to study the most distant galaxies, particularly in the mid-wavelength to near-infrared region (which is difficult to observe from ground-based telescopes because of the filtering effect of Earth's atmosphere), exoplanets, and other objects from the vantage point of outer space, as it will be placed 1.5 million kilometers outside Earth's orbit. Clery (2016a) is a brief nontechnical presentation.

The Big Data challenges raised by the new astronomical instruments is presented in a comprehensive study by STscI (2016). *STscI* stands for the Space Telescope Science Institute (Baltimore, Maryland), whose strategic goals include the science operations of the main NASA astrophysics missions; to develop advanced astronomical data, archives, and tools; and to disseminate information and engage citizens. The report by a panel has identified the computing infrastructure investment needs of the future as follows:

> We identified several key science cases that would be difficult, if not impossible, to accomplish with our current capabilities...They can be grouped into a few common challenges in astronomical "Big Data" when applied to large or complex datasets. These are: (1) computer-assisted object detection and classification, (2) multi-wavelength catalog cross-correlations, (3) time-domain event detection and classification, (4) model fitting to massive data volumes, and (5) disentangling complex datasets into their constituent systems.

Open-data astronomy also will be an ideal testing ground for the massive recourse to data storage in the cloud, such as the WorldWide Telescope sponsored by Microsoft, which enables the representation of a virtual universe by the skillful combination of astronomical observations from many sources and in a unique format. Such observations include real-time imaging from different telescopes.[43]

A systematic approach to data sharing has been developed since 2002 by the International Virtual Observatory Alliance,[44] which includes twenty-one international members and aims at interoperability of data archives, tools, and services across various centers, through agreed standards. For more on open-data astronomy, see the presentations at the United Nations organized by the United Nations Office for Outer Space Affairs and the Italian Space Agency, on behalf of the government of Italy.[45]

6.8.3   Big Data in Life Sciences

Turning to the Big Data era in biology, the change was greatly stimulated by technological advances in gene sequencing. Chial (2008) gives a very concise presentation of the sequencing approaches (including a "Quick lesson in DNA sequencing") developed by the International Genome Sequencing Consortium, the publicly funded network of laboratories, and by Celera, a private company. Hood and Rowen (2013) offers a brief history of the HGP, including the reasons for its initial lack of popularity among many biologists and its subsequent impact. It also stresses the fact that it was the first time that biologists had to collaborate on such a large scale with mathematicians, engineers, and computer scientists. Along with some government funded organizations, such as the NIH, private nonprofit institutes are now playing a leading role in biodata generation. The Broad Institute of MIT and Harvard is a major hub of genomic data analysis with a twenty-five-year history. After its involvement in the HGP, the Broad Institute has played an important role in a number of major projects, including HapMap,[46] the 1000 Genomes Project, the Cancer Genome Atlas, Comparative Reference Genomes, the ENCODE Project, the NIH Roadmap Epigenomics Mapping Consortium, the Genotype-Tissue Expression Project, and the Human Microbiome Project. The institute has studied more than 1.5 million samples from more than 1,400 groups in more than fifty countries.[47] It receives substantial support from the NIH and others.

Moreover, to confirm the close relationship between large-scale data generation and progress in the development of tools and methods for data analysis, the Broad Genomics team created the Genome Analysis Toolkit (GATK), a software for data-sequencing analysis. One GATK-based specific tool for the detection of gene variants was reported to have over 30,000 registered users to date and to have been applied over 800 million times by external users. The overall GATK code has over 50,000 users and is open source.[48]

In the last few years, the NIH (2018) has produced and made available in the public domain new tools for data compression, suites of algorithms, web-based software, application programming interfaces (APIs), public databases, computational approaches, among others.

It is interesting that similarly to CERN Openlab, there is a collaboration between the Broad software engineering group (the Broad Data Science Platform) with several major entities in data science.[49] In their words:

The life sciences are in the midst of a data revolution. Cheap and accurate genome sequencing is a reality, advanced imaging is routine, and clinical data is increasingly stored in electronic formats. These innovations—and the massive data sets they produce—have brought us to the threshold of a new era in medicine, one where the data sciences hold the potential to propel our understanding and treatment of human disease.... To bring the tools of machine learning and cloud computing to bear on problems of fundamental importance to biomedicine, the DSP collaborates with world leading technology corporations.

These entities include Amazon Web Services, Cloudera, Google Cloud Platform, IBM, Intel, and Microsoft.

A future project in neuroscience has been conceived entirely as IT research. The Human Brain Project[50] aims at a full-scale model of human brain simulation by supercomputing platforms. It will require exascale computing resources (as discussed next) and will probably cost more than EUR 1 billion.

It is not only genetics that is becoming data intensive in life sciences. There is also the National Institute of Allergy and Infectious Diseases (NIAID) at NIH. It supports and performs basic and applied research, a field where the data traditionally come from hospitals in the context of clinical trials, but also from other less statistically standardized sources of information on patients and therapies.[51]

The NIAID hosts several large-scale bioinformatics projects:[52] "The Bioinformatics Resource Centers (BRCs) for Infectious Diseases program was initiated in 2004, with the main objective of collecting, archiving, updating, and integrating a variety of research data and providing such information through user-friendly interfaces and computational analysis tools to be made freely available to the scientific community."

There are six such BRCs, each focusing on a group of pathogens (bacteria, viruses, and so on). Data-intensive programs in biosciences that may benefit from exascale computing include the All of Us Research Program (aiming at monitoring 1 million US residents) and the Cancer Moonshot; the Human Connectome project; and Brain Research Through Advancing Innovative Neurotechnologies (NIH 2018, 3). Another interesting development, discussed in this document, is the systematic integration of biodata (trials, epidemiology, genomics, clinical care, and environmental exposure) with socioeconomic data (educational records, employment, income, and genealogy).

To confirm that the benefit to researchers of large-scale bioinformatics archives can be valued in terms of time saved, a survey in 2016 (CrowdFlower, 2016) states that 80 percent of research time is spent on collecting and organizing existing data, leaving only 20 percent of the time spent on analysis and possible discovery. This is particularly true in medicine because of the still-prevailing pattern of data generation by small groups with heterogeneous research methods and data archiving.

The transfer of approaches to data analysis from different fields is increasing. According to NIH (2018, 15), "the same software used by NASA scientists[53] to determine the depths of lakes from space is being tested for use in medical-image analysis for mammography, X Rays, computerized tomography (CT), and magnetic resonance imaging (MRI) scans, as well as for ultrasound measurements."

The rise of data-intensive biology is one of the drivers of large-scale collaborations, a topic discussed in an interesting paper by Vermeulen et al. (2013). Also of interest is the emerging concept of "commons" in medicine, which draws from the institutionalist approach of Elinor Ostrom, Nobel laureate in economics; in particular, see the book *Governing Medical Knowledge Commons* (Strandburg, Frischmann, and Madison, 2017).[54]

6.8.4   The Economic Value of Open-Source and Free Software

On the economic value of open-source and free software in general, there is a wide body of literature. See Lerner and Tirole (2002, 2005a, 2005b); Kogut and Metiu (2001); Boyer and Robert (2006); Lemire (2014); the Linux Foundation value estimates (McPherson, Proffitt, and Hale-Evans, 2008) and 2015 report; Daffara (2009); April (2007); Shah and Keefe (2010); and Bitzer and Schroeder (2006).

Tresch (2008, 174–176) offers a concise but interesting discussion about the efficient pricing of software. He observes that software production is a case of global natural monopoly, as writing and testing the program amount to a fixed cost, and then distributing it over the Internet has zero marginal cost per download. In this context, the efficient price would be zero, and the entire cost must be funded by a lump sum transfer to the producer. Private software producers (as the cable industry and others) collect fees in the form of subscriptions or similar all-or-none offers to their customers. This socially inefficient arrangement creates profits for them but expropriates potential value to users. In fact, similar reasoning would apply to the production of digitized content. This natural monopoly feature of software and information provision has obvious distributive implications in the information economy era, an issue that I will discuss briefly in chapter 10.

7 Users of Science-Based Innovations

7.1 Overview

The benefits of scientific knowledge ultimately created by a research infrastructure (RI) accrue to the final users of goods and services. For example, let us consider three types of synchrotrons. A synchrotron is a particle accelerator that can increase the speed of electrically charged particles or ions (i.e., charged atoms or molecules) to velocities close to the light. This is done by high-frequency, alternating electric fields synchronized to particle passage. To keep particles on the desired circular track in the vacuum chamber, powerful electromagnets are used, and their field is adjusted to the energy of the beam of particles. Ultra-high vacuum is needed to avoid the accelerated particles interacting with gases in the tube.

The Large Hadron Collider (LHC) is a synchrotron with a special design: it is a collider managing two beams, circulating in opposite directions and made to collide. It has been designed, built, and operated at the European Organization for Nuclear Research (CERN), exclusively for basic research: the scientific service it provides is the production of experimental data on fundamental interactions of particles and forces. Direct users are scientists and students, as discussed in chapters 3 and 4. A synchrotron, however, can also be designed exclusively for medical research on carbon-ion therapy, currently at thirteen facilities worldwide, with several others under construction or planned.[1] At the National Centre of Oncological Hadrontherapy (CNAO) in Pavia, Italy, or MedAustron in Wiener Neustadt, Austria, and elsewhere, synchrotrons convey beams of protons or carbon-ions to a patient's solid cancer when other forms of radiotherapy based on photons, such as X-rays, are not appropriate and surgery is not an option. The direct users of such facilities are mainly researchers in radio-oncology and oncologists specializing in specific pathologies. However, to do so, after experiments on humans are approved, patients become the direct beneficiaries of the facility services as well. Later, if the new experimental therapy for a certain type of cancer is successful, and if it is approved by the health authority, the therapy may be adopted by other centers, and it will gradually become the mainstream treatment of that type of cancer. More lives can be saved and the quality of life of patients can be improved, and these are measurable social benefits.

A synchrotron light (SL) source, instead, is a set of accelerators (featuring a linear accelerator, a booster, and a storage ring) where electromagnetic radiation is created. The synchrotron light is produced when a magnetic field forces the electrons, moving at velocities close to the speed of light, to change direction and emit photons. This radiation has several desirable properties: It is several orders of magnitude more intense than conventional X-ray tubes and highly collimated. Its spectrum can range from infrared light to hard X-rays. It is tunable to any wavelength, highly polarized, and emitted in very short pulses (e.g., less than a nanosecond). These properties find applications in fields ranging from condensed matter physics and materials sciences to pharmaceutical research and cultural heritage. As in the previous example, the direct users are researchers using the facility, including scientists hired by research hospitals, universities, or firms. Later, if the properties of a new material or a protein are discovered thanks to the RI services, for example, further research and development (R&D) of a firm may develop this discovery into an innovative product. An ultimate benefit for consumers may arise because of the improved quality of materials or of drugs.

Frequently, the benefits of knowledge accrue to economic agents more indirectly, through a chain of downstream transmission mechanisms. Third parties such as business in various industries, medical facilities, and government agencies acquire knowledge from experiments and observations in the labs through published results, and then adapt such knowledge to their specific needs, possibly combining and developing features. This chapter discusses both the direct and indirect mechanisms of the impact of scientific knowledge creation on innovations, with examples of both types.

The evaluation of the social benefits in the first case is often straightforward. When the RI uses its equipment to provide specific services directly to external users (e.g., businesses, government bodies, other research teams, patients, etc.), any socioeconomic benefits are valued by either using the marginal cost of the services provided or by estimating external users' willingness to pay (WTP) for the service, and in some cases, estimating users' avoided costs as well. In perfect markets, the equilibrium price equals both marginal costs and marginal WTP. The market for knowledge services, however, often does not provide this efficient "textbook" equilibrium, for several reasons (externalities, information asymmetry, monopoly or oligopoly power, etc.). As it happens, the empirically estimated marginal cost is often lower than the marginal WTP by users. For instance, the marginal production cost of a new drug may be low because the upstream fundamental research and the subsequent downstream R&D are fixed costs, largely independent of the quantity sold in the market. However, patients are willing to pay much more than such a low cost of producing an additional unit of a drug. The estimation strategy is similar to what has been discussed in the previous chapters of this book.

As the focus in this chapter is on ultimate users, such as final consumers of goods, patients for medical treatments, passengers of transportation services, and citizens potentially concerned by natural hazards, just to mention some possible examples (the words *users, consumers, beneficiaries,* and *patients* will be used interchangeably as appropriate), the analysis needs to consider the existing market conditions carefully. In fact, it is not always obvious that any economic benefit arising from R&D is immediately transmitted to consumers. In these circumstances, market prices should possibly be replaced by appropriate shadow prices to compute the direct potential benefit of RIs. The approaches to quantify and value these benefits depend upon the types of new services or products. These methods, in principle, are again generally based on the WTP, marginal cost, or avoided cost approaches and are often well established in CBA.

The main problem is that a causal relation is often difficult to detect as in most cases, product and process innovations are the combination of discoveries and knowledge creation from different sources, not from only one specific RI. Substantial additional R&D expenditure is required downstream of the direct provision of scientific services by the RI, in some cases creating a new combination of very different ideas and techniques; hence, attributing and apportioning the benefits to one specific RI may be difficult.

As an example of this problem, I have mentioned elsewhere that, according to Weinberg (2012), Ernest Rutherford's discovery of the atomic nucleus (1911) was performed by a team including just one postdoc and one undergraduate. The amount of money invested at the Cavendish Laboratory in that particular experiment was minimal. Clearly, if all the economic effects of discovering that inside an atom there is a nucleus (eventually leading, inter alia, to nuclear energy production) were credited to that small project, its benefit-cost ratio would have been spectacularly high. However, proceeding in this way would be a ridiculous exaggeration because much more was needed, in terms of cumulative knowledge, to go from the Rutherford experiment to the design of the first atomic reactor that produced electricity.

Having said this, there may be circumstances when the causal chain, going from discoveries in a specific RI to indirect benefits to final users, can be identified ex post and sometimes even predicted ex ante, albeit with considerable uncertainty and due caution. This chapter focuses only on such predictable and quantifiable direct and indirect benefits and disregards any additional unpredictable effects on consumers, which will be discussed in chapter 9, where the concept of *quasi-option value* will be considered.

The structure of the chapter is as follows: section 7.2 describes several examples of the direct provision of services by the RI; section 7.3 discusses how to value such services; section 7.4 presents examples of indirect benefits to consumers and other users; section 7.5 discusses the valuation of such indirect benefits; section 7.6 is about wider effects, and section 7.7 concludes the chapter with two case studies: satellite imagery for navigation and hadrontherapy research, and with the sketch of an empirical strategy in the field of synchrotron light sources. Finally, there is the usual "Further Reading" section.

7.2   Direct Benefits of Services Provided by RIs

As previously mentioned, some RIs provide scientific services to external users, including businesses, government bodies, and other research teams. These services may include, among others, access to certain measurement equipment, machine time, computing resources, data, sample preparation, access to archives and collections, personnel training, and expert support. Some examples of these are provided in table 7.1.

In some cases, as with synchrotron light sources, a customer (external user) contacts the RI facility's manager with a request to access certain services. After reviewing the request, the manager provides the external user with a cost estimate for the time spent on the machine or for any required services. In some cases, but very infrequently, the RI's charge for the service reflects market prices. In other cases (typically when the external user is an academic team or part of such a team), there is no charge at all, or such charges cover only part of the average operation cost incurred by the facility to make the service available. In fact, in most cases, access is free of charge. Special arrangements are in place for medical research when patients are involved, and such a tariff is paid by a health organization.

The number of potential external users that may be interested in exploiting the infrastructure's equipment or services can be forecast. Typically, during the design phase of an RI, the promoters investigate external users' interests through surveys or in other ways. The data collected can then be exploited for the forecast. The design of the RI itself may pose constraints to the number

Table 7.1
Examples of services directly offered to external users by RIs

Infrastructure	Services Offered
NASA Glenn Research Center, Cleveland, Ohio	The center provides ground test facilities relating to acoustics, engine components, full-scale engine testing, flight research, icing research, microgravity research, space power and propulsion, and wind tunnels. In addition, a range of test consultation services are offered.
European Synchrotron Radiation Facility (ESRF), Grenoble, France	The facility provides synchrotron light and techniques, such as for imaging, microtomography, topography, microscopy RX, and Fourier-transform infrared (FTIR) spectroscopy and microscopy, with many industrial applications. For instance, pharmaceutical and biotech firms use synchrotron techniques to help develop new products at all stages of research, from drug design and formulation to preclinical phases. Further, the automotive industry uses synchrotron techniques to obtain more efficient catalytic exhaust converters. Synchrotron techniques are also commonly used for the study of inclusions in surfaces or the identification of a defect on silicon wafers used to produce semiconductors.
High Field Magnet Laboratory, Nijmegen, Netherlands	Access to the facility is given to all researchers who have their research proposals approved by an external review committee. Access involves the use of the installation, the use of all available auxiliary equipment, and (if necessary) the support of the local staff.
Laserlab-Europe, 22 sites in countries in the European Union	The laboratories in this consortium offer access to their facilities for European research teams. These include world-class laser research facilities and a large variety of interdisciplinary research, including life sciences, free of charge. Travel and accommodation are included. Access is provided on the basis of the scientific excellence of a research team's proposal, reviewed by an external and independent selection panel. Priority is given to new users. A typical access project lasts for two to six weeks.

Source: Florio et al. (2016), elaboration from RI websites. https://facilities.grc.nasa.gov/using
.html; http://www.esrf.eu/Industry/applications; http://www.ru.nl/hfml/facility/access_to_the;
http://www.laserlab-europe.net/transnational-access, accessed on June 6, 2018.

and frequency of access, and excess demand of beam time is widespread across the facilities.

To demonstrate the process,[2] let us consider how this is managed at the European Synchrotron Radiation Facility (ESRF), in Grenoble, France. ESRF, with thirteen member states and nine associates, is one of the world's most intense X-ray sources (100 billion times the standard in hospitals, with brilliance that will only be increased by a planned upgrade). Applicants for beam time are required to prepare an electronic application form and an experiment methods form. The first one is filed in a user portal and should indicate the required beamline, the scientific area, and the societal theme. The second one is a two-page Portable Document Format (PDF) file, with a summary of the proposal: "clear statement on essence of proposal—what are you trying to do, how you intend to do it, and why you are doing it (impact, importance of study)."[3]

In recent years, out of 2,000 proposals, 900 were rejected after examination of the applications by reviewers, mainly because of beam time constraints, even when the scientific quality was good (this proportion is similar elsewhere, such as at the ALBA synchrotron near Barcelona, Spain).[4] Applicants should show a compelling scientific case for using synchrotron radiations and are advised to consult with the RI staff to define the target measurements, based on the beamline. There is an obligation to mention the ESRF in any publication and, in any case, to deliver an experimental report to the staff, even if this contains confidential information. Eventually, the decision on the application is taken by a Beam Time Allocation Panel, after having considered the advice of anonymous reviewers. A total of 7,000 scientists are reported yearly as visiting teams of users. There is no fee for academic research, while industrial client applications are managed by a Business Development Office. Customers can either bring their samples of materials for analysis to the facility and have them analyzed by ESRF staff, or by the client's own staff. With industrial clients, a quotation for the beam time and terms of conditions, including nondisclosure agreements, are negotiated case by case. The revenues from such activities seem to be relatively modest:[5]

Over the last few years, the ESRF's commercial activities, operated through the Business Development Office, have generated over 2 M Euros annually. About 75% comes from beam time services, supplied on the basis of costs recovery, with the remainder from a growing technology transfer and instrumentation commercial programme exploiting the ESRF's intellectual property and singular pool of light source engineering know-how... in 1995, a pharma company was paying some 10,000 Euros for a day of beam time during which they might collect data on two or three protein crystal samples. The new "MASSIF" station, jointly created by ESRF and EMBL, allows us to offer a professional mail-in service at a price of 125 Euros per sample—an almost two orders of magnitude decrease in price, opening the door to any pharma or biotech, large or small, to benefit from our facilities.

The policy is no different in the United States: The Advanced Light Source (ALS), Berkeley Lab, California, is a US Department of Energy (DoE)–funded SL with a forty-beamline facility. The ALS does not charge for beam time if the users' research is nonproprietary (e.g., if the research results are published). All users, however, are responsible for the day-to-day research costs (supplies, phone calls, technical support, etc.). Users performing proprietary research pay a fee based on ALS usage cost. These users may then take title to any inventions made during the proprietary research program and treat all technical data generated during the program as proprietary (i.e., not intended for publication).[6]

Fragmentary evidence (e.g., the amount of time dedicated to commercial purposes in five European facilities) is provided in Marks (1998). He suggests that, at the existing synchrotron light facilities, the beam time allocated to industrial users is considerably lower than 10 percent of the potentially available time. This is because businesses usually do not have the in-house expertise needed to perform the experiments, or because they are not interested in research that is not yet clearly related to possible commercial exploitation. In fact, most firms just wait for academic research teams to discover something of interest and disseminate the results in conferences and publications; they then decide whether to invest their own R&D energy and capital in the most promising ones in profitability terms. Hence, firms mostly do not pay for the knowledge created at the RI in the first place, but they will profit later on by cherry-picking some of the results disseminated by others. This spillover is a clear example of a positive externality.

7.3 Valuation of Direct Users' Benefits

How can we value the services provided by the RI? The estimation of an RI service's marginal cost is a relatively simple technique to evaluate the direct benefits for users of the service. For example, one can estimate the incremental electricity and labor costs to provide a user with an hour of a synchrotron light to test new material. This would clearly be the lower bound of a possible value because, in this case, the value of previous knowledge existing within the RI (embodied in the skills of personnel assisting the external users or running and maintaining the accelerator), past investment cost, and other fixed costs are neglected. A full average cost accounting would reveal a much higher figure. The ALBA synchrotron estimates an average cost per hour for operation of a beam at around EUR 550, based on available time and yearly costs (charted only for proprietary research), but this does not include past costs that are embodied in the service.[7] Under a marginal social cost approach, this rate can be taken as an empirical proxy of the shadow price of services provided for free.

Alternatively, one can estimate the external users' marginal WTP for the service. Benchmarks with similar infrastructures are sometimes helpful, but as the case of the synchrotron light source shows, while there may be currently more than 50 facilities running worldwide, the facilities providing exactly the same service (e.g., in terms of energy and other beam characteristics) in one specific region are limited.[8]

In any case, whatever the empirical approach, it is important to avoid double-counting of benefits; thus, if the shadow value of the service (proxied either with its marginal cost or with the users' WTP) is included in the economic impact analysis, then the RI's revenues should not be considered in the same analysis. Revenue from selling the beam time to users should enter into the financial analysis of the project (as discussed in chapter 2), but not in the economic analysis (as discussed in chapter 1), where it is replaced by an estimate of marginal production cost (usually as a lower bound, for the reasons already explained); by users' WTP, if available (which usually represent the upper bound of the service value); or proxies of these variables.

7.4   Indirect Effects of Discovery

Many RIs are instrumental in creating scientific knowledge that is ultimately embodied in innovative goods and services provided to consumers by third parties, often (but not always) without any active involvement of the RI staff.

From this perspective, let us briefly recall some mechanisms identified in the previous chapters. First, scientific knowledge is mostly communicated through publications and other outputs (chapter 3). In certain cases, it is possible to track the journey from discoveries to innovations directly; for example, patents may cite scientific literature (not just other patents), and some of this literature can be tracked back to a program developed by an RI (see the "Further Reading" section at the end of this chapter). However, when this systematic evidence is not available, there is only some anecdotal evidence about how technological innovations are based on scientific results available in the public domain.

Second, the human capital effects, discussed in chapter 4, are the counterpart of the increased productivity of organizations hiring personnel who have acquired certain skills in an RI, in part captured by a salary premium. While I have insisted on PhD students and early-career researchers (ECRs), in principle, a spillover effect arises whenever know-how and acquired abilities are transferred elsewhere. Due to labor market imperfection, it may happen that scientists entering such a market can be unable to be fully compensated for the social marginal value of their labor. This creates another externality mechanism. One reason for this is asymmetric information between firm managers and employees with scientific skills, as ex ante, it is obviously difficult to forecast the economic productivity of a researcher, an issue discussed in section 3.7.

I have also discussed knowledge spillovers for firms involved in procurement contracts with RIs and for those benefiting from open access to data and software, in chapters 5 and 6, respectively. Eventually, it is the cumulative effects of all the abovementioned mechanisms, and possibly their mutual interactions, that underpin the contribution of RIs to the well-being of consumers. The knowledge embodied in publications and patents, the increased productivity of personnel leaving an RI, and the knowledge spillovers accruing to firms collaborating with RIs translate ultimately into lower production costs and market prices of existing goods, higher quality, or the introduction of entirely new goods.

The mutual interaction of mechanisms in some cases is crucial. For example, a skilled scientist in an R&D unit of a firm is needed to understand publications originating from an experiment, and in some cases, those previously trained at the RI itself can better appreciate and translate the innovation potential of such scientific publications. Alternatively, engineers of the firm exposed to a procurement relation with the RI may take full advantage of published science because their collaboration with the RI's scientists has allowed them to acquire tacit knowledge that is only implicit in a published article about a discovery. All these mechanisms are deeply related to the open-science model of many contemporary RIs and their other features, identified in chapter 1.

Eventually, it is just the dissemination of knowledge that creates the externality, without any mediation of human capital effects, procurement relations, the release of free software, and technology transfer or spin-offs. RIs' internal scientists and external users donate knowledge to the world, and somebody else elaborates on it. These types of indirect benefits for society are possibly the largest, and while it is often impossible to forecast them ex ante, or even to identify them precisely ex post, in some cases, an RI's role is clearer. An example is when the origin of a biomolecular drug design can be precisely traced to the discovery of certain genetic mutations, revealed by open data released by a genomic platform. Situations arise, however, in which the practical use of a good can be expected in principle, but it is still uncertain in the technical sense (i.e., that a probability distribution function is unknown). In the latter case, the benefit value is related to what is called a *quasi-option value*. Further elaboration on this topic is provided in chapter 9.

7.5   Tracking and Valuing Indirect Effects

If the ultimate benefits of knowledge to consumers are properly identified and attributable to a specific project, then the methods to quantify and value such social benefits in a CBA frame are generally well established. Here, I will make some illustrative examples of this idea.

Superconductivity is a physical property of certain materials at critically low temperatures. Superconductors transmit electricity, virtually without losses, as opposed to copper and other metals, which lose power in the form of heat. High-energy physics has used this property to develop superconductive magnets, accelerator cavities, and certain detectors in many particle accelerators, such as those at CERN and Deutsches Elektronen-Synchrotron (DESY) in Europe, or Fermilab and Brookhaven in the United States. Superconductive magnets are currently widely used in medicine for nuclear magnetic resonance (NMR), but superconductivity has also found applications in transportation, telecommunications, and other fields. Cities that have recently decided to experiment with superconductive cables for the transmission of electricity include Essen in Germany and Saint Petersburg in Russia. The Long Island Power Authority in the United States and municipalities in Japan and China have also developed similar projects.

Thomas et al. (2016) survey in detail the state of the art and the huge potential of the socioeconomic impact of superconducting transmission lines, building from lessons learned in high-energy physics but considering technological alternatives requiring higher temperatures to achieve superconductivity, with materials such as magnesium diboride: According to Thomas et al. (2016, 59):

Superconducting transmission lines have a tremendous size advantage and lower total electrical losses for high capacity transmission plus a number of technological advantages compared to solutions based on standard conductors. This leads to a minimized environmental impact and enables an overall more sustainable transmission of electric energy. ... The access of remote renewable energy (RE) sources with high-capacity transmission is rendered possible with superior efficiency. That not only translates into further reducing CO_2 emissions in a global energy mix that is still primarily based on fossils but can also facilitate the development of RE sources given, for instance, the strong local opposition against the construction of new transmission lines.

If, in the future, superconductive transmission-line technology becomes actually available on a large scale, it may allow savings in electricity production from fossils and consequently, sizable decreases in greenhouse gas (GHG) emissions. The improved energy efficiency benefit may be valued through a decrease in energy costs, whether incurred by the energy producer, distributor, or final user. This cost reduction is not expressed at market prices but by considering the opportunity cost (shadow price) of the avoided energy inputs, which should be calculated as the long-term marginal costs of production and (if relevant) transportation. To estimate the negative externality of GHG and polluting emissions, the usual approach in CBA is to quantify the emissions avoided because of the project and value them with a unit of economic cost (measured in EUR per ton of emissions)—that is, with a shadow price of carbon or other pollutants (European Commission, 2014).

The main difficulty here is not measurement and valuation, but rather establishing a clear attribution. The principles of superconductivity go back to the work of Heike Kamerlingh Onnes, more than a century ago (1911), with the Nobel Prize for his work awarded in 1913. Applications have gradually evolved after several theoretical and experimental works in the field, leading to further Nobel Prizes in 1972, 1973, 1987, 2003, and more recently, in 2016. While it is fair to say that the 27-km-long tunnel hosting the LHC also hosts the longest cryogenic line, the most powerful superconducting magnets, and the largest number of them in one place, it is much more uncertain to what extent one can attribute to LHC accelerator science the further R&D activities needed elsewhere to move from what has been learned at CERN with superconductivity to other applications for transmission lines, with very different temperatures, materials, and objectives. There is a certain "LHC learning effect" in applied research on superconductive transmission lines,[9] but much else was developed elsewhere.

A second example is the indirect impact of knowledge created by satellites for Earth observation. The European Space Agency (ESA) has developed as part of the European Copernicus Programme, a set of space missions, based on one or more satellites for Earth observation.[10] The data collected by the Sentinels support manifold applications, such as monitoring of land and ocean environments, thorough mapping of vegetation, soil and water bodies, inland waterways and coastal areas, measurements of sea surface temperature and water quality, atmospheric monitoring, and last but not least, climate studies, for example. This information can be further elaborated to support decision-making and forecasting, including early warnings for natural hazards such as floods and fires.[11]

Hence, these RIs have a potential impact on the context of disaster resilience and natural risk prevention and management. Social benefits include the avoided damage to capital and natural stocks, sometimes combined with effects on economic activities such as agriculture, tourism, navigation, and fisheries. In the section entitled "Case Studies: Hadrontherapy, Earth Observation, and Synchrotron Light," the economic benefits of the Copernicus Sentinels (see figure 7.2) for winter navigation in Finland and Sweden are presented in terms of cost savings and improved efficiency, from icebreakers until final consumers of goods through satellite imagery over the ice-infested Baltic Sea waters during wintertime.

For each of the benefits of the Sentinels, in principle, there are well-established CBA approaches. For example, if one could attribute to a specific

Sentinel mission the provision of data that have been instrumental in forecasting disasters (after elaborations of the environmental protection authorities or other organizations) and saving lives, then one may use valuation methods developed in CBA for health, transportation, and environmental projects and policies. Such methods, following well-established literature, are based on the concept of the value of a statistical life (VOSL), defined as the value that a society deems should be spent to avoid the death of an undefined individual (more on this in the "Case Studies" section).

Again, the most challenging issue is not the valuation of such benefits, given the stock of well-known CBA methodologies and techniques about VOSL estimation (albeit with some uncertainty), but instead is the quantification, even ex post, of the amount of social benefits that can *causally* be attributed to a specific RI, albeit indirectly. A life saved by a weather forecast is the result of a combination of data provided by a satellite, by additional data provided by other sources, the development of mathematical models to simulate complex phenomena, the professional expertise of a forecaster, and eventually, an institutional mechanism to manage a meteorological alert effectively.

Historians of science and technology are qualitatively able to detect ex post certain paths from discovery to economic progress, and this often makes for fascinating reading. After a certain exposure to that literature, however, one understands that attempts of economists to simplify the story are often misleading. It is the cumulative nature of science that explains the technological and economic change. Having said this, I claim that in some cases, it is possible to insulate some causal linkages, and, in principle, one can try to attribute to a specific scientific project the ultimate benefits for consumers. Thus, a conservative approach is to value only the effects for which strict causation can be shown, going from the RI-created knowledge to innovations. Indeed, given the innovative nature of the supplied service/product, the reach, magnitude, and extent of the actual materialization of benefits is highly uncertain. Hence, what is more challenging is forecasting the probabilities of success and the various levels of effectiveness associated with the innovation implemented. In this context, different scenarios, at least one pessimistic and one carefully optimistic, should be forecast, and each of them needs to be carefully tested through a risk assessment (see chapter 10 for more details on risk assessment). Delphi methods, mentioned in section 10.5, may be helpful to identify the scenarios.

7.6   Wider Welfare Effects from Innovations

Further enlarging the perspective, if and when it is possible to identify a broad causal link between an RI project and economic innovations elsewhere (i.e., outside the direct mechanisms discussed in chapters 4, 5, and 6), then the social value of new goods can be expressed as the sum of changes in consumer and producer surplus. This would be the preferred avenue in the standard partial equilibrium CBA. For example, a large-scale randomized medical trial, based on the international collaboration of many researchers, and involving the observation over several years of the effects on the health of a new drug, may be considered a form of distributed virtual RI. If the new drug is successful, there are direct benefits for some patients (but not immediately for those of the control group taking a placebo), and ultimately, indirect benefits to many more patients when the drug is eventually approved by the health authorities. The demand curve for the drug by final patients is often mediated by public or private organizations, acting as providers or insurers, but in principle, one can say that there should be an empirical relation between the quantity of the new drug demanded and its price, given income levels, age of patients, market substitutes, and other drivers of the demand. The integrated difference between the equilibrium/observed price and the price that patients would be willing to pay for any consumption level in a compensated demand curve is the consumer surplus, a standard concept in microeconomics.[12] At the same time, the difference between the production cost of the new drug for any level of quantity produced and the price is the familiar notion of producer surplus, which boils down to a gross profit.

A discovery that reduces the cost of producing many goods (e.g., the discovery of a new material as a result of searching and testing solutions for space explorations) may affect the production of cars or dozens of other manufactured goods, as an alternative to more costly materials. The effect may be to decrease the production costs and, ultimately, the price of cars and any other good in other industries adopting the new material. The effect is to increase the consumer surplus (or the sum of the consumer and producer surplus). A practical example is in Florio, Bastianin, and Castelnovo (2018), which gives a simple diagrammatical representation of the potential socioeconomic impact of a breakthrough in the particle accelerator technology, such as plasma acceleration, which would have effects in radiotherapy, electronics, inspection of containers, and many other fields. This analysis is framed in a partial equilibrium setting.

In general equilibrium, the sum of producer and consumer surplus is achieved by the concept and empirical estimation of the shadow profit, which is the same concept of the net present value (NPV) presented in chapter 1, when shadow prices are considered. The benefit of innovation is expressed using the incremental shadow profit expected from the change in the availability of certain goods. In particular, *incremental* means that social profits expected from the sale of new/improved products, services, and technologies generated by the project must be compared with the profits in the without-the-project scenario; and *shadow* means that market distortions should be duly considered; for instance, the shadow profit is greater than the gross financial profit if the RI is located in an area of high unemployment because the shadow wage will be lower than the observed wage. For more about this, see the "Further Reading" section of chapter 1.

Let us assume that the innovations linked to a discovery having such wide effects can be identified and attributed to firms in one or more industries. As mentioned, these firms have no direct or indirect relation with the RI where the experiments were performed in the first place. For example, these are not firms under a procurement relation with an RI or in some way related downstream of them. Thus, we are evaluating generic Marshallian knowledge spillovers that are "in the air." Given j companies ($j = 1 \dots J$) and their innovations, ($i = 1 \dots I$) (products, services, and technologies) attributed to knowledge created by an RI over time represents the expected incremental firms' shadow profits directly imputable to these innovations, and s_t represents the discount factor. Then the expected present value of developing new/improved products, services, and technologies from applied research (AR) is simply expressed as a discounted sum of shadow profits Π:

$$\mathbb{E}(AR) = \sum_{j=1}^{J} \sum_{i=1}^{I} \sum_{t=0}^{T} s_t \cdot \mathbb{E}(\Pi_{jit}). \tag{7.1}$$

The ex ante estimation of these benefits involves the following calculations.

Benefits for final consumers must be quantified by forecasting the demand for new/improved products/services/technologies over time; and then the marginal social value (shadow price) of new/improved products/services/technologies should be estimated. Next, a very crude proxy of the shadow profits is obtained by looking at potential gross profits of firms in the concerned industries and correcting them appropriately to capture the socioeconomic impact.

Information on profitability, average costs, and sales can usually be retrieved from databases available in the public domain or may be granted by data providers. Typically, in competitive industries, earnings before interest, taxes, depreciation, and amortization (EBITDA) can be used to proxy companies' shadow profits because interest and taxes are effectively transfers between agents, and depreciation is inconsistent with the discounted cash flow approach that supports the computation of the NPV. For instance, the Amadeus Database, maintained by the Bureau van Dijk consultancy,[13] which provide ORBIS and balance sheet data reported to national registries and statistical offices by European companies, has been used by Castelnovo et al. (2018) to calculate

Table 7.2
Average of companies' median EBITDA margin (%) by sector and country (2004–2013)

Industry (NACE sector)	NACE Code	Italy	France	Germany	United Kingdom
Manufacture of basic metals	24	7.6	15.3	7.1	35.0
Manufacture of computer, electronic, and optical products	26	11.2	8.3	11.7	14.4
Manufacture of electrical equipment	27	10.3	16.4	11.7	11.2
Manufacture of machinery and other equipment	28	13.1	10.3	9.8	17.6
Telecommunications	61	40.1	13.8	11.3	10.0
Computer programming, consultancy, and related activities	62	11.3	15.3	8.0	8.3

Source: Florio et al. (2016), elaborations based on ORBIS data.

average sector-specific values for profitability up to the four-digit Nomenclature statistique des Activités Économiques dans la Communauté Européenne (NACE) level. Interviews with experts can assist in conjecturing on the possible changes in the profitability of businesses in different scenarios.

For the purpose of exemplification, the ten-year average EBITDA margins, associated with companies whose primary activity falls within a selected list of NACE codes,[14] are presented in table 7.2. Data are gathered from the ORBIS world database and refer to some sectors often involved in the procurement of major RIs. By excluding outliers in the sample, the average EBITDA is between 10 to 15 percent across countries.

In some cases, instead of focusing on shadow profits, it is more convenient to consider that users avoid certain costs, given the exploitation/application of new technologies made available for free by the RI. For instance, an innovative combustion technology developed by an RI can be exploited by businesses to improve their own production processes, thereby significantly reducing their energy costs. These avoided costs represent the value of knowledge spillover for businesses.

While conceptually simple, the abovementioned empirical strategies often will be difficult in practice. A wide body of literature has linked firm- or country-level R&D expenditures, patents filed, or other proxies of knowledge production to economic performance, but here, the empirical issue is different. It should link the ultimate welfare change of agents to specific discoveries of research results, at a given time, and in one specific place. As has been mentioned, narratives of such effects are always possible and often helpful.

However, empirical testing is much more demanding, and the risk of exaggerating the benefits of a specific RI project is considerable if the cumulative nature of knowledge is disregarded.

7.7 Case Studies: Hadrontherapy, Earth Observation, and Synchrotron Light

In this section, three case studies of the benefits provided to users are provided. The first refers to the value of health benefits for patients of the CNAO hadrontherapy center in Pavia, the second to economic benefits to users of Sentinel data during Baltic navigation in winter, and then a sketch of an empirical strategy to estimate the benefits of the ALBA synchrotron light source is presented.

7.7.1 CNAO National Center of Oncology Hadrontherapy

This section heavily draws from Battistoni et al. (2016). The starting point for the evaluation of an RI in health is the consideration of the benefits associated with decreasing mortality rates and increasing life expectancy, suitably adjusted by the quality of life.[15] This is because medical research and clinical benefits are linked in many cases. In fact, ultimately, no knowledge in medicine can be created without curing patients.

Hence, there are two steps in the application of CBA to a medical RI project. First, after providing a breakdown of the various new clinical treatments arising from research, a forecast is needed for the total number of patients by each treatment. Second, a probabilistic estimate is needed of the greater effectiveness and lower toxicity levels for each treatment and for each category of patients compared to the existing treatments. Finally, an estimation of the monetary value of quality-adjusted life-year (QALY) should be used to value the total expected amount of QALYs gained by patients, thanks to the research.

CNAO, in Pavia, Italy, is a synchrotron designed to develop innovative hadrontherapy approaches for patients affected by some solid tumors, which are both radioresistant and unresectable or need high-precision beams. See box 7.1 and figure 7.1 for more about CNAO.

An important step in the CBA of CNAO was the identification of a counterfactual scenario for each type of protocol. This was needed to quantify the incremental benefit associated with the innovative therapy treatment relative to possible alternatives. The two core variables were the number of years of life gained and the percentage of successful treatment (suitably defined).

In fact, for most protocols, the counterfactual scenario is a "do nothing" option. This is because the patients included in the clinical treatment have no

Box 7.1
CNAO in a Nutshell

> CNAO, National Center of Oncology Hadrontherapy, is a nonprofit foundation
> on whose board there are several hospitals, universities, and research institutions,
> stemming from ideas originally developed in 1991 by Ugo Amaldi, a particle physi-
> cist at CERN, and by Giampiero Tosi, a health physicist of the Niguarda Hospital,
> Milan. CNAO was built between 2005 and 2009, involving around 400 companies,
> at an investment cost of about EUR 133 million. Later, the cost of a three-year
> long clinical trial phase was EUR 37.2 million. Battistoni et al. (2016) divided the
> total yearly number of patients under each treatment into six age categories, given
> the correlation of mortality with age. When this analysis was conducted, CNAO
> was authorized by the Ministry of Health to treat patients falling under twenty-
> three different protocols (each addressing a specific type of tumor in a determined
> organ), of which twelve use carbon ions, nine use protons, and two use combined
> therapies. The most innovative treatments involve the use of carbon ions.

alternative treatment, particularly after they have already been unsuccessfully
treated with other therapies. In other cases, the counterfactual includes surgery,
with or without associated conventional radiotherapy based on X-rays, and che-
motherapy. Clearly, under "do nothing," counterfactual potential benefits are
greatest, as the marginal gain is the same as the total gain.

The quantification of the marginal benefit arising from each protocol has
been estimated as follows: (1) the difference in the percentage of patients
who fully recover after treatment at CNAO compared with the counterfactual
scenario; (2) the difference in the percentage of patients who gain some additional years of life because of hadrontherapy compared with the counterfactual
scenario; and (3) the difference in the percentage of patients who benefit from
a marginal increase in the quality of life compared with the counterfactual
scenario. The quality-of-life benefit can be combined with the other ones.

As for the quality-of-life benefit, a quality score ranging from 0 to 1[16] was
based on earlier medical literature, such as the evidence on treatment toxicity. Then the economic value of statistical life estimation was based on well-
established literature.[17] In this literature, the money value of an increase in
life expectancy includes the estimation of the VOSL and the related value of
a life-year (VOLY). Battistoni et al. (2016) use the human capital approach
because of its convenience in terms of empirical analysis, but also because it
leads to more conservative estimations compared with either revealed or stated
preference approaches.[18]

With the human capital approach, the VOSL is estimated based on lost production in terms of the average annual wage or income due to an individual's

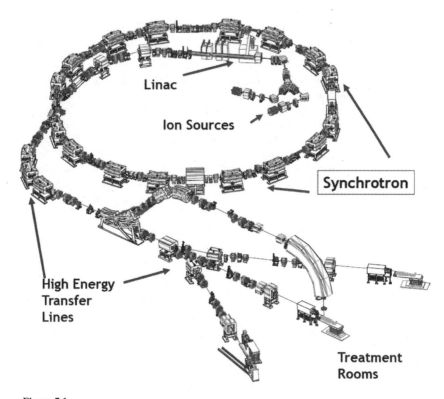

Figure 7.1

CNAO (located near Pavia, Italy) is a synchrotron composed by a set of accelerators and transport lines of particle beams (protons or carbon ions). The perimeter is 80 meters. Packages of particles are accelerated to 250 MeV for protons and 480 MeV for carbon ions. The beams are then sent to one of the three treatment rooms where a patient affected by cancer is located. The beam strikes the cells of the tumor with a precision of 200 μm (two-tenths of a millimeter) and gradually destroys the tumor in a few minutes per session. *Source*: Courtesy of ©CNAO, from Rossi (2015).

premature death. There are several variants of this approach, and a sensitivity analysis may reveal to what extent the results depend upon specific assumptions. Having established the range of variation of the three key input variables (i.e., the number of patients, marginal health improvements, and the economic value of a year of life saved), the total expected present value of the applied research benefit for CNAO's patients amounts to nearly EUR 2 billion. See Battistoni et al. (2016) for further details. Although there is a large degree of uncertainty, the health benefits are significant and repay the total discounted costs of the RIs ($\mathbb{E}(\mathrm{NPV}) > 0$), even under the most conservative assumptions (see chapter 10).

7.7.2   Copernicus Sentinel Satellites

The case study in this section draws from the Green Land and European Association of Remote Sensing Companies (EARSC) (2016) and from Tassa (2019). While this study is not a CBA, its results are, in principle, easily translated into the welfare impact on different agents of the data produced by Sentinels for a chain of specific downstream users (see figure 7.2). As mentioned

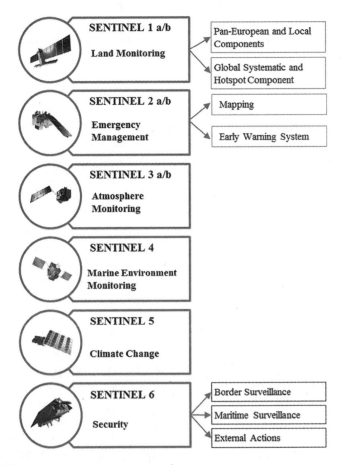

Figure 7.2

The Sentinels are the satellites supporting the Copernicus Programme. There are six families: Sentinel-1a (launched 2014) and -1b (2016) for radar imagery services; Sentinel-2a (2015) and 2b (2017) for optical imagery of vegetation, coastal and other areas; Sentinel-3a and -3b (2018) for optical and radar observations of land-surface temperature, ocean-color, any other variables. Sentinel-4 (2019) will provide data for atmospheric composition monitoring (gases aerosols). Sentinel 5 (2021) will measure atmospheric constituents such as ozone, nitrogen dioxide, sulfur dioxide, carbon monoxide, methane, and formaldehyde. Finally, Sentinel 6 (2020) will provide data for oceanography and climate studies. *Source*: Courtesy of © European Union, 2018, author's adaptation.[19]

earlier in this chapter, Copernicus, a program of the European Union (EU) and of the ESA, operates a set of satellites for Earth observation. Their data are provided for free to everybody, and at present, there are more than 140,000 registered users (June 2018). Here, I report some key figures for Finland and Sweden derived from the abovementioned EARSC study.

There are over 20,000 ships that call at the twenty-five major Finnish ports in winter. Most of these ships are not equipped to navigate in winter without icebreakers, which literally open the way to them. There are five tiers of users of radar data originated by the Sentinels. The primary service provider is the Finnish Met Office, producing daily ice charts (Tier 0); icebreakers, which without such data would need to use helicopters to find the best route in the iced sea (Tier 1); ships, such as cargos and oil tankers, using route waypoints and ice charts and following the icebreakers to minimize delays and risks (Tier 2); firms operating at ports, such as haulage and logistics, for whom time of arrival is critical (Tier 3); other business, such as paper mills or oil and gas, for which the timeliness of service is also critical (Tier 4); and the general public relying on the availability of goods at retail outlets (Tier 5). The benefits along the value chain can hence be broken down by tiers as follows: reduced fuel consumption and use of helicopters (T1); reduced fuel and operation costs of ships, including fewer ice-related damages (T2); reduced uncertainty about arrival times and related operational benefits (T3); lower probability of severe delays, and disruption of production and distribution (T4); and citizens' WTP for security of supply (T5).

The authors of the study used some generic modeling of the variables, such as time of arrival, correlated with winter ice conditions in the form of probability distribution of some parameters (however, without precise past data, and hence with large uncertainty). Satellite imagery is broadcast regularly and seen by captains of the icebreakers, who make their route decisions based crucially on it. It was estimated that yearly fuel cost savings for icebreakers in Finland and Sweden are EUR 1 million; savings on the use of helicopters per year, EUR 1.3 million; reduced costs for ships are estimated at EUR 3.3–9.4 million; economic benefits for ports are estimated in the range of EUR 4.2–21 million; the value of the lower probability of extreme delays is estimated at EUR 3.6–36 million per year, based on certain hypotheses on number of days that ports are iced, and a proportional share of gross domestic product (GDP) per capita. Finally, the WTP of citizens for service continuity is conjectured as EUR 1 per year (without a stated preference survey), which would lead to EUR 2.10 million per year. The report concludes that the sum of the economic benefits generated in Sweden and Finland is in the range of

EUR 24–116 million per year (the wide range reflecting model uncertainty) because of the free availability of satellite radar images of the iced surface of the Baltic sea in winter.

To compute a benefit-cost ratio, what would be needed is the marginal cost of providing these data by the Sentinels. This is the *total cost change* because of the specific data provided, which is probably much lower than the *average total cost* (the latter can be estimated in terms of the per capita contribution of EU taxpayers to the Copernicus Programme, against multiple services such as the abovementioned one). In the specific case of the Baltic Sea navigation, the only other relevant marginal cost component is the personnel and instrument expenditures of the meteorological services. Against such costs, there are the net benefits of the data provided by the satellites. This would be an example of the avoided cost approach, as the benefit is estimated in this way, rather than WTP by all the agents involved in using the data. A NPV is not provided in the report, but in principle, it could be estimated. In spite of several empirical shortcuts, this study suggests that in some cases, tracking the economic effects along a specific value chain is feasible, and it is a matter of adequate research design.

7.7.3 ALBA Synchrotron Light Source

The case study in this section[20] presents a possible evaluation strategy of the innovation impact of ALBA (see figure 7.3), a third-generation SL source located near Barcelona, Spain, which is 50 percent funded by the Spanish government and 50 percent by the Catalan regional government. Other third-generation sources include ESRF in Grenoble, an international project, and national facilities such as Diamond (Didcot, United Kingdom), Soleil (Saint-Aubin, France), Elettra (Trieste, Italy), and DESY (Hamburg, Germany). According to https://lightsources.org/about-2/ (access date: March 19, 2019), "In Europe, in the past 5 years alone, light source facilities have welcomed 24,000 users, who have had an impact on a wider network of 35,000 researchers, with 23,400 unique articles published in peer-reviewed journals." The operations of ALBA started in 2012, and the number of beamlines was initially just seven, out of a potential of around twenty, because of severe financial constraints. Four new beamlines will be built in the next few years, and additional ones will follow as soon as funding is available, at a cost for each beamline in the range of EUR 5–7 million, according to its technology. Each beamline is, in fact, experimental equipment with different characteristics tailored to the needs of different communities of users. The techniques, according to the existing beamlines, include X-ray powder micro-diffraction, X-ray scattering, X-ray absorption and emission spectroscopy, infrared microspectroscopy, photoemission, macromolecular crystallography

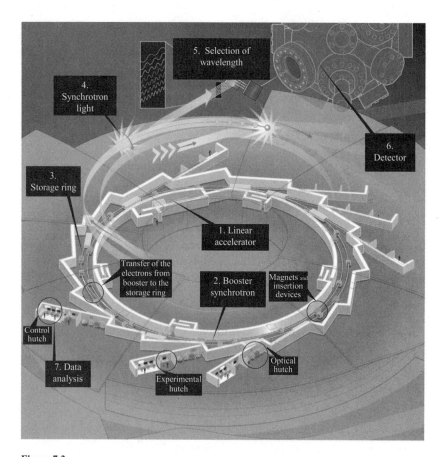

Figure 7.3

The ALBA Synchrotron (Cerdanyola del Vallès, Barcelon) is a synchrotron light source based on a combination of a linear accelerator of electrons and a booster placed in the same 270-m-perimeter tunnel as the storage ring. The X-rays emitted by the 3 Gev electon beam allow for studying the atomic structure of matter, mainly for bioscences, condensed matter, and materials science research.

1. Linear accelerator: The electrons are generated and subjected to an initial acceleration. **2. Booster synchrotron**: The electrons accelerate once again in the inner ring until reaching speeds close to the speed of light. **3. Storage ring**: The electrons are stored in the outer ring, guided by magnetic fields. **4. Synchrotron light**: Upon passing through the magnetic fields, the electrons emit energy in the form of synchrotron light, which is sent to the beamlines. **5. Selection of wavelength**: The synchrotron light contains numerous wavelengths, and a monochromator is used to select the most suitable one for each experiment. **6. Detector**: The sample to be analyzed is illuminated, and a detector captures the interaction of the sample with the light. **7. Data analysis**: The data are stored and analyzed. *Sources*: Courtesy of ©ALBA Synchrotron, author's adaptation, https://www.cells.es/en.

Box 7.2
Users of ALBA Beam Time

The vast majority of users of ALBA beam time are academic teams and research institutes, including hospitals. These nonprofit users are granted free beam time if their research project has a proven scientific validity. A minor share of beam time is given to business for proprietary R&D. These users pay around EUR 550 per hour (average operating costs). Calls for applications are organized twice a year, and the ALBA users' office manages all the selection procedure. At ALBA, a national facility, 65 percent of users are Spanish. The remaining 35 percent are users from abroad, a share that is in line with other synchrotrons. Half of the applications (around 430 in 2018, for around 1,440 users, including over 180 remote users) per year are accepted on average, but there are wide differences across beamlines (e.g., CIRCE, the photoemission spectroscopy and microscopy beamline, is in high demand, and thus more than two-thirds of the proposals cannot be accepted). This does not necessarily mean that the beamlines in higher demand are the same as those supporting the most advanced research such as MISTRAL, a beamline specializing in cryo-nano-tomography for biological applications. This is considered cutting-edge technology, with possible applications in biosciences. Another example is BOREAS, a beamline dedicated to fundamental, as well as applied, polarization-dependent spectroscopic investigation of advanced materials, an important technology for magnetic research.

(particularly proteins), magnetic dichroism, soft X-ray microscopy, magnetic reflectivity, and resonant scattering.

As with all other SLs, ALBA basically makes available to users both the light itself and well-trained staff to support researchers in their experiments. A fraction of beam time is booked for internal ALBA researchers for their own studies (12.5 percent); 20 percent is reserved for testing and buffering (including reserve capacity), while all the remaining beam time (over two-thirds) is for external users. Box 7.2 describes the typical ALBA users.

The selection process of the proposals is based mostly on their scientific quality and feasibility. Each proposal is reviewed by a team of international experts who assess their merit, also taking technical and safety aspects into consideration.

Experiments using the beamlines can be classified by scientific field: chemistry (e.g., structural characterization of solid samples at the atomic level); advanced materials (e.g., metals, ceramics, superconductors); nanotechnology (e.g., structure, characterization); pharmaceutical [inter alia, structural biology, interaction between drugs and therapeutic target at atomic level, detection of impurities, three-dimensional (3D) reconstruction of cells]; health products (including cosmetics and personal care); food and agriculture (e.g., toxicity, fertilizers, etc.); environment (e.g., analysis of polluted water, air, and soil); automotive and aerospace (including catalysts for reducing emissions oils and lubricants), energy (batteries, solar cells); and cultural heritage (ancient materials, paintings).

In fact, except for the very small share of beam time allocated to proprietary research, the vast majority will generate experimental data, supporting publications in scholarly journals in different domains. In recent years, almost 200 publications in scholarly journals per year are tracked by the ALBA Scientific Division, and each publication can be precisely attributed to one specific beamline, as the experimental data are duly cited with such reference. It is expected that the yearly number of publications will increase over time because of the combined effect of the new beamlines and the cumulative effect of past research.

A possible strategy to evaluate the economic impact of the knowledge created by ALBA (or similar SLs) could take advantage of this feature. A minimalist approach would be to evaluate the citation impact of the ALBA-related literature (more than 700 articles by the end of 2018), in the ways discussed in chapter 3. However, this would miss the innovation impact. In principle, the latter may be tracked with the citations of the ALBA-related literature as nonpatent literature (NPL) in filed patents, and then value such patents according to what has been discussed in chapter 5. A practical example on the implementation of a data-mining approach for patents citing biodata is Bousfield et al. (2016). This strategy, while informative, is constrained by the fact that NPL references are unsystematic and, importantly, by the fact that many innovations are not patented.

A more promising strategy would be a periodical survey of external users, as it is highly likely that they will track the applications of their published research. At ALBA, around 30 percent of the applicants for beam time declare on the online application form that they are already aware of the possible interest of industry for the results of their experiment. Hence, this awareness is declared at a very early stage of the project, possibly because some academic teams in fact have already established some linkages with corporate R&D at the time of the application or will create such linkages with industry soon after the experimental data are available, or after publication of the results.

Over the years, thousands of articles will be published, based on the scientific services provided by ALBA, and a regular survey of the authors can reveal whether they are aware of product or process innovations related to such literature. The last step would be estimating the value of such innovations with the methods described in this chapter, pointing to shadow profits. For the small share of industrial users, those firms who directly buy beam time, a similar survey after some years may reveal the contribution of experiments with the SL to their R&D and innovations.

A shortcut strategy for any SL such as ALBA could point to the identification of a threshold value of innovation benefits against costs. Potentially, ALBA shares the CBA model of equation 1.2 (repeated here for convenience) with any other RI:

$$NPV_{RI} = [SC + HC + TE + AR + CU] + B_n - [K + L_s + L_o + OP + EXT]. \tag{7.2}$$

Benefits to scientific users, measured by the value of publications (SC) using ALBA data, to ECRs (HC) in the experimental teams, to firms through technological learning (TL), and (CU) to users of cultural goods (e.g., ALBA organizes open days for the general public and other outreach activities) are all qualitatively similar, for any of the three types of synchrotrons mentioned in section 7.1.

Costs of capital, scientific and operative labor, other operative expenditures, and negative externalities ($K + L_s + L_o + OP + EXT$) are also qualitatively similar. The main difference between the LHC and ALBA, or between an RI for basic science and one for applied research more in general, potentially lies in two terms: B_n, the nonuse benefits, which include the WTP of citizens for just knowing that LHC and ALBA make certain research; and AR, the benefits of innovations deriving from services provided. The former are probably much higher for LHC than for ALBA (see the discussion in chapter 9), but the latter are possibly higher for ALBA than for the LHC, given the applied research of an SL. Setting conservatively:$B_n = 0$ for ALBA, then $NPV_{RI} > 0$ if $AR > [K + L_s + L_o + OP + EXT] - [SC + HC + CU]$. As all the terms on the right side of this disequation can be determined, the minimum value of AR leading to $NPV = 0$ can be determined as $min(AR) = [K + L_o + OP + EXT] - [HC + CU]$, and $L_s = SC$.

Taking stock of all the ALBA-related literature as N_t at any given time t, the NPV test is passed if the average paper generates an economic value of innovations in terms of shadow profits $> min(AR)/(N_t)$. Perhaps most scientific publications may not generate innovations and profits at time t, but some of them will lead to innovations (see the example of the Daresbury SL discussed in the "Further Reading" section), and their economic value may be estimated. Hence, through a survey of both scientists and firms or by other empirical methods, it would be possible to focus on a subsample of the ALBA-related literature and conservatively test whether the measurable cumulated benefits are greater than the required $min (AR)$ value per published paper times the number of papers:

$$\sum_{j=1}^{J} \sum_{i=1}^{I} \sum_{t=0}^{T} s_t \mathbb{E}(\Pi_{ijt}) > \left(\frac{minAR}{N_t} \right) N_t, \tag{7.3}$$

where j are companies ($j = 1 \dots J$) and i their innovations ($i = 1 \dots I$) (products, services, and technologies), and t is the time.

This would require an adequate economic impact-monitoring mechanism over time, beyond simply counting papers and citations by each beamline.

7.8 Further Reading

This section briefly reports additional material on the economic impact of innovations made possible, thanks to some RIs. These include Copernicus satellite data for forestry in Sweden, an ex post evaluation study of the Daresbury synchrotron light source, a report on RIs in Australia, an evaluation of RIs in the UK, a report on the economic impact of the Human Genome Project (HGP), and some references about the NPL.

A study of the economic impact of Sentinel satellite data on Swedish forestry, similar in terms of objectives to the abovementioned one on winter Baltic navigation, is provided by the Green Land and EARSC (2016, 6). This report estimates the benefits that arise from the control by the Swedish Forestry Agency (SFA) of illegal cutting of plots, through the data provided by the satellites:

Through the use of clear-cut maps, i.e. maps showing where forest has been cleared for harvest, the SFA can check whether this clearing was allowed under law and can take action where appropriate. But most importantly, the forest owners know that the SFA can monitor their land which has improved compliance with the law. As a consequence of the availability of the imagery, the area of forest cleared "illegally" has fallen from around 10% of harvested forest each year (in 1998) to less than 0.5% (according to a 2003 study carried out internally by the SFA). … The gathering and use of the imagery and the clear-cut maps cost very little (€64k) whilst the benefits are … between €16.1m and €21.6m per annum.

An unusually detailed (218 pages) ex post impact report on a specific large-scale RI is STFC (2010b), as it covers the entire life of the now-ceased British Daresbury Synchrotron Radiation Source (SRS), the first in the world of the so-called second generation of multiuser X-ray facilities. While the report shares the limitation of not being able to provide an estimation of a socio-economic impact with most of this kind of literature, it is very informative.

For the purpose of the discussion in this chapter, several aspects are of interest. First, the SRS has been a learning laboratory not just for its British successor, the Diamond Light Source, but also for the development of many existing similar machines, with staff trained at the SRS also subsequently working elsewhere (for example, at ALBA). This is an example showing that discoveries in one RI may be, in part, credited to know-how transferred from another RI in various ways, but particularly through formal or informal collaborations.

Second, impacts on the quality of life of citizens are reported, including the development of new drugs, new materials for electronics, clothing, new detergents, the taste of chocolate, the safety of aircraft, and iPod memories. Moreover, industries, governments, hospitals, and other organizations are cited, which have benefited from the SRS through more than 200 proprietary research projects on the properties of materials and structure of pharmaceuticals, chemicals, and other substances. A particularly interesting example is the X-ray analysis of the structure of the Foot and Mouth Disease virus (FMDV), which is responsible for a cow disease. This research at the SRS was instrumental in the development of a vaccine. It is worth reporting a sketched attempt at CBA in this report (STFC 2010b, 48):

It is not possible to give an accurate estimate of the potential savings resulting from FMDV vaccines. However, the magnitude of the opportunity can be assessed by examining the costs of the 2001 outbreak in the UK.

The costs of the 2001 outbreak are estimated by DEFRA [Department for Environment, Food, and Rural Affairs] at £8.4 billion, with the following split:

Agricultural producers (£355 million).

Food industry (£170 million).

Tourism (£3,000 million (average)).

Indirect costs for supply companies (£2,100 million).

Expenditure on FMDV by Government (£2,800 million).

The figures given relate to those which are easily measurable, human and other costs have not been measured... If the knowledge gained through SRS and associated work could contribute even 1% to the costs presented above, this would represent a financial impact of £ 84 million. In contrast, the work undertaken to determine the X-ray crystal structure using SRS was estimated to take some hundreds of scientist days, valued at tens of thousands of pounds.

This example suggests that, at least in some cases, a causal link between the RI services and ultimate social benefits can be established and empirical CBA estimates provided (after adding the cost of developing the vaccine, not mentioned in the above citation), and with some uncertainty on the figures.

A report by the Australian Academy of Sciences (2016), after reviewing the state of play of applications of particle accelerators worldwide, lists the stock in the country as follows: 1 synchrotron, 9 electrostatic, 17 cyclotrones, and 182 linacs, mostly for cancer therapy. Cited research projects include analysis of air pollution, data on soil erosion, desertification, rainfall changes, groundwater quality, aging, sustainability, new polymer-based photovoltaic, and several other topics. It concludes with recommendations to establish a national center for hadrontherapy (see the CNAO case study earlier in this chapter) and to support new and existing RIs. This report is similar to several others, as it is basically intended to show qualitatively potential impacts on consumers and firms and on researchers and students, with no attempt to compare benefits and costs. It is, however, highly informative about the potential impact in terms of innovations.

Not all RIs are an unconditional success in terms of benefits to users, and understandably, the literature on their downside is scant. However, as an example of planning problems, one may read the British National Audit Office report (NAO, 2016), which shows the evaluation of several projects in a color-coded chart.[21] For example, an RI on graphene, an innovative form of carbon, which is very strong and conducts electricity and heath well, was criticized by this report. Graphene can revolutionize fields such as solar cells, light-emitting diodes (LEDs), touch screens, and others. Two Nobel Prizes were awarded, in 2010, to Manchester University scientists Andre Geim and Konstantin Noveselov for their studies on it. However, the National Graphene Institute (NGI), promoted by Manchester University,[22] a 61-million-pound investment, and other graphene projects by the NAO lacked an adequate assessment to 2012, in terms of: "Alternative options assessed; Running costs estimated and funding/affordability confirmed; Sensitivity analysis of costs and benefits; Plan to tracking and assessing benefits realisations; return on investment estimated."

This amounts to five "reds" against just one "green" (estimation of demand) cells in the aforementioned figure. In an article in *Nature*, Peplow (2016) discusses controversies about the objective of the NGI as a link between the university and industry, and about intellectual property protection, due to the different approaches adopted by businesspeople and researchers.

As previously mentioned, a possible empirical approach to track the impact of an RI is through citations of scientific literature, generated by it in patents filed. These may be part of the so-called NPL citations (see section 5.4 in chapter 5). The US and European patent offices (USPTO and EPO) require patent applicants to provide the NPL that has had a role in their invention, including articles in scholarly journals; see Jaffe, Trajtenberg, and Fogarty (2000), or Gittelman and Kogut (2003), and OECD (2009, 2013), which suggest that NPL citations are a proxy of scientific influence. For applications, see McMillan, Marin, and Deeds (2000) and Callaert et al. (2006).

Simmonds et al. (2013), while mainly focused on the impact of the innovation of large-scale research facilities in the United Kingdom, review seventy papers on the topic, both theoretically and empirically. However, the cited literature deals with three types of impacts: the economic impact of money spent on the RI (mainly from a macroeconomic perspective, with input-output or other aggregate models); knowledge spillovers from procurement; and local effects. Hence, these are the mechanisms discussed in earlier chapters, with only limited coverage of the impact on final users discussed in this chapter. An exception is the Tripp and Grueber (2011) report on the HGP. Surprising figures are mentioned (see Simmonds et al., 2013, 38, 93, for a summary):

Between 1988 and 2010 the human genome sequencing projects and associated research and industry activity, directly and indirectly generated: (1) $796 billion in U.S. economic output, (2) $244 billion in personal income for Americans, and (3) 3.8 million job-years of employment. A federal investment of $3.8 billion ($5.6 billion in 2010 $) was judged to have made possible more than $796 billion in economic output...This is a return on investment (ROI) analysis, which compares the US government's investment of $3.8 billion (£2.5bn) in the HGP project across the 10-year period 1993–2003 with the combined effects of that expenditure on the US economy and on the US genomics industry. Overall, the report estimates a return on investment of 141:1 for the federal investment in the programme.

The economic impact figures are striking. Unfortunately, the methodology seems to be affected by the attribution issues discussed in this chapter. Simmonds et al. (2013, 39) seems to be aware of the problem, as they state: "The principal methodological weakness, common to all impact types, is the perennial challenge faced by any research evaluation, which are the questions of attribution and additionality." I agree, and I have tried to stress this issue in this chapter, albeit proposing with due caution what can be done.

7.A.1   Appendix: Valuation of Social Benefits from Applied Research

Box 7.A.1.1 on page 220 presents, as illustrative examples among many possible ones, the evaluation methods referring to typical benefits in the environmental, energy, and health sectors, which may benefit from science created by RIs.[23]

7.A.2   Appendix: Cosmo Skymed and the Effects of Earth Observation

Data from satellite Earth observation have been increasingly used to track the evolution of a variety of environmental issues, ranging from marine pollution caused by oil leakage, to terrestrial desertification or deforestation problems, to atmospheric issues like GHG concentration of the thinning of the stratospheric ozone layer (Young et al., 2008).[33] See, for example, the Resources for the Future consortium for the study of the socioeconomic benefits of Earth observation (www.rff.org).

Box 7A.1.1
Examples of Social Benefits Derived from Applied Research

Evaluation Methods	References

Improved energy efficiency

The improved energy efficiency benefit is valued through the *decrease in energy costs*, whether incurred by the energy producer, distributor, or final user. The cost reduction is not expressed at market prices, but by considering *the opportunity cost* (shadow price) *of the avoided energy sources*, which should be calculated as the long-run marginal cost of production and (if relevant) transportation.

European Commission (2014), Chapter 5, on energy sector.

European Investment Bank (2013) contains a chapter on energy efficiency and district heating.

ENTSOE (2015).

World Health Organization (2006).

Clinch and Healy (2001).

Note that producing electricity from a renewable source could be, at least initially, more expensive than from other sources. In fact, emerging renewable technologies are typically not competitive with fossil fuel alternatives (HM Treasury 2006). Thus, the project would produce a cost and not a benefit. However, this cost would be (partly or fully) compensated by higher benefits from reduced GHG and pollutant emissions.[24]

Reduction of GHG and air pollutant emissions

To estimate the externality of GHG and pollutant emissions,[25] the usual approach consists of *quantifying the emissions avoided because of the project* (measured in kg/ton of waste) *and valuing them with a unit economic cost* (measured in EUR/kg of emissions). However, when a new ecofriendly technology is developed and sold to enterprises, for example, the selling price could already incorporate the environmental benefit. In such a case, the externality should not be estimated to avoid double-counting.

IMPACT study (European Commission 2008b), which lists unit cost values for the main relevant air pollutants (in EUR/ton) on the basis of HEATCO[26] and CAFE CBA[27] reports;

NEEDS Integrated Project,[28] which provides unit damage costs for air pollutants from emerging electricity generation technologies. NEEDS also provides reliable cost factors for ecosystem and biodiversity damage from air pollution.

An Extern-E study[29] provides the unit values of air pollutants produced by energy infrastructures in EU member-states.

Teichmann and Schempp (2013).

Reduction in vulnerability and exposure to natural hazards

When an RI infrastructure project is aimed at developing tools and disaster management systems to facilitate disaster resilience, risk prevention, and

The World Bank (2003) contains Chapter 3, on natural disaster risk and CBA.

management for natural risks, a benefit from *avoided damage to capital and natural stocks* is expected.[30]

The cost of the avoided damages is estimated using information and data contained in risk and hazard maps and modeling. A shortcut consists of adopting the market insurance premiums available for different typologies of risks to proxy the value of the avoided damage to the capital stock. Instead, for assets for which an insurance market does not exist, averaged calculations on the basis of the avoided costs of the public administration for civil protection

activities, compensation paid to citizens, relocation of buildings, and other activities should be carried out and added to the economic analysis (European Commission 2014). Alternatively, the people's WTP for decreasing the vulnerability and exposure to a natural hazard could be estimated. Finally, when the project addresses natural assets, additional effects should be evaluated in terms of increased use or nonuse values. Regarding use values, the typical additional effects to be considered are increased recreational value, typically valued through the travel cost method (TCM), and preservation of productive land, typically valued through its opportunity cost. Regarding nonuse value, the preservation of a natural asset in good condition must be estimated by eliciting its existence value (typically, through contingent valuation or benefit transfer).[31]

Improved health conditions

Changes in human mortality and morbidity rates can be triggered by a RI infrastructure[32] with different aims, such as improving the health conditions of the people affected by a certain disease by producing a new drug or an innovative treatment technology; improving the health safety of people (or a group of people), such as through food and transport; remediating a polluted environment (e.g., a radioactive dump or a site contaminated by chemical waste); and mitigating the risk of natural

Guha-Sapir, Santos, and Borde (2013).

MMC (2005).

Benson and Twigg (2004).

Kunreuther and Michel-Kerjan (2014).

On VOSL, see, for instance, Landefeld and Seskin (1982); Viscusi and Aldy (2003); Abelson (2003, 2008); and Viscusi (2014).

On VOLY, see, for instance, Johannesson and Johansson (1996) and Desaigues et al. (2011).

For a major meta-analysis of VOSL estimates, derived from surveys that asked people around the world about their WTP for a small reduction in mortality risk, see OECD (2012) and Lindhjem et al. (2011).

(continued)

Box 7A.1.1 (continued)

Evaluation Methods	References
disasters. In such cases, the project's marginal benefit is the reduction in mortality or morbidity rates or improved health conditions. Following the literature, these reductions can be valued using VOSL, The QALY may also be used. The preferred approach to value changes in health outcomes is to calculate the WTP of people affected by the project. This calculation can be done using the stated preference methods (surveys) or revealed preference methods (hedonic wage method). However, in practice, the human capital approach (for mortality) or the cost of illness approach (for morbidity) is more frequently used. Each method has its benefits and drawbacks.	For a meta-analysis of VOSL estimates, derived from revealed preference studies, see Mrozek and Taylor (2002). On human capital approach, see, for instance, Landefeld and Seskin (1982) and Brent (2003, Chapter 11). On the cost of illness approach, see, for instance, Rice (1967); Rice, Hodgson, and Kopstein (1985); Byford, Torgerson, and Raftery (2000), and World Health Organization (2006).

Source: Florio et al. (2016).

Cosmo-SkyMed (Constellation of small Satellites for Mediterranean Basin Observation), developed by the Italian Space Agency (*Agenzia Spaziale Italiana,* or ASI) in cooperation with the Ministry of Defense in Italy, is an Earth satellite observation system designed for dual purposes, civil and military, based on a constellation of four identical satellites equipped with synthetic aperture radar (SAR), working in the X band, and fitted with particularly flexible and innovative data acquisition and transmission equipment. The system is completed by dedicated, full-featured ground infrastructures for managing the constellation and granting ad hoc services for the collection, archiving, and distribution of acquired remote sensing data. The system is capable of taking up to 450 shots of the Earth's surface—equal to 1,800 radar images—every 24 hours.

The microwave sensors observe the Earth in a completely different way from optical systems. This is due to the possibility of exploiting the various spectral characteristics (e.g., electromagnetic reflectance) of different types of objects on the Earth's atmosphere, land surface, or sea. Cosmo-SkyMed is able to observe the Earth day and night and, most important, almost independent of atmospheric conditions. Various information can be extracted from these remote sensing data, such as illegal leakage of hydrocarbons into the sea or the observation of floods or volcanic activity. Moreover, the comparison of SAR images acquired at different times also allows the measurement of surface deformations (e.g., land subsidence/uplift) or changes in water levels. The continuous availability of precise measurements of vertical of horizontal ground movements has applications in such fields as civil engineering, oil and gas industries, or mining.

COSMO-SkyMed can be used in a very flexible way. The radar eye can focus on an area of a few square kilometers and observe it with resolution up to a single meter. It can observe a continuous strip of land surface or cover a region of 200 km. The same point can be monitored at a time interval shorter than 12 hours. Radar images of some of the most serious natural disasters in 2008, such as Cyclone Nargis in Burma, an earthquake in China, and Hurricanes Hannah and Ike on Haiti, were used by the United Nations and humanitarian organizations involved in aiding the affected populations.

COSMO-SkyMed has activated a complete interferometric mapping service of the Italian national territory every sixteen days in acquisition mode StripMap-HIMAGE. Time series of images are used for interferometric investigations aimed at analyzing the phenomena of landslide and endogenous risk of the territory itself (landslides, subsidence, seismic phenomena, volcanoes) and regularly populating an intensive, specific interferometric historical archive of national geographic references.

Even though all this information has not been yet fully integrated and automatized in the official monitoring and decision processes, it has been used occasionally to prevent environmental disasters, and more intensively to manage them and to intervene rapidly when they occur. Application fields included the monitor and control of the state of the coasts and their erosion.

The ability of COSMO-SkyMed satellites to utilize both horizontally and vertically polarized signals allows better land classification and crop monitoring during the growth cycle, which also is useful in order to optimize crop production. The high geometric accuracy of COSMO-SkyMed images and the high spatial and temporal resolution make it a powerful tool to monitor the presence of new settlements or works and to control all the situations of lowering of the ground or subsoil, which are frequent causes of structural failure and collapse.

A systematic analysis of the socioeconomic benefits stemming from the availability of this new information and their fields of application is the focus of a three-year research project developed by the University of Milan and the Italian Space Agency, scheduled to be completed in 2020.

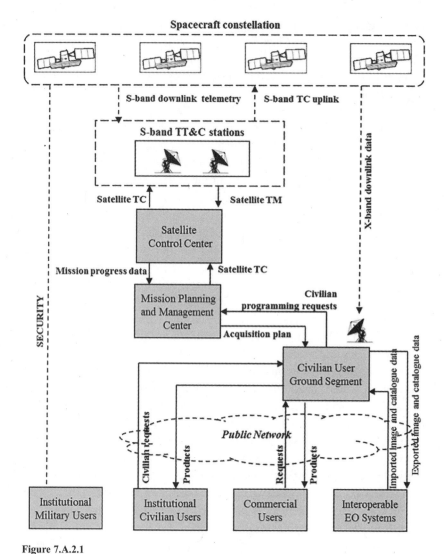

Figure 7.A.2.1

COSMO-SkyMed. *Sources*: Courtesy of ©Italian Space Agency, author's adaptation, https://www .asi.it/en.

Note: S-band: 2–4 GHz of the electromagnetic spectrum; Satellite TM: telemetry of the satellite; Satellite TC: remotes to the satellite; TT&C: Telemetry, Tracking, and Command (to and from the satellite). Only two civilian Earth observation systems are shown.

7.A.3   Appendix: From the Individual Accelerators to the Industry

At CERN and elsewhere, research is continuing on radically new technological concepts, such as plasma wakefield acceleration, which would dramatically decrease the size and the cost of accelerators (Wing, 2018).[34] The technology underlying particle accelerators has been evolving for more than a century since the first experiments at the Cavendish Laboratory in Cambridge.[35] Accelerators have found several applications beyond basic research, from medical to industrial uses. Nowadays, probably more than 40,000 accelerators are operated,[36] from the Mev- to the Tev-energy scale, from the meter to the kilometer length, and in the thousands-to-billions euro cost range (not counting the small X-ray equipment).

Chernyaev and Varzar (2014) review the literature on the accelerator business, and based on a metadata analysis, state that as of 2014, there were 42,200 accelerators worldwide: 27,000 (64%) in industry, 14,000 (33%) for medical purposes, and 1,200 (3%) for basic research.

Figure 7.A.3.1 represents a sketch of the impact of a possible technological breakthrough in the market for accelerators. In each panel, the horizontal axis represents the standardized quantity of accelerators produced and purchased per year, while the marginal WTP of buyers (e.g., hospitals, industry, universities) and the price required by producers appear on the vertical axis. The latter

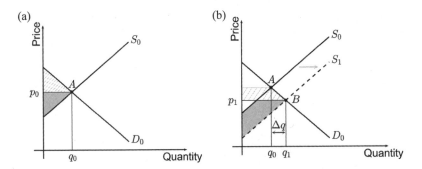

Figure 7.A.3.1

Socioeconomic impacts of a breakthrough in the technology of accelerators. (a) Initial equilibrium; (b) post-breakthrough equilibrium. *Source*: Florio, Bastianin, and Castelnovo (2018).

Note: Panel (a) shows the market equilibrium, consumer surplus (*CS*, hatched area) and producer surplus (*PS*, shaded area) before the technological breakthrough. Panel (b) shows the increase in consumer surplus (hatched area) and producer surplus (shaded area) after the breakthrough.

reflects marginal costs, including R&D, for the new technologies. The starting point is panel (a), where the market is at equilibrium point A.

When the equilibrium price is p_0 and the equilibrium quantity is q_0, we identify the consumer surplus with the hatched area and the producer surplus with the shaded area, while we use the term *economic surplus* to indicate their sum (after Boulding, 1945). The consumer surplus is given by the difference between the maximum price that a consumer is willing to pay and the equilibrium price, p_0. It can be represented as

$$CS = \int_0^{q0} D(q)\,dq - p_0 q_0. \tag{7.4}$$

Similarly, the producer surplus is the additional private benefit to firms, in terms of profits, when the market price is greater than the minimum price at which they are be willing to sell:

$$PS = p_0 q_0 - \int_0^{q0} S(q)\,dq. \tag{7.5}$$

Figure 7A.3.1(b) shows the impact of a technological breakthrough due to the appearance of a new generation of accelerators produced at a lower unit cost (upon a standardized measure of their performance), and hence sold at a lower price: The new equilibrium is at point B, with a lower price p_1 and a higher quantity q_1. The hatched and shaded areas now represent the change in consumer and producer surplus, respectively; a technological breakthrough yields an increase in economic surplus for the whole society. These first-round effects, however, are just a share of the total increase in benefits due to the new technology. In fact, there are some second-round effects on other parties because more accelerators are now available (Δq), and this increase in the stock will generate additional economic surplus.

8 Outreach and Benefits to Users of Cultural Goods

8.1 Overview

The NASA Kennedy Space Center (KSC), at Cape Canaveral in central Florida, is probably the most popular worldwide site for science tourism. Its Visitor Complex offers a variety of attractions, such as the Rocket Garden, a three-dimensional (3D) movie theater, exhibits of artifacts and robots, a memorial dedicated to astronauts, visits to the *Atlantis* space shuttle, activities simulating astronaut training, and a close-up view of the NASA rocket launches. The complex, one of Florida's most popular tourist destinations, hosts more than 1.5 million visitors per year from all around the world, despite the cost of a one-day ticket being $57 for adults and $47 for children (for an additional price, people can even have lunch with an astronaut).[1]

The mandate of the National Aeronautics and Space Administration (NASA)—namely, to communicate to the general public its mission and achievements—is written into legislation: "The Administration in order to carry out the purpose of this act, shall … provide the widest practicable and appropriate dissemination of information concerning its activities and results thereof" (the National Aeronautics and Space Act, 1958 as amended in 1961). In fact, worldwide, NASA is the most successful research organization in terms of media exposure, including in social media. In 2015, it had 11 million followers on its Facebook account, leaving the White House a distant second (4 million), and the US Navy and US Air Force in third place (about 2 million each).[2] On February 23, 2018, the NASA Facebook page had 20,911,149 "likes" out of 20,937,006 followers (who are the ones who dared not to "like" it?). NASA is also currently far ahead of the White House in the number of Twitter followers, with 28,6 million and 16.7 followers, respectively, although the Donald Trump administration is working relentlessly to improve its Twitter track record.

Most RIs, particularly but not only large-scale ones, have developed their own outreach strategies to attract the interest of the general public. Such strategies include, inter alia, the organization of on-site permanent or temporary exhibitions, open days, and guided tours, typically free of charge or at only a modest price. Moreover, many RIs sponsor social media and websites for the general public and support communications with traditional media: newspapers, magazines, and television and radio broadcasting. The CERN Communications Strategy 2017–2020[3] states its objective as follows:

Thus, the overall objective of the CERN Communications Strategy is: to help ensure the long-term future of CERN's mission and share it with society. This overall objective may be broken down into the following goals:

1. Contribute to maintaining and increasing support from current Member States.

2. Contribute to attracting new Member and Associate Member States (in line with the current strategy for scientific and geographical enlargement).

3. Maintain high public awareness and engagement with CERN's activities.

4. Foster community-building efforts within both CERN and the international particle physics community.

5. Raise awareness of and provide information about CERN's societal impact.

6. Enable CERN to serve as an effective voice for fundamental research in relevant multilateral debates and with the public.

A previous document[4] concisely stated: "The purpose of this communications strategy is to generate and secure sustained political, financial, and popular support for CERN's scientific and societal missions from all its stakeholder groups. In capitalizing on its currènt visibility, CERN will build the communications foundation to engage with many aspects of society and thereby contribute to embedding science firmly in mainstream culture."

Some of these communication strategies are indeed successful and attract the interest of large numbers of citizens from various backgrounds. Overall, every year, news of the scientific knowledge produced by RIs reaches millions of people. When important discoveries are announced, such as the Higgs boson in July 2012 and gravitational waves in February 2016, science makes prime-time news. In these instances, the audience may be more than 1 billion, despite the esoteric features of particle physics and general relativity. After the paper on gravitational waves was published in *Physics Review Letter* (PRL)—not exactly a popular magazine—the journal editor sent this internal message to the writing team: "The stat that really struck me was that in the first 24 hours, not only was the page for your PRL abstract hit 380K times, but the PDF of the paper was downloaded from the page 230K times. ... Hundreds of thousands of people wanted to read the whole paper! This is just remarkable" (Collins, 2017, 230).

In fact, hundreds of *millions* of people listened to television broadcasts on the discovery of gravitational waves and read the news in the press. Some of them even issued tweets about it, including an enthusiastic one from President Barack Obama ("Einstein was right!"), citing the Laser Interferometer Gravitational-Wave Observatory (LIGO), the RI in which the signal was first detected. Collins (2017, 246), however, compares the Google Trends data about gravitational waves with those about Kim Kardashian (not known as a scientist)[5] and finds that the former were just 2 percent of the latter in terms of Google hits. This is just a sober reminder that, after all, science is much less popular than reality shows. RI communications offices need to work hard to make discoveries and research visible in the competitive media arena.

On-site visits at the RIs involve smaller numbers of people than media exposure, but they are particularly significant for those who enjoy permanent or temporary exhibitions, guided tours of the facilities, special events, open days, lectures, and workshops designed for the general public. Even relatively lesser known facilities attract several visitors. For example, according to STFC (2010b), since 1995, the Daresbury synchrotron committed an increasing volume of resources to public outreach activities at the local and regional levels. This produced a flow of about 3,000 visitors annually. Additionally, 3,000 school students per year have been involved in ad hoc programs and activities, either in the laboratory or in schools. The ALBA synchrotron light source near Barcelona reports similar figures: Each year, ALBA organizes an open day addressed to the general public (especially families), which can accommodate around 2,000 visitors, and the demand for the free admission tickets exceeds the full capacity. Open-day events are internally designed and organized, with the scientists of each division collaborating to present their activity to the visitors. Interactive presentations are structured in a way that is interesting and funny. In addition to the open day, ALBA organizes visiting programs addressed to students (mostly at the high school and university levels). Around 5,000 students visit the facility each year and are briefed on experimental methods and results.

Turning to websites and social media, it seems difficult to find an RI that has not invested at least some resources in competing for attention on the Internet to attract virtual visitors and possibly to engage some of them via social media (Kahle, Maureen, and Dumbacher, 2016). Some of these websites are particularly well designed and sometimes offer special experiences. For instance, the European Organization for Nuclear Research (CERN) website offers a virtual guided tour *inside* the Large Hadron Collider (LHC), which is impossible during normal operations for safety reasons. As mentioned in chapter 6, some telescopes allow virtual visitors to access the view of the sky as if they were on site, even in real time, and they often encourage various forms of citizen-scientist engagement, scientists ask citizens to report potential discoveries of new objects after a visual inspection of a large number of pictures of the sky in a website.

Coverage of discoveries by the press, television, radio, online news, and blogs happens regularly, but it is enhanced when scientific papers are published in highly visible journals, such as *Nature*, *Science*, or the *New England Journal of Medicine*, and when specific media events are organized by the outreach departments of the RI. In some cases, fictional and non fictional books, movies, and other cultural goods are produced, drawing on the science produced at the RI. Examples may include *The Right Stuff*, a film on the first astronauts, or *Hidden Figures*, a 2016 film on black female matematicians at NASA in the 1960s. From this perspective, an RI is creating science and culture at the same time. The relation between the former and the latter is potentially a common feature of any research enterprise, even often in little science environments, as many scientists love to achieve a wider audience beyond recognition from peers and may consider it their civic duty to spend some time giving nontechnical explanations of their discoveries. Moreover, for managers of the RI, outreach is also a necessary complement to their strategy to justify the cost of their enterprises, particularly when they are funded by taxpayers (as the previous citations from CERN documents suggest).

In this context, how is it possible to measure the social value of the cultural goods produced by RIs? In this chapter, I present some ideas, drawing on the cost-benefit analysis (CBA) of outdoor recreation, but also of cultural and heritage goods.

The core idea of the travel cost method (TCM) in the CBA of outdoor recreation is that benefits obtained from visiting a place in leisure time cannot be inferior to the costs sustained to reach the site, including the opportunity cost of time. This perspective has interesting conceptual and empirical ramifications in this context. Most of this chapter makes the additional assumption that the value of leisure time is related to the opportunity cost of labor. Feather and Shaw (1998) examine this assumption in detail and compare empirical approaches. While their discussion is related to the demand for outdoor recreation, I do not see any important difference between time spent "virtually" traveling through content on the Internet or other media and time spent physically traveling to reach an exhibition at an RI site (or a museum). Obviously, the implicit willingness to pay (WTP) would be different according to age, employment position, income, and preferences, as widely discussed by the abovementioned paper and by the TCM literature.

The LHC case study, presented in section 8.5, suggests that for highly visible RIs, the social benefits deriving from reaching the general public and from the media can be substantial, as they are assessed ex post. Making forecasts in this area may be difficult, but in principle, it is no more difficult than assessing any large-scale cultural project or event.[6]

Another aspect related to culture not covered in this book, but which is often mentioned by scientists,[7] is the value of the RI as a meeting point and melting pot of different cultures across nations. In such a way, contemporary RIs may promote the concept of science as a global place where scientists from various countries, possibly having otherwise hostile relations, learn how to cooperate and coexist. It may be unusual to see the United States and Iran collaborate on a project, but this has happened with one of the LHC experiments. Another example of these wider cultural effects is cited in Benjamin (2003, 4): "The impact of seeing the Earth from space focused our energies on the home planet in unprecedented ways, dramatically affecting our relationship to the natural world and our appreciation of the greater community of mankind and prompting a revolution in our understanding of the Earth as a living system."

There may be other good examples, but while these cultural effects may be important, I do not know how to measure their socioeconomic impact, and they are best left aside from a quantitative evaluation framework.

This chapter has the following structure. Section 8.2 addresses the social value of on-site visits. In accordance with standard CBA approaches in cultural economics,[8] the expected marginal social value (MSV) of on-site visits is the expected visitors' implicit WTP for such visits. Benefits to virtual visitors are then detailed in section 8.3, a discussion that extends to the traditional media covered next, in section 8.4. Citizen-scientist engagement is presented in section 8.5. Two case studies, very different in the scale of cultural activities [CERN and the National Centre of Oncological Hadrontherapy (CNAO)], are presented in section 8.6, before the usual "Further Reading" section suggests some literature on the economics of media and culture.

8.2   Outreach: A Generic Model and the Value for On-Site Visitors

There are standard CBA approaches to evaluate cultural tourism in museums or other recreational activities, such as visiting a natural park. These methods usually rely on the direct or indirect estimation of the WTP for tourism or other cultural goods and services and can be exploited for various types of outreach benefits of RIs as described next.

Let us denote the WTP ($WTPx = WTP_1$, $WTP_2 \dots WTP_X$) for each type of outreach activity ($x = 1, 2 \dots, X$) by the type of user in the general public ($g = 1, \dots, G$), such as high-school students, retired people, and working-age laypeople. Such activities may include visiting the RI project; consulting its website; using social media sites related to the RI; accessing other media, including exposure to broadcasting; and seeing news and reports in the press. Hence, we can express the cumulated intertemporal benefit of outreach activities of a specific RI simply, as follows:

$$CU = \sum_{x=1}^{X} \sum_{g=1}^{G} \sum_{t=0}^{T} \frac{1}{(1 + SDR)^t} \cdot WTP_{xgt}. \qquad (8.1)$$

Each element requires ex ante forecasting or ex post assessment and valuation of the WTP. From an empirical perspective, the approach to estimating the WTP may be rather different according to each of the goods.

In the rest of this section, I focus on on-site visits. As previously mentioned, the standard CBA approach to estimate WTP for recreational sites is to use the TCM.[9]

The TCM was suggested by Hotelling (1947) and developed by Clawson and Knetsch (1966) to assess the value of environmental resources and recreational sites.[10] TCM has also gained popularity in cultural economics, particularly regarding cultural heritage.[11] The method attempts to place a value on a nonmarket good by drawing inferences from the expenditure incurred to consume it, even when there is no entrance fee; the value includes the cost of the trip (e.g., gasoline for cars, and train or airplane tickets), plus the opportunity cost of time spent traveling, on-site expenditures, and accommodation costs. In particular, two types of TCM exist: the "individual demand approach" and the more common "zone of origin approach" (Anex, 1995). The latter is the simpler and less expensive approach; it is applied by collecting information on the number of visits to a site from different representative origins of the trip. This information is used to construct the demand function and estimate the economic benefits for the recreational services of the site. The individual demand approach, instead, uses survey data from individual visitors in the statistical analysis rather than the average data from each zone. Although widely adopted, the TCM is affected by a limitation related to the apportionment issue. This issue arises whenever it is reasonable to assume that a trip is made for multiple reasons (a multipurpose trip) rather than to visit an RI specifically.

For example, the full travel cost of people going to Florida to visit the KSC, among other destinations, should not be entirely imputed to the KSC, given there are a number of other attractions in the area, such as Disney World, LEGOLAND, and SeaWorld. Thus, a tourist may travel to the Orlando area and visit several places. An apportionment assumption is then necessary to account for the RI-related cultural impact to estimate, as far as possible, the relative contribution of the RI on the total flow of visitors.

The zone of origin approach is simpler. Instead of being based on empirical evidence and the level of individual visitor, it assumes that there is a statistical visitor from a geographical area and estimates the weighted average travel cost from the different origins.

In an economic analysis, WTP estimated in this way replaces any possible revenues from visitors, such as revenues from selling entrance tickets. These revenues should be included in the financial analysis (chapter 2) but excluded from the economic analysis to avoid double-counting. For example, the revenues from tickets to access the KSC Visitor Complex are collected by Delaware North Companies Parks & Resorts, a hospitality private company, with a long-term contract until 2028. The Space Shop sells 40,000 items and generates additional revenues, to be added to the cost of restaurants and other services not included in the price. Given the annual intake of visitors of around 1.5 million, as well as the fact that the company claims that no taxpayer money is involved in the management of the complex, one may think that the revenue is substantial. However, it would be double-counting to add these revenues to the WTP of visitors.

It is often mentioned (e.g., COST Office, 2010) that some economic opportunities are likely to arise around the tourism demand created by the existence of the RI. Commercial and accommodation activities and other business opportunities near the infrastructure could benefit, in fact, from the higher flow of customers. In part, however, these effects are already considered in the TCM and should not be counted twice. For instance, the price of accommodations in hotels may be part of the costs paid by visitors of the KSC; hence, it is within the scope of the benefit to them, as assessed through the TCM. This is also revenue for the hotel business, but it should not be counted again. In the gray literature on the local impact of RIs, there is often a tendency to account for these effects in ways that are not acceptable from the CBA perspective.

Some nontechnology-driven spillovers such as these are what are generally called *pecuniary externalities* (e.g., externalities operating mostly through price adjustments in goods, properties, and land). It is usually difficult to find a precise and direct causal relationship between a project (of any type) and price adjustments so that this kind of wider effect is generally not accounted for in a social CBA. Also, due to the limited relevance of this effect for the majority of RI projects, my suggestion usually would be not to value it, in spite of the fact that RI managers may want to argue that there are visible benefits to local business. This may be true, but market price mechanisms are double-sided, and if, for example, the prices of hotel and restaurant services in the area surrounding the RI go up, there are beneficiaries (the tourism business) and losers (the customers). It is usually not worth investigating the consequences of such reshuffling of welfare effects in detail. An exception is when there are important redistributive implications, such as the risk that apartment price increases because of tourism around an RI will affect the welfare of the poor disproportionately, but this seems to be uncommon.

From the CBA perspective, there occasionally may be a hidden social cost of tourism if congestion is created around the location (as it is well known to the inhabitants of such high-demand destinations as Paris, Rome, and Venice). The social value of such congestion, should it occur because of a particularly aggressive strategy of an RI to increase its on-site visits by the general public, should be valued in terms of the economic consequences of pollution and local traffic, including time delays. Clearly, this would be an issue mainly for large-scale RI sites.

8.3   Virtual Visitors

In addition to visits in person, participation in activities on social media and visits to websites are further indicators of the size of the cultural impact produced by the RI, also to be included in term WTP_{xgt} of equation (8.1). These can be forecast through proper techniques commonly used by marketing specialists. Cultural impact is here defined, similar to the impact of scholarly literature, as the number of users of the cultural goods produced by the RI.

Revealing the tacit WTP for social network sites has been receiving increasing attention. Westland (2010) stated that when a social network reaches a certain critical mass, a WTP for network membership arises. Han and Windsor (2011) found that the trust generated from social activities favorably affects trust in business transactions on social network sites, thereby influencing users' WTP. Vock, Van Dolen, and De Ruyter (2013) modeled the willingness of social network users to pay a premium fee to benefit from upgraded services, compared to regular membership for free, and found that social capital and the perception of people as being bonded together in a coherent unit[12] resulted in specific values for members, which in turn positively affected their WTP (if any). Indirect evidence of these effects is the market value of listed companies managing social media, such as Facebook.

A direct extension of the TCM is to focus on the marginal cost of accessing and using the media. This is mostly given by the opportunity cost of users' leisure time, given that most web resources are free. Thus, information on the income of users and the time they spend enjoying outreach activities by the

RIs would provide a reasonable estimation of the benefits when the explicit WTP is not available through survey data. In practice, what is needed ex ante is a forecast of the number of distinct virtual visitors, the time spent by each of them on RI websites and related resources, and the MSV of leisure time. One may think, however, that the cultural benefit of virtual visits should be considered only beyond a certain minimum threshold of the duration of the visit because of the high frequency of visits that are interrupted after the initial access. The opportunity cost of working time may be higher than for leisure time. Figure 8.1 shows some reference figures for the social value of 1 minute of leisure time for selected countries.

Data on virtual visitors per type can be forecast through extrapolation techniques that use the recorded volume of web traffic, registrations on the official RI website, the number of tweets or followers on Twitter, posts or pages on Facebook, the number of views of a video posted on YouTube, the estimated number of blog conversations, and so on. The most appropriate valuation approach in this context is the estimation of the WTP for a virtual

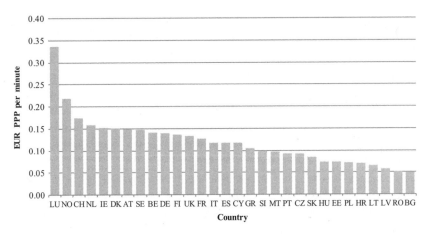

Figure 8.1

Values of leisure time. *Source*: Author's elaboration based on Wardman, Chintakayala, and de Jong (2016).

Notes: Wardman, Chintakayala, and de Jong (2016) report the most extensive meta-analysis of values of time yet conducted, covering 3,109 monetary valuations assembled from 389 European studies conducted between 1963 and 2011.

Legend: European Union countries plus Switzerland and Norway. AT: Austria, BE: Belgium, BG: Bulgaria, CH: Switzerland, CY: Cyprus, CZ: Czech Republic, DE: Germany, DK: Denmark, EE: Estonia, ES: Spain, FI: Finland, FR: France, GR: Greece, HR: Croatia, HU: Hungary, IE: Ireland, IT: Italy, LT: Lithuania, LU: Luxembourg, LV: Latvia, MT: Malta, NL: Netherlands, NO: Norway, PL: Poland, PT: Portugal, RO: Romania, SE: Sweden, SI: Slovenia, SK: Slovakia, UK: United Kingdom.

visit or activity. Vock, Van Dolen, and De Ruyter (2013) provide a statistical analysis of determinants of WTP in two social networks: one among former classmates and friends and the other among professional users. The authors take advantage of the option to pay for a premium service against the standard one provided for free. They conclude that *entitativity* (the perception that the social network is a coherent entity) and *social capital* (the value of a resource for collective action) have an impact on the WTP by social network members in terms of fees, with variability across types of networks.

An alternative to the opportunity cost of time approach is the contingent valuation of cultural goods, which consists of directly asking people to state either the maximum amount that they would be willing to pay to obtain a good or the maximum amount that they would accept as compensation to give away a good (willingness to accept, or WTA), contingent upon a given scenario. However, empirical studies show that when consumers are accustomed to receiving an online service or content free of charge, their WTP is very low or nil. For instance, Chyi (2005) reports the results of a survey of 885 Hong Kong residents and a review of earlier literature in this area, pointing to very small WTP for online news. Wang et al. (2005) suggest that the most important determinant of WTP is the perceived added value and quality of the content provided (see the "Further Reading" section for more recent surveys in the United States and elsewhere).

Difficulties in obtaining WTP values through contingent valuation have also been experienced in the cultural sector.[13] In this context, choice experiments or conjoint analysis methods are often considered more useful than traditional contingent valuations. Using stated preferences, these techniques imply asking a sample population to choose or rank various combinations of attributes of the same good (e.g., a museum or an archaeological site), for which the price is included as an attribute. This method requires the uncovering of preferences in terms of the WTP for each attribute and for the entire set of attributes, to be as effective as possible. The same techniques could be usefully exploited to attempt to value the public interest for an RI, focusing on certain attributes of the visit.

Given the soaring number of web users, including social network site visitors, the importance of estimating the social benefits of this form of outreach activity cannot be exaggerated. Two additional issues can be mentioned in this context: how to value the indirect impact of outreach, such as when a blog takes advantage of the information provided by outreach activities of the RI (or of any communications) to generate its own content; and how to value negative comments. The answer is the same as what applies to any kind of indirect effect of information, if measurable. Citations of scientific papers, even by other papers that criticize their results, are part of their impact (chapter 3).

In the same way, citation of material from the NASA website in a blog is part of its cultural impact, even if the blog suggests that no men ever landed on the moon (conspiracy theories are still around fifty years after Buzz Aldrin and Neil Armstrong's *Apollo* 11 walk).[14] While it may be discomforting for scientists, there is no way for an applied economist to discriminate quantitatively between bad and good cultural impacts. As a matter of fact, RI communication management may monitor the sentiment expressed on the Internet, but this is a different issue.

8.4   Impact on Traditional Media

The cultural impact of RIs on traditional media (broadcasting, newspapers, magazines, books, and other media) is often significant; hence, this issue cannot be ignored in terms of the RI management strategy and evaluation of the socioeconomic impact of scientific projects. In this regard, the valuation strategy needs to be conducted with prudence because, as opposed to web resources, some cultural goods and services in the traditional media are priced in the final consumer markets, but such markets have unusual features. For example, advertising repays the production and distribution costs of some media much more than does charging users.

Similar to what has been proposed for virtual visitors, it can be argued that the social value of leisure time spent reading, listening, and viewing content related to RI outreach in the traditional media is a reasonable proxy for the benefits to users. This valuation strategy seems more useful than looking at revealed preferences through the lens of standard demand curves, because of the limited role of prices paid by the audience to access content in the media. In fact, respondents to a survey on their WTP for a newspaper may have a reference point given by the price they actually pay, which is only part of what in fact they are indirectly willing to pay. If advertising is effective, the average reader would pay another hidden price for the consumption of the newspaper, in the form of the cost of advertising embodied in the price of an advertised good or service he or she is probably going to buy.

There may be more conservative valuation strategies, however, based on the cost of producing the service of the media. Unfortunately, this information is usually not available in the public domain, while information on the audience of broadcasting and readership of newspapers and magazines is more frequently available for research (and analyzed in depth by specialists in advertising). In some cases, it is also possible to have some information on the time spent on an individual piece of information, but special surveys may be needed to understand the attention paid to content originated by the RI outreach.

A more problematic issue is the valuation of indirect effects, such as the mention of CERN in popular movies such as *Angels and Demons*, or of NASA in the docudrama film *Apollo 13*. It seems prudent to ignore such further effects of the RI in terms of social benefits, except perhaps when the content is the result of direct coproduction with the RI outreach departments.

8.5   Citizen Science

An interesting way for RIs to engage with the general public is to involve them actively in online experiments and observations, and possibly in games based on actual scientific data, usually in visual form. While these activities need to be much simpler than professional research, they remain valuable in terms of outreach, and occasionally as a complement to mainstream scientific work. Online resources have created new opportunities which, in some cases, are directly supported by teams working in some RIs. I give here just some examples, among many, in life sciences, astronomy, and particle physics.

EyeWire[15] is an online game, initially launched at the Massachusetts Institute of Technology (MIT) and now managed by the Princeton University Computational Neuroscience Seung Lab, supported by grants from the National Institutes of Health (NIH), the National Science Foundation, and from others. Players are required to construct a 3D neuron map, starting with the retina of a mouse. To do so, real images from an electron-microscope are used. About 250,000 players from more than 145 countries have signed up to EyeWire and contributed to solving visual puzzles and, in some cases, to identify interesting patterns.

Open-data astronomy is a field where a combination of artificial intelligence and voluntary computing or other citizen-scientist projects is increasingly popular. According to some astronomers, the contribution of citizen science may actually provide discoveries.[16] Reported examples include 1.3 million classifications of events in the first five days of activity of the Muon Hunter project, hosted on the Zooniverse.[17] This online platform displays several projects and has offered the possibility for millions of amateur scientists to analyze data in various domains, from astronomy to particle physics, from biology to Earth observation. The classification of observed events was done through a visual inspection of images with VERITAS telescopes in Arizona by around 4,000 volunteers (a number that grew to around 6,000 in January 2018).[18] The project aims at screening muons from outer space against gamma-rays and other particles. The project is presented on the website to citizen scientists (more detailed instructions are given to volunteers on a subsequent webpage)[19] as follows: "Astronomers using the VERITAS telescopes to detect some of the highest-energy photons in the Universe need your help! ... We need your help to identify camera images that contain muon rings so we can teach computers to better identify such images and efficiently filter out those pesky muons that are masquerading as gamma rays."

This statement also shows the possible interplay between citizen scientists and artificial intelligence ("we can teach computers"). According to STscI (2016), citizen science should not be seen as outreach, but as a complement to automated procedures of image recognition. For example, light echoes (LEs) are a physical phenomenon of reflection from a light source to a cosmic dust cloud. The report suggests that because LEs are rare and faint, and shape and size greatly vary, it is computationally inefficient to use standard automatic recognition procedures, while citizen scientists can perform the task very quickly. Similarly, Beaumont et al. (2014) show how a combination of citizen-science methods with machine learning was instrumental in building a new classification of certain patterns in the Milky Way galaxy.

The NIH takes seriously the potential role of citizen science in the context of their stated objective to "engage a broader community." NIH (2018, 19) writes:

NIH encouraged development of new or significantly adapted interactive digital media that engages the public, experts or non-experts, in performing some aspect of biomedical research via crowdsourcing. NIH will work to find additional ways to engage the public and healthcare providers in making use of biomedical data and data-science tools. Doing so will help to expand the biomedical "sandbox" to researchers without access to large-scale computational resources, such as non-research academic organizations, community colleges, and citizen scientists.

How can we value the social benefits of citizen science in a CBA framework? The benefit for citizen scientists is related to the free consumption of a cultural good. By definition, the time devoted by an amateur astrophysicist to identify a muon through the Muon Hunter project is not professional time. Thus, it is not meaningful to think that the benefit of such projects is mainly in terms of research time saved as, in this context, the value of the time *spent* represents the benefit. The fact that in some cases, the citizen scientist may contribute to mainstream science at the RI is an additional benefit, and this should be valued in terms of time *saved* by scientists; but in practical terms, if a student or a retired person spends 1 hour per week looking at images released for free by a scientific project and accepts performing certain simple tasks of image interpretation, this can be valued in terms of the opportunity cost of such an hour. Additionally, if this work saves, let us say, 1 hour per month of a scientist, the value of such hour is an *additional* benefit.

Theobald et al. (2015) study 388 biodiversity projects and estimate that between 1.36 million and 2.28 million citizen scientists participated with an in-kind contribution of time whose worth they estimate to be in the range of $667 million to $2.5 billion per year. Another example is Sauermann and Franzoni (2015), who, after analyzing six months of activity by more than 100,000 participants in seven Zooniverse projects in 2010,[20] estimate that around 130,000 hours of voluntary work was offered. They suggest that at an estimated $12 an hour (which, according to the authors, is a research scientist's minimum wage), the total value of this contributed time is more than $1.5 million, with an average of more than $200,000 per project (with a large variance). From my perspective, however, while the scientific projects may benefit from this in-kind donation (in the form of voluntary work), I would often be more conservative because it is unlikely that 1 hour from these citizen scientists replaces 1 hour from a professional scientist. Hence, the MSV of leisure time of citizen scientists (which, in turn, is probably less than the MSV of scientific time because many volunteers are students, or retired people.)[21] multiplied by the time spent empirically measures the direct benefit of consumption of such a cultural good. In special cases, where there is a clear contribution to research by amateurs, the time saved by scientists should be added. Given the massive volume of data generated by large-scale RIs, their support for this form of outreach to create a cultural good seems to be justified.

8.6   Case Studies: CERN and CNAO Outreach

This section presents two case studies on the valuation of the social benefits of outreach. The first one is about CERN, a large-scale case in terms of impact. The second is about CNAO, the Italian national center for hadrontherapy, which is on a much smaller scale; this may be relevant for many RIs, such as synchrotron light sources, genome campuses, marine biology observatories, and so on.

8.6.1   CERN as a Cultural Attraction

Some statistics[22] may help to characterize CERN as a creator of cultural goods and services. In six months, from January–June 2016, CERN was visited by 320 journalists; there have been 80,000 press cuttings of news and articles in the printed media, 630,000 mentions on social media, and 1.764 million visitors to the CERN website. Twitter mentions of CERN in these six months have amounted to 429,000, Instagram 111,900, and Facebook 108,600. One single tweet on CERN by the *Economist* reached 11.4 million followers, and three tweets by CNN got 26.3 million. Moreover, CERN is visited on site everyday by many people with a general interest in science, particularly secondary school students and teachers. Media can be excited by strange facts, not only by news about discoveries. In April 2016, a curiosity-driven but unfortunate weasel, a tiny mammal living near the LHC, was killed after chewing a 66,000-volt cable in a high-voltage transformer.[23] The accelerator had to be shut down for several days, and there were around 5,500 global press cuttings about this unusual accident.

The rest of this section draws heavily from Florio, Forte, and Sirtori (2016). The key social groups to be considered in the valuation of LHC cultural benefits are (1) on-site visitors to the facilities in Switzerland and France; (2) visitors to CERN's traveling exhibitions in different locations (one is currently at the Museum of Science and Technology in Milan); (3) people reached by the media reporting LHC-related news (including the weasel story); (4) visitors to the websites of CERN and its collaborators; (5) users of several LHC-related social media (YouTube, Twitter, Facebook, and Google) directly sponsored by CERN; and (6) participants in two voluntary computing programs. Note that there is also a citizen-science program, Higgs Hunters,[24] but it is not considered here.

Benefits for on-site visitors are determined using the revealed preference method (Clawson and Knetsch, 1966), with the MSV of the time spent in traveling obtained from Harmonised European Approaches for Transport Costing (HEATCO) country-level data (see figure 8.2).[25] The forecasts to 2025 are extrapolated by a constant yearly value based on the trend observed in the previous years. An estimated 80 percent overlap between visitors to LHC experiment facilities and permanent CERN exhibitions (Microcosm and Universe of Particles, in the Globe of Science and Innovation) is estimated; moreover, only 80 percent of visitors to CERN are attributed to the LHC.

The valuation of the benefit is based on the segmentation of visitors in three areas of origin, with increasing distance from CERN (see figure 8.2), and by average travel costs for each zone, based on seven origin cities taken as cost benchmarks. For each zone, a transport mode combination and length of stay are assumed (see figure 8.2).[26] Based on the distribution of visitors by country and mode of transportation, Florio, Forte, and Sirtori (2016) have estimated an overall probability distribution of visitors.

For CERN traveling exhibitions, the number of past visitors as provided by CERN (between 30,000 and 70,000, for the period 2006–2013) was used, assuming a constant number of 40,000 visitors per year from 2014 to 2025. The WTP was prudentially assumed to be just 1 EUR per visitor (assuming local transportation was used).

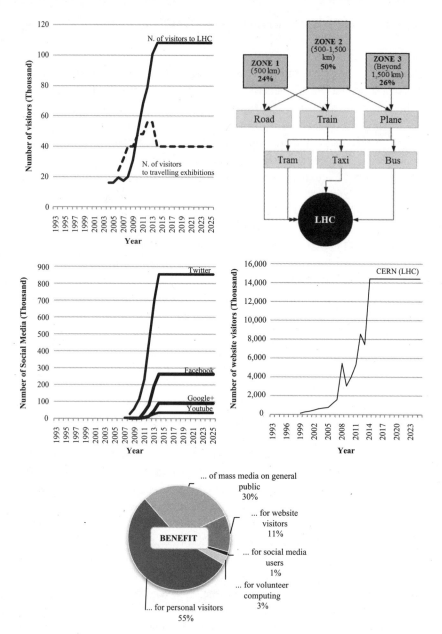

Figure 8.2

Cultural impact of CERN: Onsite visitors. Number of visitors (upper panel left); valuation through the TCM: origins (upper panel right); number of social media users (bottom panel left); number of website visitors (bottom panel right); share of benefits by type (pie figure).

Source: Florio, Forte, and Sirtori (2016).

Note: Origin zone 1, 2, and 3: Radius distance from CERN—share of visitors. Activity types (cumulated impact to 2025): Benefits to personal visitors, social media users, and LHC website visitors. Forecast 2014–2025 are taken as the 2013 level to be conservative. Constant values are assumed after 2014.

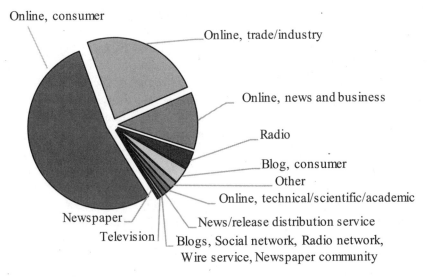

Figure 8.3

CERN visibility through specified media. *Source*: Author's elaboration on CERN data (2014–2017).

For the benefit of LHC coverage in the media (see, e.g., figure 8.3), the authors have conservatively considered only the news spikes on September 10, 2008 (first run of LHC), and July 4, 2012 (announcement of the discovery of the Higgs boson). Based on interviews with experts, the average time per head devoted to each LHC news item during such spikes was estimated at 2 minutes. The value of time of the target audience was estimated, based on current gross domestic product (GDP) per capita in the average CERN member-states and the United States [for 2013, using International Monetary Fund (IMF)], and the number of working days per year (8 hours times 225 working days), with mean equal to EUR 17.

The number of website visitors was estimated on the basis of historical data on hits until 2013–2014.[27] The forecast is conservatively based on the (unrealistic) assumption that the value at the last available observation remains constant. The benefit comes from the number of minutes per hit from users of the websites, with estimated mean equal to 2 minutes.

Further benefits come from LHC-related social media and website visits, with the MSV of time of the general public proxied by the hourly value of per-capita GDP.[28] Historical data were used until 2014, and for the subsequent years, the previous year's data were taken as constant. The average stay time is assumed for all social media to be equal to 0.5 minutes per capita, on average. Time was then valued as described here.

Finally, some CERN projects exploit computing time donated by volunteers to run simulations of particle collisions, with WTP revealed by time spent. Two such LHC-related programs are SIXTRACK and TEST4THEORY, where outsiders donate the machine time and capacity of their own computers to CERN and then are able to access some data and join a social network. The stock number of volunteers in 2013 is provided by the CERN PH department;[29] based on this information, a constant rate of increase from the program start years (2007 and 2001, respectively) was assumed. A forecast of the future volunteer stock was given to 2025 by the same source; again, a yearly rate of change over 2014–2025 was assumed. The opportunity cost is the time to download, install, and configure the programs (15 minutes per capita *una tantum*) and the time spent in forum discussions (15 minutes per month per capita). Again, time was valued as described here. The total mean value of the abovementioned cultural effects is EUR 2.1 billion. This value contributes to around 16 percent of the LHC total cost to 2025.

8.6.2   CNAO Case Study

This section heavily draws on Pancotti et al. (2015). Since it become operational in 2011, CNAO, the National Hadrontherapy Center in Pavia, Italy, has organized guided tours for students from high schools, universities, the general public, and for outside researchers. The application of the TCM consists of two steps. First, the number of visitors must be estimated. The historical number of visitors to CNAO guided tours, for the period 2011–2014, was made available by the CNAO communication office. Additionally, the number of visitors from 2014 onward was estimated, considering that free tours usually take place four times per year, when the accelerator is off for regular maintenance;[30] hence, a maximum of 8 days per year is available for organizing guided tours, with a maximum number of 1,800 visitors per year. A second step consists of defining origin-destination matrixes. Considering data on the historical data of visits, possible areas of origin, located at different distances from Pavia, are identified: for example, Zone 1 (less than 20 km from Pavia), with 28.6 percent of visitors; Zone 2 (distance 20–500 km), with 67.8 percent.

The average travel cost associated with each travel zone was estimated, including the cost of lunches/accommodation, the cost of the trip, and the opportunity cost of time spent traveling. A subdivision between visitors remaining for one day only and those staying over one night have also been made according to historical data. Specifically, almost 97 percent of visitors were supposed to remain for one day only. Considering that all CNAO guided tours are offered for free, the cost of their stay includes lunch only, ranging from EUR 8 to 12.[31] The remaining almost 3 percent of visitors, which includes those traveling from more than 500 km to come to CNAO, were assumed to stay one night, at a cost ranging from EUR 60–80,[32] plus the cost of two lunches in Pavia, amounting to a total cost ranging from EUR 40–60 per person.[33]

Depending on the origin zone, transport modes such as airplane, train, bus, car, and taxi were taken into account when estimating the cost of trips for visitors to CNAO. Also, according to the origin zone, the total travel length was in one or more sections, corresponding to the use of various transport modes. The result of this segmentation process is a number of combinations of origin and transport modes, which includes bus, car, taxi, train, and airplane. For example, the average travel cost and time for Zone 1 is estimated as the average travel cost from Pavia city center to CNAO; the average travel cost and time from Turin is considered for Zone 2; and so on. Representative cities of origin were selected among those placed in countries from which significant shares of visitors usually come (according to available data). Then different options and travel costs were put into the model according to each zone. The share of passengers per transport mode and the joint probabilities between the combination of transport modes and the share of passengers were estimated. For example, the baseline for Zone 2 was 24 percent of visitors using train and bus, 6 percent train and taxi, and 70 percent bus and car.[34]

The opportunity cost of time for different categories of visitors was estimated based on the HEATCO travel time values related to working or leisure trips. Specifically, if visitors were the general public or high-school and university students, the value of time for leisure was used. Conversely, for visitors coming from other research centers, the value of working time was used. The estimated 2013 adjusted values of time are presented in table 8.1.[35] In the risk analysis, these values were assumed to take a normal probability distribution, with a standard deviation of 0.3 with respect to the mean values, as presented in table 8.1.

Table 8.1
Estimated values of time (per passenger hour)

	Air	Bus	Car, Train
Working time	35.80	20.90	26.00
Nonworking time; short distance	13.81	6.66	9.26
Nonworking time; long distance	17.76	8.55	11.90

Note: 2013 Adjusted values of time (EUR).
Source: Pancotti et al. (2015), based on HEATCO data.

Finally, after running a Monte Carlo simulation, the total expected present value of the CNAO cultural outreach resulted in nearly EUR 6.2 million in the time span considered.

8.7   Further Reading

This section provides some suggested readings on (1) cultural goods in general, (2) value of content in social media, and (3) value of news and traditional media services, particularly broadcasting time. These materials (very selectively cited) provide some examples that may or may not be relevant for RI outreach, but taken together, they suggest that the value of time of exposure to any cultural good in the media can be empirically estimated from different angles and is often the most practical empirical approach to the evaluation of cultural benefits.

8.7.1   Cultural Goods

UNESCO (2009, 9) defines culture as "the set of distinctive spiritual, material, intellectual, and emotional features of society or a social group, that encompasses, not only art and literature, but lifestyles, ways of living together, value systems, traditions and beliefs ... Whereas it is not always possible to measure such beliefs and values directly, it is possible to measure associated behaviors and practices." The *UNESCO Framework for Cultural Statistics* considers cultural and natural heritage, performance and celebration, visual arts and crafts, books and press, audiovisual and interactive media, and design and creative services. From this perspective, RIs directly or indirectly produce cultural goods, including exhibitions, music performances, fine arts and photography, books (including fiction), newspapers and magazines, films, video, Internet, architecture, and tourism and sports.

From a broader perspective, Throsby (1999) introduced the wider concept of cultural capital, to be added to physical capital, human capital, and natural capital. He also makes a distinction between the cultural value and the economic value of a cultural good, the two ideas being correlated, as the cultural value can enhance the economic value. He also mentions the possible interest of using CBA in this field (p. 9): "if there does exist a distinct phenomenon that can be called cultural capital, with some of the attendant characteristics of more conventional forms of capital, perhaps it would be possible to apply to it the sorts of investment appraisal techniques that are used in other contexts, such as capital budgeting and cost-benefit analysis (CBA)."

In this chapter, it was assumed that an RI can increase cultural capital (beyond the stock of knowledge and human capital) by producing certain goods and services. A possible starting point for a review of the literature on CBA of cultural goods and services is O'Brien (2010), a report prepared for the British Department for Culture, Media, and Sports. The author shows that some confusion in the field of economic valuation of culture arises from the fact that to many, there are cultural values that are simply immeasurable. O'Brian (2010, 23) quotes this view:[36] "I know of no economic theory that comes remotely close to expressing the 'intrinsic' value of a great performance of Bach's 'St Matthew Passion', or for that matter of Bernstein's 'West Side Story', or their capacity for enriching, even changing lives." This type of rhetorical argument was also very common in discussions of environmental valuation: that it is simply inappropriate to value an intangible object such as the environment using economic value (Pearce, 1993).

Here, the confusion arises from the fact that the notion of socioeconomic value in the context of CBA is unrelated to the concept that anything has an intrinsic value. Even if CBA in some circumstances mentions an "intrinsic value" concept, as I will discuss in the next chapter, this is something different: a measurable *social* attitude or preference, not an immeasurable absolute judgment (possibly by an individual opinionated expert of Bach's performances). This is similar to the confusion, on the one hand, about the absolute value of scientific discovery, which is obviously not a quantitatively measurable concept, as it is a social construct depending upon the consensus of peers; and on the other, its measurable socioeconomic impact.

8.7.2   Value of Content in Social Media

There is some literature on the value of content in social media and online news. These are often available for free, and yet they have a utility to users of these services. Most of this literature is of the gray type, such as reports prepared by marketing consultants about the WTP, based on surveys of users such as Boston Consulting Group (BCG 2009) or the report "Basic Willingness to Pay for Journalistic Online Content in Germany 2013–2014,"[37] and many other similar ones by Statista, a statistics portal.

In some cases, it is helpful to convert the word count of news into reading time. Reading 1,000 words may require 5 minutes (you can check this yourself). Some data analytics on the web take the opposite route and would tell the reader how much time an online visitor has spent on a page. Donatello (2013) contains a survey of the relevant literature and empirical results for the United States (around 900 people). See also Hermansson (2013), who discusses different media and studies an online news provider in Australia. Dou (2004) reports interesting empirical results of a survey of around 800 users of a clip art website, comparing those who pay for the full service with those who use the basic content provided for free. According to Punj (2015), based on survey data, while the actual payment for digital content is related to income and education, WTP is more related to age and gender.

On the value of social media, a consultant at Deloitte Digital (Heuer, 2012) writes that "the stream of customer opinions expressed on popular social media sites about a product or service represents a flow of knowledge that can be extremely influential. What does this mean for measuring the value of social media?...These flows can now be identified, measured, and converted into financial equivalents, thus enabling organizations to aggregate the disparate forms of returns into a more traditional view." The paper proposes an empirical approach to such computation of returns.

8.7.3 News and Traditional Media

Among the many reports on traditional media, "The Economic Value of the BBC," about the prestigious British public broadcast corporation, claims that its annual spending of 4 billion pounds generates gross value multiplied by a factor of 2 (BBC, 2013). It also claims that this a conservative estimate, as it does not include its wider effects on the creative sector (the promotion of open standards, training, encouraging the development of creative clusters, and its joint ventures and partnerships).

There are many reports on the value of broadcasting time. For example, Broughton (2016), an Ampere Analysis report, suggests the revenue from commercial broadcasting and related services per hour of television viewed in 2015 is roughly $0.23 per viewer, measured across eight countries in Western Europe and North America (with large variations across countries).

These values are obviously much lower than the MSV of leisure time for consumers, which is the preferred valuation approach in this chapter. Another way to value television broadcasting would be to consider the WTP of companies buying advertising time. For example, according to Nielsen data cited by AdAge (2015), the mean price of 30 seconds of advertising on prime-time television broadcasting was $112,000 in 2014. However, commercials during the popular and acclaimed series *The Big Bang Theory* (on the CBS network), with 16.7 million viewers (including several physicists I personally know), was $344,827. This is equivalent to about $0.04 per minute, or $2.40 per hour. There is huge variability in these data, and further research is needed about to use them to value outreach.

9 Taxpayers: Science as a Global Public Good

9.1 Overview

The Gran Telescopio Canarias (GTC) is the world's largest optical telescope by aperture (10.4 meters).[1] It is located in an exotic place—the volcanic peak of Roque de los Muchachos (on La Palma, one of the Canary Islands, in the Atlantic Ocean). The island belongs to Spain, even though it is more than 1,200 kilometers from its coast. The median Spanish taxpayer is probably not an amateur astronomer and has never tried to read a scientific paper in astronomy. Furthermore, this taxpayer does not plan to visit the Roque de los Muchachos GTC site. Like anybody else, this citizen is occasionally informed through the media that a new object in the sky has been observed, but cannot guess when (or even if) the knowledge arising from the discovery of an asteroid dangerously approaching Earth, or of an exoplanet similar to the Earth will be of some practical use to humankind in the future.

Even so, any citizen de facto contributes to the funding of scientific knowledge creation every day, through a small share of any collected tax on sales, income, or property. For example, each Spanish and European taxpayer would have contributed to the construction cost of the GTC—about Euro 130 million in total.[2]

Space exploration is costlier than space observation. The 2019 budget for the National Aeronautical and Space Administration (NASA) includes about $20 billion of federal funds (NASA, 2018a). Considering only US citizens 18 years and over, each one directly or indirectly contributes more than $95 per year on average to space exploration and other NASA activities. In a cost-benefit analysis (CBA) framework, the question is: If a research project (possibly) leading to a discovery currently has no predictable use value, is there nevertheless a value to the taxpayer of such a project? Is research at NASA worth the $95 per year it costs each US citizen? What is the benefit for the Spanish taxpayer of the GTC?

In this chapter, I argue that there may be a nonuse social value of public projects in science. In the words of Johansson and Kriström (2015, 24–25): "A resource or a service might be valued even if it is not consumed. Such values are referred to as non-use values, but sometimes they are labelled passive-use or intrinsic values. ... If the project being evaluated affects non-use values, this should be reflected in the cost-benefit analysis ... among these are existence values."

The existence value (EXV) concept[3] plays an important role in the CBA of environmental goods, such as the protection of biodiversity or natural landscapes. People are often willing to pay to protect whales, pandas, tropical forests, bald eagles, and many other goods they are not directly using. Most people will never eat whales or visit the Amazon rain forest. They are just happy to know, however, that certain species and environments are preserved. Millions of citizens fund conservationist organizations by donating money or, in some cases, by offering their time as volunteers. For example, in the fiscal year 2017, the World Wildlife Fund (WWF) attracted around $104 million in individual contributions, and twice this figure in other forms of support, including $49 million of grants from foundations and corporations and $78 million of in-kind support.[4]

If properly asked through well-designed special surveys (contingent value experiments), citizens may reveal their willingness to pay (WTP) to protect the existence of environmental goods. A similar preference has been detected for the preservation of cultural heritage goods, such as archeological sites, monuments, libraries, and opera houses. There is wide empirical literature on the EXV of such goods. This chapter suggests that there are similar social preferences for scientific research and their discovery potential. Many citizens seem happy to know that somebody is exploring the outer sky, the microscopic world of quarks, the depth of oceans, or the molecular dimensions of life, even if the future use of these observations is unpredictable. Thus, the analytical framework established in the earlier literature for environmental and cultural goods valuation can be extended to research infrastructure (RI) activity.

I suggest, in fact, that there is a measurable public good value of science. As taxpayers ultimately pay the bill of government-supported RI, they may have a tacit WTP for the RI's discovery potential. Such WTP, if empirically found to be positive, suggests that taxpayers perceive research and potential discoveries as having an intrinsic value. This is a *nonuse value of discovery as a public good.*

My claim in this chapter is that this social preference for curiosity and knowledge per se, regardless of whether one knows whether it may have some future use, is empirically testable, either by stated preference techniques or occasionally through revealed preferences. The objective is to elicit the tax-payers' WTP for *knowing that something exists*, which is how a discovery is defined, regardless of its actual or potential use, instead of the WTP to *preserve something that is already known to exist* (the usual definition of *EXV* in CBA).

Another interesting concept related to the value of information in the CBA frame is the *quasi-option value* (QOV) of potential discovery. The way I use this idea, which goes back to Arrow and Fisher (1974), is as an adaptation or reinterpretation of the original concept. The classic example is the value of a forest area that can either be developed (meaning that the original forest is cut down and the land used for agricultural production) or not (the area is protected). The social net benefits of either choice are known with precision for the current period of time, but are uncertain in the more distant future. Perhaps in the distant future, the scarcity of forest areas will make the idea of protecting forests more attractive than clear-cutting them and producing agricultural goods. If the clear-cutting option is irreversible (i.e., the natural forest is destroyed forever), it may be worthwhile to acquire new information and make the decision later. In the words of Boardman et al. (2018, 339), a standard CBA textbook:

It may be wise to delay a decision if better information relevant to the decision will become available in the future. This is especially the case when the costs of returning to the status quo once a project has begun are so large that the decision is effectively irreversible. If information revealed over time would reduce uncertainty about how future generations will value the wilderness area, then it may be desirable to delay a decision about irreversible development to incorporate the new information into the decision process. The expected value of information gained by delaying an irreversible decision is called quasi-option value.

I will suggest, in this chapter, that investing in an RI is a decision leading to *endogenous learning*: You acquire the information at a later stage through the investment in discovery. Hence, you can either wait and see (perhaps somebody else will discover something, and you learn from that) or you can invest time and effort to learn something yourself. I will argue that in practical terms, the QOV in this context should usually be ignored because it is too uncertain.

It is important to stress that the QOV and intrinsic value of investing in discovery activities are two distinct concepts, even if they can be empirically difficult to disentangle. The first is related to unknown future use value, and the second to known nonuse value.

Sample average values of WTP (conditional to individual characteristics) can then be compared to what the taxpayer actually pays for a specific RI, and implicitly for its discovery potential. For example, is the French taxpayers'

WTP for research at European Organization for Nuclear Research (CERN) greater or less than (or by chance, equal to) the EUR 2.7 per year they actually pay (implicitly through their government contribution to the CERN budget)? This is an empirical question that can be answered by careful investigation (an answer will be given at the end of the chapter).

This chapter elaborates on these ideas and provides examples of possible experiments to detect the public good value of scientific discovery. Section 9.2 expands the discussion about the difference between use and nonuse value that was initially proposed in chapter 1 and in this brief overview; section 9.3 then discusses the various concepts of option value and QOV in the context of scientific research; section 9.4 addresses the public good value of science; section 9.5 discusses empirical approaches, based on contingent valuation (CV) of WTP or on revealed preference; section 9.6 discusses case studies, and further reading material is suggested in section 9.7.

9.2   Use and Nonuse Value of Discoveries

In the framework of the model of chapter 1, the estimation of use benefits (B_u) should often be sufficient to justify the worthiness of the investment in applied RIs. Hence, $NPV_u > 0$, where NPV_u is the economic net present value (NPV) when the benefits are mostly for producers and, ultimately, for consumers of goods embodying innovations (chapter 7). However, with regard to basic RIs, a crucial impact on social welfare is related to their discovery potential. Fundamental research experiments (in a broad sense, including exploration and observation) are associated with a broad set of all possible outcomes, defining a "probability space." Each outcome, as seen ex ante, has a probability of occurring (see figure 9.1) in such a space.[5] When research hypotheses are not confirmed, discoveries do not occur, but the results are nevertheless valuable, as they add something to our knowledge (e.g., we can reject the hypothesis that a drug is helpful to cure a disease). In this figure, probabilities are purely illustrative and could just as easily be unknown or uncertain.

But how can we attach a value to these probabilities? Some helpful concepts can be borrowed from environmental economics, in which any good or natural resource can be assigned a total economic value.[6] Such value can, in turn, be broken into two general but different classes.

The *use value* refers to the direct or indirect benefits arising from the *actual use* of an asset or its known potential or *option use*. These benefits have been discussed thus far in chapters 3 to 8 of this book. On the other hand, the *nonuse or intrinsic value* often denotes the social value of preserving a natural resource compared with not preserving it. The nonuse value includes a *bequest*

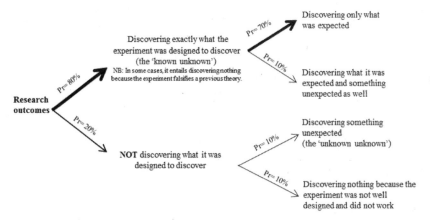

Figure 9.1

Illustrative example showing probabilities associated with various research outcomes. Note: Pr = probability. *Source*: Florio et al. (2016).

value, which arises from the desire to preserve certain resources for the benefit of future generations;[7] and an EXV, related to knowing that a good simply exists, even if it has no actual or planned use for anyone.[8] In addition, situations could exist in which further information, available at a later stage, can reveal the practical use of a good; however, if and when such new information will be acquired is often unknown or very uncertain ex ante. In these cases, a good's total value includes the QOV (considered by some authors[9] as a correction of the NPV). Later in this chapter, I elaborate further on these issues.

In the previous chapters, the use benefits (B_u) and costs have been discussed, but I have left aside B_n, the nonuse term in equation (1.1) in chapter 1. In most cases, for applied research forecasting, (B_u) and costs are all that are needed to justify a well-designed RI (or in any case to assess its NPV adequately). However, for infrastructures for basic research, NPV_u would grossly underestimate the whole societal impact. Scientists tend to be confident that something useful will arise sooner or later from what they do.

Nevertheless, when we guess that $NPV_u > 0$, in most cases for practical evaluation purposes, there is no need to go further with the analysis. One should just assume that B_n is nonnegative (as I argue later in this chapter), and no more is needed for decision-making (obviously, along with the scientific case and all the other ingredients, as discussed in chapter 1). If $NPV_u < 0$, it should be considered that NPV_u is only a part of NPV_{RI}; thus, a negative NPV_u does not necessarily mean that society has lost from the RI, but we cannot be content with just a guess that B_n is nonnegative: We need to estimate the EXV, a public good, as suggested in chapter 1 (see equation 1.10).

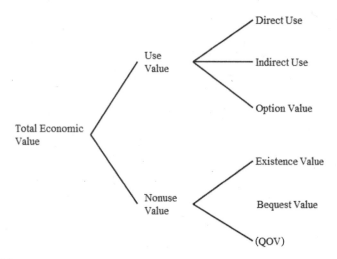

Figure 9.2
Total economic value. *Source*: Adapted from Pearce, Atkinson, and Mourato (2006).

An EXV is often related to the benefit of cultural goods, arts, or sports, to which citizens attach a utility. Some people get utility from the mere knowledge of the existence of a cultural good, such as from the pride that they attach to their heritage, despite not having any practical use for it (DCMS, 2010, 23–24). Similarly, a scientific discovery could benefit people who have a preference for knowledge. We are not referring here to scientists, but to so-called ordinary people who, even if they do not fully grasp the meaning and implications of a discovery, are happier simply because they know that discovery occurred or can occur. The focus in this chapter is exclusively on this nonuse value. I start, however, by the more esoteric concept of QOV, and how it differs from option value in this context.

9.3   Option Value versus Quasi-Option Value

In the context of environment CBA, Pearce, Atkinson, and Mourato (2006, 21) define QOV as the "difference between the net benefits of making an optimal decision and one that is not optimal because it ignores the gains that may be made by delaying a decision and learning during the period of delay." Instead, Pearce, Atkinson, and Mourato (2006, 146) define the option value as "the difference between the option price, namely the maximum WTP expressed now for something uncertain, and the expected value of a consumer's surplus (given a probability distribution)."[10]

In this section, I elaborate on the two concepts and focus on the QOV in the context of the evaluation of RIs. The abovementioned QOV definition clearly has a general scope. Any investment project uses some resources today, without perfect information about the future. Planning a new particle accelerator for basic research (such as the proposed Circular Electron Positron Collider in China, perhaps operative around 2030,[11] or the Future Circular Collider at CERN, possibly operative around 2040)[12] is often based, first, on a scientific case given the present understanding of fundamental forces, and second, on a forecast of the future development of existing technologies, without knowing if something will be discovered in the meantime. Or, to give a more applied example, a hadrontherapy synchrotron is built now without knowing if different, perhaps less expensive, therapies for cancer will be available in the future, perhaps taking advantage of progress in genetics or immunology. Hence, there would possibly be a value of waiting and learning from experiments elsewhere, but there would be a cost as well, because the proposed synchrotron would have generated technological and scientific learning in itself, and lives could have been saved in the meanwhile. Thus, what is the value of waiting for information? Is it a benefit or a cost in the ex ante evaluation of an RI?

To answer this question, it is important to clarify the difference between the QOV and the option value. The latter is a concept initially advocated by Weisbrod (1964), who suggested that some people would consider the expected value of future utility of a good, given a probability distribution function, and would be willing to pay something to preserve the option to use the good in the future. This concept points to a use value (albeit elsewhere in previous literature, it has been classified otherwise), as it has to do with the preference for a known, expected future use.

The article by Weisbrod (1964) was written in response to Friedman's (1962) advocacy of a policy of closing a national park if the commercial value of timber or minerals exceeded the WTP for recreational use. Weisbrod argued that the consumer surplus of current users of a park understates the park's value to society because many individuals would also be willing to pay for an option that guarantees their future access, even if they do not currently plan to use the park.

Let us suppose that a genomics RI may launch an ambitious and expensive program of experiments that may or may not lead to breakthrough medical discoveries. Alternatively, the RI managers may plan a less costly program leading to relatively modest incremental improvements of existing therapies. The commercial value of the former is highly uncertain, while there is much more information about the latter. What is the social value of retaining the uncertain option versus the more certain one? The RI managers do not know what the probability is of discovery and the economic outcome of a new therapy, if and when such discovery actually happens, and pass all the trial procedures from the laboratory to the commercialization of a new therapy. However, they may have an ex ante WTP for the new breakthrough knowledge created in the laboratory, based on imperfect current information and some conjectures. This WTP is the option price. An investor, such as a venture capitalist, can offer to fund a laboratory, provided that a contract offers a return on the discovery outcomes: this person actually buys the option. Equity crowdfunding may be another mechanism of raising money for a risky project.

In social terms, the current value of the less ambitious, less risky, and less expensive research program is, instead, the mean of all the probable outcomes, in terms of the current WTP of patients (or the organizations representing their interests), for the incremental innovation of therapy when it becomes available. Such aggregate consumer surplus, to be empirically gauged, needs some knowledge of a demand curve by patients (e.g., based on the price of current therapies) and complete information on the probabilities that the discovery occurs (e.g., based on statistics on similar experimental programs).

In contrast, the option price on the more ambitious program is a bet on investing the resources of the laboratory with possibly a higher payoff for its uncertain discoveries, given the current information on consumer surplus, but also with a risk of failure. Such a divergence between the option price and the expected consumer surplus is the option value of the ambitious research project. On the other hand, the option value is possibly zero for the less ambitious project if information on future outcomes is perfect for all economic agents, including patients (and all parties discount the future in the same way). It is obvious from this example that the option value is related to a future, albeit risky, use of discovery against current known uses: thus, it is a use value (and, in principle, it can be observed in venture capital markets, or, albeit imperfectly, in the value of the shares of science-based listed companies). From this perspective, the option value is related to the discussion of expected economic benefits of innovation (chapter 7).

I can now turn to the concept of QOV as introduced by Arrow and Fisher (1974), ten years after Weisbrod (1964), when studying how the uncertain effects of some economic activities could be irreversibly detrimental to future environmental preservation. This concept describes the impact of a development intervention in one period on the expected costs and benefits in the next; namely, the expected net benefits in future periods that are conditional on the realized benefits in the present period. Arrow and Fisher (1974) make a clear opening statement that uncertainty would in general *decrease* the net benefits of an activity with environmental costs. After referring to Weisbrod (1964) and option value, they argue that (1) if a project involves an irreversible choice, (2) and consequently the loss of other perpetual benefits because of such choice, (3) then the net benefits are reduced if the expected values of costs and benefits of the various choices are conditional on information available at a later stage. In this context, the QOV works in the same direction as risk aversion (even under the risk neutrality of decision-makers). In this model, the crucial point is that expected values in the second period are conditional on what happens in the first period. Arrow and Fisher (1974, 319) had in mind irreversibility and uncertainty creating a *cost*: "Essentially, the point is that the expected benefits of an irreversible decision should be adjusted to reflect the loss of options it entails."

Going back to the abovementioned definition of QOV by Pearce, Atkinson, and Mourato (2006), there is an opportunity cost of not delaying a decision to invest because, in a precise sense, our ignorance of the future makes any decision that we make now suboptimal compared with a decision that we may make if we were able to know the future perfectly. The payoff of investing now, t_0, given the information we have, is, say, $\mathbb{E}(NPV)|_{t_0} = 100$. Let us wait one year, $t+1$, and perhaps some innovations will be available such that the costs are lower and expected $\mathbb{E}(NPV)|_{t+1} = 120$ (after considering the loss of one year of benefits because of the delayed start-up of the project). In such a case, $QOV = 20$. This is the value of waiting and learning, or the value of the additional information, which is lost forever when we make an irreversible decision now.

In the context of investment in a scientific project, the reasoning is more complex. After all, an RI almost always creates knowledge, even when it discovers "nothing," because a failed but well-designed experiment also produces knowledge. For example, it may discard previous expectations that "something" should be there (as suggested by figure 9.1). In fact, Arrow and Fisher (1974, 319) conceded that the QOV is a general notion that may be applied outside environmental economics because it is linked to uncertainty, information, and irreversibility issues that affect decision-making in general. From this perspective, the RI context is one of endogenous learning (Boardman et al. 2018): the investment itself can create information.

Thus, what could be the QOV in the case of investing in science, particularly through a costly, large-scale RI? I think that there may be different cases, with different results.

A delay in knowledge creation may have a negative welfare effect because learning is postponed. This is the opposite of the Arrow-Fisher example of the natural forest, mentioned in the overview to this chapter. It could be expected that there are potential uses for knowing the very large fraction of the human genome for which, at the time decisions were made (and even today), no specific function is identified. Opponents of the Human Genome Project (HGP) were initially against it, considering it premature. They proposed to delay the start-up of the full-scale project and to learn where to search, or to dispose of better technologies.[13] These concerns included inter alia the risk that genetic testing and genome editing is used for new eugenics (i.e., selection of genes perceived as favorable), beyond therapeutic objectives. It was feared that the insurance companies, employers, or governments required access to information about certain genetic characteristics of individuals. One may also think of risks related to the design of new biological weapons based on genetic engineering of bacteria or viruses. Several problems of this nature may have been settled by the ethical protocols subsequently developed and, in many cases, embodied in legislation, but there is still uncertainty about the boundaries of what should be considered a too-risky genetic manipulation, which would be forbidden for precautionary or ethical reasons.

Ex post, it seems safe to say that it would have been wrong to postpone the HGP because it stimulated a cascade of learning mechanisms (particularly in gene-sequencing technology), which were possibly greater than in a counterfactual scenario of delayed start-up. One also may conjecture that there was nothing irreversible about the HGP in the Arrow-Fisher meaning. The HGP could have been stopped by the core funding agencies, the US Department of Energy (DoE) and National Institutes of Health (NIH), as happened to the Superconducting Super Collider (SSC). However, there is always something irreversible in any implemented RI decision, as certain resources are committed for many years and knowledge is created forever. It is usually impossible to erase or forget knowledge, once it has been created. Hence, if such knowledge is found to be beneficial, there is a cost of delaying its creation, and not a benefit. From such a perspective, while the Arrow-Fisher QOV has the same effect of risk aversion, for scientific research, QOV may have the opposite effect. But is this always true?

Leaving aside for a while the difficulty of giving an empirical estimate to such value ex ante, the question arises if one may always be confident that the QOV of any RI project must be positive. This would amount to saying that any knowledge created is beneficial to society, and the sooner the better. Most scientists would subscribe to this view (at least when they consider their own field, and their own project). Many others would strongly disagree. First, if the future uses of knowledge are totally uncertain (and the costs, as appraised today, surely very high), how is it possible to always guess a positive QOV in order to balance a negative NPV_u (users benefits minus costs)? In some cases, it would be a good strategy to start with small-scale experiments, proof-of-concept, and wait before risking a large-scale investment.

Second, in certain cases, the social preference may be *not to know something until the means to make good use of such information is understood.* For example, some legislation forbids certain types of research on humans, for either ethical reasons or as a (often vague) precautionary principle about the good use of science. One may think that such concerns are excessive in some cases, but it is easy to see that total freedom of research on anything by any means may lead to unacceptable projects. The HGP had to invest up to 5 percent of its budget in studying ethical and legal issues and laying down guidelines that support the prohibition of certain experiments.[14]

Of course, preoccupations about black holes or strangelets created by the LHC are absurd, but they attracted a lot of attention (e.g., Posner, 2004). There may be people who think that sending radio messages into deep space[15] may attract the attention of hostile aliens. The list of arguments against science is long and in most cases, not worthy of serious consideration. There are, however, cases where the scientific community itself thinks that it is better to refrain from expanding our knowledge. An example is research that potentially leads to new and lethal biological weapons.[16]

According to this discussion, the potential social costs or benefits of the "wait-and-learn" approach in the case of scientific knowledge are totally unknown, including their sign, because of potentially positive and negative effects. Hence, I suggest that conservatively setting QOV = 0 as a default rule seems reasonable. However, this assumption may be considered excessively prudent. It amounts to saying that, in CBA terms, we need to be neutral about the unpredictable value of scientific discoveries per se.

Restating something that I have already stressed several times in this book: This is not an argument against the value of science, which should be assessed as the core aspect of peer review evaluation of any research project. It is an argument against giving, in general, a monetary value in the CBA of an RI to some future *potential use* that can be positive, occasionally dangerous, or definitely negative. In any case, we simply do not know anything that can be included into a quantitative forecast in terms of the benefit or cost of *waiting and learning* or, alternatively, of *doing and learning.*

Conversely assuming QOV = 0 as the default rule is consistent with the first principles of CBA. For example, building a new highway may have far-reaching effects in the future. It may create new opportunities to connect distant people, but it can also divert traffic and disrupt a local economy. New unpredictable, positive and negative, cultural and economic circumstances may arise. We may wait to build the highway to learn these circumstances, and impatience may lead to a suboptimal decision. However, our present knowledge about these wider effects is too uncertain, and QOV is not usually included in the CBA of a motorway; namely, it is implicitly set to zero.

Hence, in my interpretation, differently from Arrow and Fisher (1974), the QOV could be either positive or negative, producing potentially either an increase or a decrease in the RI project's NPV based on the best possible bet one can make, given the available information. Setting this CBA component to zero means taking a neutral attitude about unknown future uses of new knowledge.

This assumption may seem strange, and perhaps may be unpopular among scientists, as it suggests that discovering gravitational waves, *while having scientific value*, has zero socioeconomic value per se because we do not know what the use of such knowledge will be. Readers are invited to be patient and not to jump to the conclusion that I am throwing away the true essence of science. In fact, what I am proposing is to rely on a different evaluation strategy, leading to the empirical estimation of the intrinsic public good value of the gravitational waves discovery, in terms of social preferences for knowledge.

9.4   The Value of Discovery as a Public Good

In economics, a public good, according to the famous definition by Paul Samuelson (1954), is a good for which exclusion of any person from its consumption is impossible or too costly and there is no rivalry in consumption among consumers. Joseph Stiglitz (1999) suggests that knowledge is a global public good. The knowledge by somebody that gravitational waves have been detected by LIGO, an observatory, does not prevent the use of this information by anybody else in the world. The open model of science typically associated with RIs would make it contradictory to exclude somebody from such knowledge, in contrast to the esotericism of the ancient alchemists or the top secrecy of their heirs, such as the Manhattan Project's nuclear scientists.

As already mentioned, an *existence value* can, in principle, be attributed to an RI's discovery, reflecting a social preference for pure knowledge per se. In other words, EXV, in the model of chapter 1, refers to the value of knowing the object of discovery, regardless of the fact that it may have some (good or bad) use sooner or later. Knowledge acquired along with the discovery in an RI context is also a global public good. There may be exceptions, and perhaps occasionally some of the applications are never made available to the public for security reasons, but this is probably an unusual occurrence outside military research. Contrary to the QOV, which is unknown, current preferences for the public value of discoveries can be observed in principle. Hence I claim that the existence value of knowledge is the value of a public good.

The standard way of estimating such nonuse values is by stated preference techniques, based on answers given by a representative sample of citizens to derive the respondent's tacit WTP for a good. Hence, I suggest estimating the WTP of taxpayers for the pure discovery potential of a government-funded RI. Conceptually, the issue is no different from estimating the WTP of the general public for climate change policy or for biodiversity conservation. Most people have only vague ideas about these very complex issues, but to a certain extent, when they are given a piece of information, they express their attitudes in appropriately designed surveys. Some criticism has been raised against the legitimacy of nonuse values.[17] However, the importance of the EXV as a component of the total economic value of goods is even reflected in some legislation, such as the US federal preservation regulations.[18]

In this chapter, I refer to taxpayers who are members of a specific jurisdiction, such as the United States for NASA, even if the public good of discovery is such that virtually all human beings have standing. In practice, any RI in basic science is supported by a state or a coalition of states, not by the universe of potential beneficiaries. Thus, it seems convenient to focus on those who, in fact, will fund the RI through government transfers, even if there is free-riding from other countries.

Consistent with environmental and cultural economics, the empirical estimation of EXV as a benefit of RIs is in line with the Boardman et al. (2006, 229) prudent approach:

Should existence values be used in CBA? The answer requires a balancing of conceptual and practical concerns. One the one hand, recognizing existence values as pure public goods argues for their inclusion. On the other hand, given the current state of practice, estimates of existence values are very uncertain. This trade-off suggests the following heuristic: Although existence values for unique and long-lived assets should be estimated whenever possible, costs and benefits should be presented with and without their inclusion to make clear how they affect net benefits. When existence values for such assets cannot be measured, analysis should supplement CBA with discussion of their possible significance for the sign of net benefits.'

Our model can be seen as an operationalization of these ideas in the field of RIs. The CBA test when $NPV_u < 0$ is

$$EXV_0 - |NPV_u| > 0. \tag{9.1}$$

The pure subjective public good value of discovery may or may not counterbalance the negative net present use value. In other words, the RI is deemed to have a positive measurable social impact if the positive EXV is greater than the net costs (i.e., negative NPV_u of measurable use-components).

9.5   Empirics of Intrinsic Values: Stated versus Revealed Preferences

The "Questions and Answers" area of the Earth Observation Environmental Satellite (Envisat) website replies to a question on the cost of the project by stating: "Envisat cost 2.3 B EUR (including 300 EUR million for 5 years operations) to develop and launch (launch price tag: 140 EUR million). This is equivalent to 7 EUR per head of population across all the ESA member states, or about one cup of coffee per year spread over its 15 year lifecycle." Are taxpayers happy with this answer? To figure this out, we need to estimate the maximum amount of money that taxpayers should be asked to pay. We can proceed in three ways: stated preference techniques, revealed preference techniques, and benefit transfer.

Under the first approach, a CV on a representative sample of taxpayers should test the WTP, an amount of money equal to or greater than the threshold necessary to get a positive NPV_{RI}.

Stated preference techniques involve soliciting responses to hypothetical questions regarding the value that people place on goods. Thus, these techniques are based on answers given by a representative sample of the population of interest to derive a respondent's WTP for a good. Within the class of stated preference methods, two main alternative groups exist: choice modeling and CV. The latter seeks measures of WTP through direct questions such as: "What is the maximum that you are willing to pay for this telescope project?" and "Are you willing to pay x for this telescope project?" The former seeks to secure rankings and ratings of alternatives from which WTP can be inferred.

The use of a CV methodology has found widespread application in public services, and particularly in environmental economics for estimating the economic value of ecosystem services[19]. In the same vein, one could attempt to grasp the WTP of taxpayers with preferences for RIs and their discovery potential, regardless of a discovery's actual or possible use.

The literature on CV debates numerous issues. Reviews of these debates can be found in Mitchell and Carson (1989), Portney (1994), and Carson, Flores, and Meade (2001). A possible objection is that CV experiments are rather costly. Nevertheless, the typical cost per capita of a well-designed CV in Western Europe is in the range of about EUR 20–40 per interviewee, and it would be a very modest fraction of the overall cost of the RI in the first place, particularly for the large ones. A sampling of around 1,000 respondents per country is regularly accepted for many professional surveys, and much of the academic literature on CV is based on smaller samples.

Another possible objection is that asking individuals their WTP for the mere knowledge of the existence of any good may not be easy and may result in bias due to a number of individual, cultural, and socioeconomic circumstances.[20] To address these issues, the evaluator should take into account a number of recommendations developed since the early 1990s by a panel of distinguished economists for the US National Oceanographic and Atmosphere Administration (NOAA) (Arrow et al., 1993), including indications about the modalities and structure of the interviews.

A slightly adapted version of the NOAA panel guidelines is proposed here:

• *The target population of the RI should be identified in terms of potential interest (those who have standing in CBA):* The entire population of the country (or region) in which the RI is located, or a defined group of people in a reference geographical area (world, nation, region). Because new knowledge can be a global public good, in some cases, all of humankind is the potential beneficiary, but only taxpayers of some countries would in fact support the project—such as the member-states of the European Molecular Biology Laboratory (EMBL) or the European Space Agency (ESA).

• *The CV experiment (random) sample type and size should be identified:* This should be the closest practicable approximation to the target population (e.g., all taxpayers in a region, all students from certain universities, and all subscribers to certain magazines).

• *Careful pretesting should be done because the interviewers may contribute to "social desirability bias" in the event that face-to-face interviews are used to elicit preferences:* Pretesting is also essential to verify whether respondents understand and accept the context description and the questions. This is particularly important for scientific projects, as most respondents will be unfamiliar with them.

• *The purpose of the questionnaire should be stated, and an accurate description of the RI, its mission, and research potential should be provided:* This ensures that respondents understand the context, are motivated to cooperate, and are able to participate in an informed manner. The use of pictures or videos could help.

• *The payment vehicle (i.e., the manner in which the respondent is hypothetically expected to pay to sustain the infrastructure's research activity/mission) should be described:* One example of this is an increase in taxes or donations to a fund.

• *The elicitation format, for which open-ended, bidding game, payment card, and single-bounded or double-bounded dichotomous choices are the most broadly used formats, should be carefully selected:* Dichotomous (i.e., referendum-like) valuation questions that allow for uncertainty by including a "don't know" option, are recommended by the NOAA panel. See the section entitled "Case Study: Willingness to Pay for Discovery," later in this chapter.

• *Follow-up questions should be inserted:* These are essential to understanding the motives behind the answers to WTP elicitation questions.

• *Questions allowing for cross-tabulations and multivariate analysis should be inserted, including a variety of other questions that assist in interpreting the responses to the primary valuation questions:* For example, income, education level, prior knowledge of the infrastructure, prior interest in the issue, and attitude toward RI are items that would be helpful in interpreting the responses.

• *Both sample nonresponse and item nonresponse should be minimized:* A reasonable response rate should be combined with a high, but not forbidding, standard of information.

• *The conservative approach should be preferred:* When the analysis of the responses is ambiguous, the option that tends to underestimate WTP is preferred. Similarly, the reliability of the estimate should be increased by eliminating outlier answers that can implausibly bias the estimated values.

Johnston et al. (2017) update the stated preference approach based on the experience gained from more than twenty years of research after the NOAA panel, including technological progress with online surveys.

Finally, some experimental economists would prefer a setting in which participants are incentivized with actual money. Academic research with this methodology usually involves a relatively small number of participants (often students) and tests specific behavioral hypotheses rather than estimating the WTP. It is not obvious that the advantages of a laboratory approach in the CBA of an RI would be clearly more convincing than a well-designed CV survey.

To overcome the difficulty of explicitly stating a WTP in some circumstances, valuation methods based on revealed preference can be employed. These methods assume that intrinsic value can be determined through the observation of economic behavior in a related market, such as voluntary contributions to organizations devoted to the preservation of a public good. For instance, in the RI context in some countries, several scientific institutions are supported by taxpayers who can name a charity or a research body to which a percentage of their taxable income should be donated. Additionally, universities regularly receive donations for research from firms and individuals.

A third valuation method is the benefit transfer approach, in which a meta-analysis of CV studies on the intrinsic value of goods produced by other projects is used to establish a benchmark median value or a range of values, after considering some controlling variables.

Unavoidably, in practice, WTP, as elicited by a CV approach, is usually a blend of perceived QOV and EXV. Some respondents make conjectures or mental associations about the possible future use of potential discoveries. This fact per se is not disturbing, provided that the design and interpretation of the survey results are careful. Moreover, if QOV is set to zero, there should be no concerns about double-counting pure preference for knowledge and uncertain conjectures about future use values.

To conclude this discussion, as always in applied welfare economics, an estimated value cannot be taken as anything more than a structured conjecture based on some empirical evidence. WTP is no exception. It cannot be measured with the precision typical of hard sciences. However, one can try to guess its value as a stochastic variable, drawing from well-designed experiments as far as possible in social science. Examples are given next.

9.6   Case Study: Willingness to Pay for Discovery

Two CV experiments have been performed about the LHC and future CERN investment in new particle accelerators—one with students in four countries and the second with a representative sample of French taxpayers. The first was a pilot study (Florio, Forte, and Sirtori, 2016), based on a survey of university students in a wide range of disciplines, including the humanities, in France, Italy, Spain, and the United Kingdom. The data were further analyzed by Catalano, Florio, and Giffoni (2018). The students were taken to represent a sample of future taxpayers with tertiary education. Referring to students in experimental economics and political science is common practice (e.g., Druckman and Kam, 2011). However, this approach should be considered with due caution when extrapolating taxpayers' preferences. The second CV experiment was performed in January 2018 with a survey of a representative sample of taxpayers in France; this was probably the first of its kind in the field of the public good value of scientific research.

As for the survey of students, a brief description of the LHC was provided to the interviewees: namely, a shortened version of the Wikipedia entry "Large Hadron Collider," including five photographs.

This first survey was conducted between June 2014 and March 2015; 1,022 valid questionnaires were filled in by students from five universities located in four European countries: University of A Coruña (Spain), University of Exeter (United Kingdom), University of Milan (Italy), and University Paris 7- Denis Diderot and Sciences Po University (France). They were enrolled in more than thirty university degree areas: around 65 of the respondents were in the social science and humanities area, and the remainder were getting science degrees;

857 students were aged between 19 and 25, while the remainder were over 26. A total of 83 percent of the interviewees stated that they had an interest in scientific research; 85 percent recognized that funding RIs was at least important. The LHC was known by 52 percent of the interviewees. Their source of information mainly included the Internet, magazines, and television. The Higgs boson discovery was familiar to 61 percent of respondents, and around 9 percent had already visited CERN.

While 48 percent of the respondents declared that they did not know whether they were willing to contribute to the cost of LHC, 73 percent of respondents revealed a positive WTP. The type of curricula did not discriminate between the available options: *Yes*, *Do not know*, and *No*. Among the source of information about the LHC, online news and television programs were the most quoted by the interviewees, while only 20 percent of the sample indicated that they had learned at their university. Students in social science and humanities faculties are willing to pay for basic science, at least as students enrolled in scientific curricula. The variables representing knowledge of research at LHC, personal values, attitude, and interest toward science are, as expected, strong drivers of being willing to pay.

While the survey of students was not representative of the WTP of the current taxpayers, the subsequent CV experiment in France aimed at a representative sample of 1,005 citizens over 18 years of age.

The survey was carried out in February 2018 after a long design and pre-testing phase. The questionnaire, in French, was structured as four sections. Section A investigated interviewees' interests and their opinion about the importance of scientific research in general. Section B focused on CERN and its research activity. A two-page description of CERN and a two-minute video, visualizing what particle physics research at CERN consists of, were provided to interviewees. Section C contained questions aimed at eliciting respondents' WTP for a scenario of new CERN investment in particle accelerators against a scenario of no additional investment. Finally, Section D included questions on the respondents' demographic and socioeconomic characteristics: sex, age, level of education, occupation, income, family size, and region and zone (urban-periphery or rural) of residence.

The main findings were, on one side, the estimated maximum average unconditional WTP, and on the other, the estimated conditional double-bounded average WTP. The unconditional average maximum WTP for CERN investments in particle physics research amounts to EUR 13.5 per person per annum. The WTP was elicited through the question: *What is the maximum annual amount you would pay for supporting investment in Scenario A?* Such a scenario was this: *CERN Member States decide to invest in a new particle accelerator in the next decade. It will make discoveries on phenomena that cannot be explained today. This new accelerator will be operated for at least twenty-five years.* The alternative Scenario B was the following: *CERN Member States decide not to invest in a new particle accelerator. The research activity with the existing accelerator, the LHC, will gradually decrease over the next twenty years. The possibility of finding answers on unexplained phenomena will remain limited.*

It was clarified that the respondents' contribution would be made through tax increases. The unconditional WTP value is an average between a slight majority of respondents who have a positive WTP and those who say that they have no money or have other spending priorities. Moreover, the above-mentioned unconditional average WTP does not control for income and other characteristics and is influenced by some outliers. Hence, this should be considered an upper bound of the maximum WTP. Instead, the results of the referendum-like, double-bounded questions are conditional on a range of controlling variables for the respondents' characteristics. This dichotomous CV experiment was designed, as far as possible, according to the standards of the NOAA panel and the Johnston et al. (2017) guidelines. The conditional WTP was elicited by asking the question: *Would you agree to pay the amount of EUR X per year as a taxpayer for the construction of a new particle accelerator at CERN as described in Scenario A? In this question,* the variable X is an amount of money preprinted in the questionnaire. The bounds work as follows: If the respondents answered that they were (not) available to a pay a certain amount, they were asked whether they would accept paying a given (lower) higher amount. There were five subsamples, each with different initial and final bids. Obviously, the respondents could also reject the lower bid.

The resulting estimated average conditional WTP is slightly more than EUR 4 per person per annum. As previously mentioned, this is not the maximum WTP, as it is constrained by the bid values and calculated by taking into account socioeconomic characteristics (income, gender, education, employment, and age) and interests, as well as opinions with a multivariate econometric model. Gender and age play no role if income, employment, and education are considered, while income and education are positively correlated with the WTP of respondents.

These results can be compared with the actual contribution that French taxpayers currently pay to CERN in the form of taxation (the information was not reported in the questionnaire). This contribution (budget year 2017) amounts to EUR 2.7 per person per annum. It was concluded that even when adopting the most conservative assumptions, the WTP of French citizens to support future CERN investments in particle physics research is higher than their current tax contribution of a factor of 1.50 when the bounded conditional average WTP is considered. For details, see Florio and Giffoni (2018).

9.7 Further Reading

This section suggests some additional material from a very extensive body of literature on the following topics: option value, QOV, CV, revealed preferences, and EXV.

9.7.1 Option Value

Greenley, Walsh, and Young (1981) use the Weisbrod concept of option value to measure the social value of water quality for recreation in the South Platte River Basin, Colorado. The option value should be added to consumer surplus, as mineral and energy development may lead to degradation of water quality. Brookshire, Eubanks, and Randall (1983) study the option price and EXV of grizzly bears and bighorn sheep in Wyoming through a CV where individual discount rates for nonmarketed goods are also estimated. They also discuss the difference between the concepts of expected consumer surplus and option value, on the one hand, and option price on the other, the latter being the sum of the former (without double-counting). Outside environmental economics, in transportation project CBAs, the option value is the WTP for continued availability of a line of nonusers of rail service.[21]

There is some terminological confusion here, as a nonuse value (or even an EXV) is associated with future use, while, in this chapter, I prefer to consider any *known* future use, albeit partially uncertain, as a use value. Also see Johansson (1987) for a discussion of the concepts. Conrad (1980) shows that the Weisbrod (1964) option value is the expected value of perfect information.

9.7.2 Quasi-Option Value

In the same vein, one can say that QOV is the value of imperfect information in a multiperiod setting and with irreversibility of investment decisions. My point in this chapter, about endogenous learning and the sign of QOV in the case of the RI, is supported by a short note by Freeman (1984, 292):

It is not difficult to imagine situations in which the relevant information to guide future decisions can be gained only by undertaking now at least a little development. In such cases there can be a positive quasi-option value to development. ... I will show that quasi-option value is a neutral concept ... Whether quasi-option value exists and whether it is positive or negative ... depends on the nature of the uncertainty, the opportunities for gaining information (reducing uncertainty), and the structure of the decision problem.

Fisher and Hanemann (1987), however, insist that QOV is usually positive (meaning that it has a cost). In the RI field, the predicted negative effect of the Arrow-Fisher frame (decreasing net benefits) can be reversed: Investing now, while possibly having an irreversible cost effect on the one hand (e.g., should it be very costly to decommission a large particle collider, while a rival particle accelerator may disrupt the scientific case for it), also immediately produces some new knowledge through its experiments. This is an irreversible benefit per se. It is the peculiar feature of research that probably often makes the sign of the QOV a benefit, thus acting in the opposite direction of risk aversion. For more recent applications of the QOV concept, see Basili and Fontini (2005) and Messina and Bosetti (2003). Strazzera et al. (2010) presents an interesting attempt to compute QOV from survey data with two choice experiments. Choice experiments have been considered as occasionally being more revealing than CV. For an application to water supply options with mixed methods, see Powe et al. (2004).

About endogenous learning, a particularly interesting paper is Stange (2012), who studied the concept of the value of information in the context of sequential schooling decisions and estimated that 14 percent of the total value of attending a college for the average high school graduate is related to the information acquired about one's own abilities.

In turn, the QOV is in some way related to the real option approach developed by Dixit and Pindyck (1994). Sanchez et al. (2012, 265), in a much simpler but still interesting study, suggest that there is an option value of innovative medical care of chronic myeloid leukemia. They estimate that there is an additional value for patients to receive some innovative drug that enables them to survive until an even more effective drug is available in the future. Again, the value here arises from the fact that new information helps to bring new opportunities. Another example is Sanderson et al. (2016), a good analysis using real-option modeling of agricultural investment decisions in South Australia, under the uncertainty posed by climate change effects on alternatives between using land for wheat production versus for livestock and grazing.

9.7.3   Contingent Valuation

The studies on CV are very numerous indeed. Carson (2012), in the *Journal of Economic Perspectives*, is my preferred suggestion as a starting point, as it summarizes, in a nontechnical way, the research originated by the *Exxon Valdez* accident in Alaska, which caused an oil spill in 1989, followed by important legal controversies about the consideration of damages in terms of the EXV of the landscape, estimated at $3 billion, beyond the use value for fisherfolk and tourists, estimated at just $4 million. One important point mentioned by Carson is that while many economists assume that the WTP in CV studies may be exaggerated, the view of Samuelson (1954) about stated preferences for public goods was in the opposite direction because of the "free-riding" argument. Carson (2012) also mentions that—contrary to what would have been expected by some skeptical economists—in empirical terms, revealed or stated preference approaches tend to converge, or even to show that CV is more conservative than revealed preference, as observed by Carson et al. (1996), where a meta-analysis of eighty-three studies with 616 comparisons of CV estimates to revealed preference estimates found that CV estimates were on average lower by 11 percent than revealed preference estimates, while correlating (R = 0.78) with them.

Carson (2012) concludes that while CV is not perfect, the alternative would often be given no value on public goods or other goods for which there is a demand, but no market. He claims that a body of evidence in support of appropriately done CV as a tool to gauge the WTP for public goods is now considerable. I agree, and I think this conclusion is highly relevant for the evaluation of RIs.

A specific point of debate in the CV literature is the incentive and information issue, or how to ask respondents the questions in a credible way. See Carson and Groves (2007) for a neoclassical framework (as opposed to behavioral economics) approach.[22] They discuss what makes a questionnaire "consequential" from the perspective of the respondents, and to what extent various elicitation mechanisms are incentive compatible. For earlier criticism, see Diamond and Hausman (1994).

Many empirical CV, choice experiment, and other stated preference studies are available in the Environmental Valuation Reference Inventory (EVRI),[23] a searchable storehouse of empirical studies on the economic value of environmental benefits and human health effects. The EVRI was developed in the 1990s by Environment Canada (a government body), in collaboration with a number of international experts and organizations, as a tool to help policy analysts use the benefit transfer approach to estimate the economic values of changes in environmental goods and services. Currently, the database makes available more than 4,000 valuation study records from more than thirty fields.

In some cases, stated preference valuations of EXV may include choice experiment or conjoint analysis methods, already mentioned when addressing the issue of the value of an RI's cultural effects. CV has been developed as a method for eliciting market valuation of damages to environmental resources, but it also has been used to value a wide range of nonmarket goods and services, such as museums,[24] cultural heritage,[25] and local soccer clubs,[26] even for small projects. For example, via a survey, Jura Consultants (2005) estimated that the museum, library, and archive services of the community in Bolton, United Kingdom, were worth 10.4 million pounds, of which 3 million pounds were related to nonuse value.

9.7.4 Revealed Preferences

Turning to other empirical strategies to value public goods, an example of the experimental economics approach that may be potentially relevant in the RI context is Tonin and Vlassopoulos (2010). They wanted to disentangle warm-glow altruism and pure altruism in prosocial attitudes by measuring the level of effort by groups of students enrolled in a short-term data entry job.

In this experiment, the charities mentioned were, inter alia, Amnesty International, Red Cross, Cancer Research UK, Greenpeace, and others. Cancer Research UK was selected much more frequently, and as a consequence, it received much more money than the others. In principle, one may think to design a similar experiment tailored to see the destinations of the donations to various scientific organizations, but I do not know how this could be helpful for inference on the social WTP for an RI project, as these kinds of laboratory experiments in economics have different objectives.

In terms of observing actual behavior, in some countries, there are tax deductions for donations to charities. In Italy, 0.5 percent of taxable personal income can be donated tax-exempt to a specified list of charities, including for scientific research. The list of recipients (more than 50,000 in 2016) is in the public domain.[27] The lion's share of these contributions (in money terms) was for the AIRC, the Italian Association for Research on Cancer (with more than 1.67 million choices). The following top recipients were also in the field of medical research or health-related charities. There were also many universities and research centers, most of them with a small number of donors. In 2016, the total donation, by more than 16.2 million taxpayers, was EUR 491 million, with the average donation being close to EUR 30. A systematic analysis of tax data in countries where similar opportunities are available may be interesting in order to reveal the WTP for scientific research (or at least the ordering of preferences across themes).

For an application in the cultural sector, see Fujiwara (2013).

9.7.5 Existence Value

Probably the first paper that proposed the idea of an existence ("sentimental") value was Krutilla (1967, 781): "There are many persons who obtain satisfaction from mere knowledge that part of wilderness North America remains even though they would be appalled by the prospect of being exposed to it."

For a general discussion, see chapter 13 of Boardman et al. (2018). The terminology is rather unstable. Some authors prefer *active* versus *passive*, meaning using the good in the former case and enjoying it mentally (a nonuse value) in the latter case. The terms *intrinsic value* and *existence value* are often synonymous in the literature. Altruistic, EXV, and bequest value are nonuse value, the only difference among them being the utility arising from knowing that other agents now enjoy the good as opposed to future generations doing so.

Whitehead and Blomquist (1991) make the important point that EXV is based on some information about the good, and this may have observable implications in terms of behavior. They present a simple model where the utility of households is based on goods produced by a function that includes purchased goods, natural resource, time input, and a vector of other goods. Information is introduced as a factor that allows utility from the natural resource, either through on-site visits or off-site ones (e.g., through magazines, television, and other remote sources). Empirically, they find that the WTP for the EXV of the Clear Creek wetland in western Kentucky is three times higher for CV respondents who had acquired prior information about the area.

There are authors who strongly disagree with the EXV concept. One is Weikard (2005), who provides a good discussion of the terminology in previous literature, but then claims that knowledge of the existence of a good per se should not be used in CBA because of what he perceives as a logical contradiction as, in any case, some information, such as that provided by a CV survey, must be given to the respondent. Nelson (1997) attacks the concept of EXV as being a way for some economists to advance a progressive theology of nature, mixing a religious argument with a scientific one. Other critical voices are discussed by Carson (2012).

10 The (Expected Net Present) Value of Investing in Discovery

10.1 Overview

What is the bottom line of the evaluation framework proposed in the previous chapters?

Trying to assess quantitatively the multifaceted socioeconomic impacts of a radio-telescope, a genomic platform, or a satellite for the study of climate change can be seen as an impossible task. Some earlier contributions are based on narratives, case histories, and mixed (quali-quantitative) methods with multiple indicators. For example, Martin and Irvine (1984b), in a discussion of the expected scientific performance of the Large Electron-Positron Collider (LEP), the collider that preceded the Large Hadron Collider (LHC) in the same 27-kilometer underground tunnel at the Swiss-French border, mentioned thirteen criteria: costs relative to other projects; accessibility of funding resources; degree of technical difficulty; technical track record of the laboratory; relative increase in energy; period of world leadership; event rate (a measure of relevant experimental observations); number of experimental areas; variety of possible experiments; ease of interpretation of the data; scientific track record of the community of users; potential spin-off to accelerator physics; and flexibility. Drawing on previous studies in radio astronomy and particle accelerators, Ben Martin (1996, 346) suggests that the key dimensions of basic research include "a) scientific contribution to the stock of knowledge; b) educational-contributions in terms of skills and trained personnel; c) technological-contributions to the development of new or improved technologies; and d) cultural-contributions to the wider society." These four dimensions are broadly the same as in the cost-benefit analysis (CBA) approach presented in this book, and Ben Martin should be given the credit for having qualitatively identified some of the core issues many years ago.[1]

This multiple indicators approach has been refined and extended several times.[2] While some of these indicators may be informative, nevertheless, given the lack of a common metric (you cannot sum the number of patents, the number of papers, and the percentage of scientific personnel, just to mention some indicators proposed in earlier literature), they cannot provide an absolute or comparative evaluation of the socioeconomic impact of research infrastructure (RI). In principle, one may perhaps build something as an aggregate score (such as in a multicriteria analysis frame),[3] but such aggregation would be arbitrary. Hence, this approach cannot answer the simple question: Are the measurable social benefits of a specific RI project greater than its costs?

In this chapter, I suggest that ambitious long-term RI projects, even in fundamental science, may often (but not always) confidently pass a net present value (NPV) test, (i.e., showing from a long-term perspective a flow of total benefits exceeding its costs, at present value). Finally, I explain why such a test, while informative on some aspects of the socioeconomic impact, should not be misinterpreted as a "blind" stopping rule. In other words, NPV < 0 does not necessarily mean that an RI project is not worth being implemented. In fact, what I want to show is that at least a (sometimes significant) share of the total costs is almost *always* already paid back to the society in the form of predictable technological, educational, innovation, and cultural benefits. Further, the magnitude of this share can be empirically estimated.

The model presented in chapter 1 provides a framework for the assessment of such socioeconomic benefits and costs associated with an RI, measurable with a common metric. After identifying such impacts, valuing them in a numeraire and discounting them with a social discount rate (SDR), the resulting net effect is expressed in money terms as an economic NPV (ENPV). This is not a financial indicator (see chapter 2), but rather a way to represent social welfare effects. In some cases, the communication of results to decision-makers may be easier by reporting absolute numbers, such as the internal rate of return (IRR) or the benefit/cost ratio (BCR), see section 1.A.1.

From an ex ante perspective, the probability of an error being related to each forecast and estimate included in the analysis should be carefully considered. To address this issue, a full-fledged quantitative risk assessment is required, meaning that costs and benefits become part of a probabilistic model. A risk assessment requires assigning each critical variable entering the model a specific probability distribution. As a result, the probability distribution of the outcome of interest, (i.e., NPV, BCR, or IRR), conditional on the distributions of the critical variables, assesses the project performance. A Monte Carlo simulation approximates such probability distribution functions of the performance indicators, their cumulative distribution functions, expected values, standard deviation, and other statistics.

While crucially, the case for investing government or charitable funds in a research project depends on its own scientific merits, there is also an obvious interest in recognizing its net socioeconomic impact from the perspective of governments and citizens that support science. Clearly, a positive net benefit can only reinforce the scientific case, as a signal that project's costs to society are more than recouped by its benefits. I am confident that for many (even if not for all) well-conceived RI projects, the social cost is lower than the measurable total benefits.

The chapter unfolds as follows. After this brief overview, section 10.2 discusses the time horizon of the project; section 10.3 deals with uncertainty and risk; section 10.4 suggests ways to present the results to decision-makers; section 10.5 briefly discusses some distributive issues; section 10.6 suggests caveats and further research needs; section 10.7 presents two case studies: the LHC and National Centre of Oncological Hadrontherapy (CNAO) expected NPV in socioeconomic terms; and section 10.8, "Further Reading," presents additional case studies and materials.

Readers more familiar with applied welfare economics will see that what follows is a fairly standard way to present the results of any CBA of an investment project. I regard this as a clear advantage because the usual applied welfare economics toolkit can be used to evaluate a large-scale RI—a toolkit that has been tested worldwide, over several decades, by hundreds of professional teams, in different fields, for thousands of projects, and is widely used by governments.

10.2   Time Horizon

Discussions between the US and Russian governments about the future International Space Station (ISS) go back to 1992. Construction began in 1998 and was still ongoing in 2017. Decommissioning is expected in 2024.[4] This would amount to an around a thirty-year life cycle of the ISS, from initial construction steps to decommissioning.

Since 2013, European Organization for Nuclear Research (CERN) has been discussing the feasibility of a Future Circular Collider (FCC), whose construction would perhaps start around the year 2040 and begin operations after 2050. Given what we have learned from the study of the LHC, should the FCC ever be funded, some long-term effects of such a project would extend to the twenty-second century. This very long time frame may be extreme, but there are other similar examples. The discussion of the feasibility of the Human Genome Project (HGP) started in the mid-1980s, and then the HGP began operations in 1990, (near) completion was announced in 2003, and its direct and indirect socioeconomic impact will probably be evident for decades.

What is the appropriate time horizon for an RI? Time enters into the NPV function (as discussed in chapter 1); hence, it is crucial to have a dynamic representation of the project. Some of the benefits produced by the RI last and even materialize well beyond the projects' operational phases (i.e., after their useful technical or economic life). Indeed, in some cases, benefits are permanent because knowledge is cumulative. For instance, when a telescope detects a new galaxy, an accelerator observes a new type of particle, or a clinical trial confirms a new therapy to cure a disease, such knowledge exists forever and is potentially transmitted generation after generation to future researchers, without any clear end point. Thus, knowledge accumulation has a longer time horizon than the RI operational period. Possible similarities can be found in environmental economics and, in particular, the economics of climate change, in which a few hundred years are frequently considered to evaluate the impact of a policy intervention. For example, the Stern Review (HM Treasury, 2006) sets a time horizon of 200 years.

Another extreme example is suggested by Boardman et al. (2018). These authors, to emphasize that the benefits of some projects may continue to flow for many years, even if these projects are complete from an engineering or administrative perspective, mention the case of roads: the roads originally constructed by the Romans more than eighteen centuries ago remain the basis of contemporary motorways. Dalgaard et al. (2018) study some economic effects of Roman roads to contemporary times, using econometric methods.

Should we assume that the time horizon of an RI is infinite, given that knowledge lasts forever once created? I think this would be wrong. A reason to adopt a finite reference period is related to the obsolescence process of tangible equipment, human capital, and knowledge itself over time. Empirically, this obsolescence is observable in the trend of citations of scientific publications and patents. Indeed, in retrospect, past discoveries and inventions are known to have lost some of their scientific/technological value and have been surpassed by new knowledge and technology. As a result, a long but finite time horizon for an RI seems reasonable. Such time horizon is project specific and changes greatly in accordance with the scientific domain.

Another problem related to the adoption of an infinite time horizon is that this would often lead to paradoxical results. Suppose that the SDR declines to zero or close to zero after some years;[5] large investment costs, which are spread throughout a finite range of years, may always be less than the sum of a series of constant (even if small) benefits spread over an infinite time horizon.

When a finite time horizon is selected for a specific RI, the residual value of assets and benefits should be computed at the end of the period of analysis in order to consider that these assets may still have an expected value.[6] The

European Commission (2014) recommends calculating the residual value in the same way as the remaining market value of fixed capital, as if the capital were sold at the end of the time horizon. Another method consists of estimating the amount that an entity would currently obtain from the disposal of assets net the estimated cost of disposal. This can be negative if there are substantial decommissioning costs of radioactive components (to cite one example). A simple, albeit imprecise, commonly used method is the straight-line depreciation method, in which the residual value is equal to the nondepreciated amount of the asset and the concept of remaining service life is exploited.

In the RI context, where some benefits may continue after the infrastructure decommissioning (such as technological spillovers to firms and human capital effects on former students and young researchers), the calculation of the residual value includes the discounted value of the benefits that exceed a project's time horizon.

10.3   Uncertainty

Given such a long-term time horizon to account for the uncertainty that characterizes the future, a risk assessment of an RI project can take advantage of established methods.[7] In such assessment, costs and benefits become part of a probabilistic model. As a result, the probability distribution of the performance indicators of interest is estimated. In general, ex post deviations from baseline predictions can be related to *endogenous* errors (e.g., errors of prediction incurred ex ante or during project implementation, which could have been avoided with a more accurate forecasting exercise) and *exogenous* errors (e.g., changes to the project context caused by an unpredictable event). The difference between the two types is, in a sense, just a matter of accuracy thresholds, as an unpredictable event is in fact usually something with a very low ex ante probability of occurring, but usually not with zero probability (i.e., impossibility).

According to Flyvbjerg, Bruzelius, and Rothengatter (2003), three broad categories of explanations exist for endogenous errors in project appraisal: technical (errors and pitfalls in forecasting techniques), psychological (optimism bias), and political–economic (the deliberate overestimation of benefits and underestimation of costs by a project's promoter in order to get funding). All these factors can potentially affect the appraisal exercise and provide the decision-maker with a biased outlook of the expected project outcomes.

For a number of contributions in this area, see Priemus, Flyvbjerg, and van Wee (2008), a collection of papers on the appraisal of megaprojects, including chapters on CBA in this context. In a chapter on managerial risks associated with megaprojects, De Buyn and Leijten (2008) point to some technical characteristics, such as the lack of solidity of the technical design, unproven technology, indivisibility of the functional elements, tight coupling among subcomponents, unavailability of fallback options, an excess of multifunctionality, and radical versus incremental implementation. They also mention the risks arising from social complexity, including third-party blocking power and long transformation time.

With regard to RIs, the uncertainty of CBA results can be greater than the uncertainty associated with traditional infrastructures when, for example, the experimental environment is unique and based on technological breakthroughs; thus, no previous experience is available, or suitable enough, to forecast future trends. Another source of uncertainty typical of the RI is related to the probabilities of discovery itself, which, in some cases, cannot be estimated based on previous experience. Other risks are the same mentioned previously for megaprojects, and even for any capital-intensive infrastructure project with a long time horizon.

An overall risk assessment is traditionally split into three steps. The first is a sensitivity analysis. The impact of each variable entering the analysis on the predefined outcome measure (such as the NPV) is assessed by changing each "best guess" value in absolute terms or by arbitrary percentages, one by one. The European Commission (2014) suggests focusing on a neighborhood of 1 percent of variation around the best guess. In other words, a 1 percent change in the value of the independent inputs of the CBA is assessed, and the variables leading to a greater than 1 percent change in the outcome measure are considered critical. However, a good practice is to consider a number of percentage variations around the best estimate on a continuum scale ranging from, say, −10 percent to 10 percent (Florio, 2014). This practice allows detecting the nonlinear and nonsymmetric effects of the variables on the project outcome.

Scenario analysis (Vose, 2008), also called "what-if" analysis, is a specific form of sensitivity test. Whereas with standard sensitivity analysis, the influence of each independent variable on project performance is tested separately, scenario analysis studies the combined impact of sets of values assumed by the critical variables. In particular, combinations of the optimistic and pessimistic values of a group of variables could be useful for building extreme scenarios and calculating the extreme limit of project performance indicators. To define optimistic and pessimistic scenarios, the extreme values defined by each critical variable's distributional probability should be used.

After the sensitivity analysis, a range of variations and a specific probability distribution function are assigned to each identified critical variable. Probability distributions depend greatly upon the specific type of RI project under evaluation and may be guessed from various sources of information, including experimental data, distributions found in the literature and adopted in similar projects, and time series or other types of historical data. When insufficient data exist to construct probability distributions based on past experience, the range and likelihood of possible values ultimately rest on project promoters' and evaluators' judgment.

In some cases, Delphi methods can be helpful (Linstone and Turoff, 1975), which are based on gathering an international panel of experts and eliciting their views about likely future scenarios. This information is used to estimate the probability distribution function of various technological scenarios, conditional on existing information. The Delphi method involves multiround forecasting challenges, in which experts provide initial forecasts and then adjust them based on feedback they receive. This process is iterated until a satisfactory level of consensus is reached and final forecasts are constructed from the aggregation of individual forecasts. See the section entitled "Further Reading," at the end of this chapter, for more information.

Finally, the project's riskiness is assessed using the Monte Carlo simulation method, which enables estimations of the integral corresponding to the probability distribution function of the project performance indicator of interest (e.g., the NPV). By extracting one value of each critical variable from the respective distribution function and plugging it into the CBA model, the associated NPV is computed. This process, if repeated over a large number of iterations, leads to the probability distribution of the project's NPV. In other words, through the law of large numbers, which implies the convergence of the NPV empirical distribution with its true counterparts, the CBA result can be considered in probabilistic terms and the NPV's minimum, maximum, mean, and standard deviation values can be computed (with a Monte Carlo error associated with the fact that draws are not infinite).

10.4   Presenting the Results

How should these results be given to decision-makers? The NPV cumulative distribution function returns the probability that the outcome is less than or equal to any given value in the range of the variation in the considered performance indicator. Thus, the performance indicator can be directly exploited to observe the cumulated probability that corresponds to some threshold. Namely, if $\Pr \{NPV \leq 0\} \approx 0$, the RI project can be judged as almost certainly desirable in terms of its socioeconomic impact. When considering the IRR, if $\Pr \{IRR \leq SDR\} \approx 0$, the project can be judged as socially desirable. The range of variations consists of the window of values, from minimum to maximum, within which the NPV and the IRR vary. The range of variations provides a picture of the project's riskiness. In general, a project with a narrow range of variability in its performance indicators is preferable, ceteris paribus, but the scientific case may be associated with higher risk and may counterbalance such preference.

The mean values of the NPV and the IRR are the key estimates for a risk-neutral funder and are interpreted as the outcome expected to occur over a large number of potential project realizations. Thus, the expected mean values provide an immediately readable synthesis of the indicator of the most likely discounted social value of a project. The standard deviation consists of the variation around the mean values of the NPV and the IRR. Scientists are accustomed to seeing values expressed in the form of +/− 1 standard deviation (see the way it is done in table 10.2, later in this chapter, for instance), and it seems advisable to use this convention whenever possible. In general, no rule exists for interpreting the standard deviation (and in general, the shape of the distribution) as high or low risk in absolute terms. However, this synthesis indicator can provide useful information if compared with those of similar projects.

For decision-making, it is often informative to assess the probability that the NPV is lower than its baseline. Another useful synthetic ratio is the coefficient of variation (e.g., of the NPV); namely, the ratio of standard deviation to the mean, which is a dimensionless number and thus, in principle, comparable among projects.

In practical terms, whereas the probability distribution function summarizes the likelihood of occurrence of all outcome values randomly extracted during the Monte Carlo simulation, the cumulative distribution function returns the probability that the outcome is less than or equal to any given value in the range of the variation in the considered performance indicator.[8] Thus, the latter can be directly exploited to observe the cumulated probability that corresponds to some feasibility threshold.

Finally, for communication purposes, presenting CBA results in a disaggregated manner could be useful, such as by breaking down each discounted benefit according to social groups (firms for technological spillovers, students and ERC for human capital effect, scientists for publications, general public for cultural effects and for public good value, etc.; see, for example, figure 10.2 later in this chapter). In some cases, this is also important to assess distributional or territorial impacts. Hence, it is suggested showing to whom the benefits accrue in terms of involved stakeholders (see section 10.6).

All these elements from the CBA results of the RI are used to judge the investment's social worthiness and riskiness, along with other criteria. Having said this, how likely is it that $\mathbb{E}(NPV) < 0$ for any RI? To ask this question in a more precise way: Suppose that a government considers a portfolio of infrastructure projects in any field, such as transportation, environmental protection, education, health, energy, telecommunications, and eventually science. Do we expect that CBAs of scientific projects, taking the specific form of RIs, are more likely to end up with a negative economic NPV relative to the more traditional infrastructures of similar size in terms of total costs and of other project-specific characteristics (such as location)?

The answer can be based only on conjecture because of the limited number of RI projects for which a full-fledged CBA has been available until now. In contrast, the project portfolio of the World Bank includes more than 3,000 implemented projects,[9] the European Commission and the European Investment Bank have probably seen the CBAs of thousands of projects in total, and several thousand other CBAs have been performed for other international institutions and national governments, such as those of the United Kingdom, the United States, France, and Canada, just to mention a few countries included in a recent survey by the Organisation of Economic Co-operation and Development (OECD) on the use of CBA (OECD, 2015a).

Perhaps what, until now, has hindered the development of CBA methods for RIs is the perceived difficulty to value ex ante the main benefit—the creation of scientific knowledge. Hence, as costs are more predictable while the main benefit remains unpredictable, many have concluded that CBA should lead to a negative NPV for large-scale investment in science.

However, going back to the benefits classification by Martin (1996) and the discussion in the previous chapters, at least four groups of RI socioeconomic benefits can be identified and valued: human capital benefits, technological spillovers, and cultural benefits, including the public good value of scientific research. Moreover, in some cases, benefits for direct users of RI services (e.g., of medical research) and some indirect benefits of science-based innovations, as well as at least a small part of the contribution to scientific knowledge in terms of the impact of publications, can also be predicted and valued. My guess is that while the total costs of many RIs are lower than most traditional infrastructure projects (such as railways and highways), they usually have limited negative and large positive externalities as a side effect of their operations. In many cases, these effects will be greater than costs, even without a full consideration of the value of knowledge created per se. Finally, while the decision-making process for traditional RI may be severely distorted, particularly in countries with weak institutions, where CBA may be little more than a way to justify political choices, the peer review process in science and the concern of scientists for their reputation may tend to contain the generation of very bad projects, known as *white elephants*.

Table 10.1
Major RI projects in the context of EU regional policy between 2008 and 2013

Country	No. of projects	Investment Cost (a)	ENPV (a)	ERR (b)	SDR (b)
Czech Republic	1	104,200	3,998	13.09	5.5
France	2	88,440	81,192	12.16	4.0
Hungary	2	229,350	192,600	16.20	5.5
Italy	1	83,000	50,236	22.39	3.5
Lithuania	2	76,818	21,878	9.64	5.5
Poland	8	79,018	93,129	24.31	5.4
Slovenia	1	111,100	71,561	15.26	7.0
Spain	2	225,864	21,281	7.35	5.5
United Kingdom	5	65,608	296,344	25.85	4.0
Total	24	104,143	124,435	19.38	5.0

Source: Author's elaboration of data provided by Directorate General for Regional and Urban Policy. Sector classification was taken and reelaborated from the Commission Implementing Regulation (EU) No. 215/2014 of March 7, 2014, at http://eur-lex.europa.eu/legal-content/EN /TXT/?uri=CELEX:32014R0 215. See Florio, Morretta, and Willak (2018) for details.
Notes: Data do not include R&D investment in firms. (a) average thousand EUR, (b) average %.

The only way to confirm my guess is to apply the CBA methods in practice on a sample of RIs and look at their estimated NPV either ex ante or in retrospect. From this perspective, it would be important that national and international institutions move on with the empirical analysis. As stated by Flyvbjerg (2006, 219): "A scientific discipline without a large number of thoroughly executed case studies is a discipline without systematic production of exemplars, and ... a discipline without exemplars is an ineffective one."

There is, in fact, already some initial evidence from the major RI projects submitted for funding to the European Commission, in the context of EU Cohesion Policy between 2008 and 2013. Table 10.1 shows a list of projects for which CBA was performed by the member-states of the European Union (EU) with expected NPV > 0 and an economic rate of return (ERR) greater than the SDR. See also the "Case Studies: The Net Benefits of Two Synchrotrons" and "Further Reading" sections later in this chapter for additional evidence on RI projects for which a CBA with encouraging results is available.

From the perspective of RI managers, who obviously mainly focus on scientific excellence, the main message of this book is that some side effects (also known as co-benefits) may generate sufficient socioeconomic benefits to justify the investment of taxpayer money. Hence, the RI strategy should carefully consider how to maximize the synergy between science creation on one side and human capital creation, technological spillovers, and the production of cultural goods on the other.

10.5 Distributive Issues

Any major project affecting different segments of the society may alter—to a certain extent—the distribution of social welfare. This topic is still unexplored for RI projects, but in some cases, it may be important. To see this, let us consider a slightly modified version of the CBA model of equation (1.2) in chapter 1:

$$NPV_{RI} = [a_1 SC + a_2 HC + a_3 TE + a_4 AR + a_5 CU]$$
$$+ B_n - mcpf[K + L_s + L_o + OP + EXT], \tag{10.1}$$

where $a_1, ... a_5$ are welfare weights (Florio, 2014; Johansson and Kriström, 2015) *inversely* correlated to the welfare of each group, such as scientists to whom the RI benefits accrue: scientists (SC), students (HC), investors of supplier firms (TE), investors in other firms and consumers of innovative products (AR), and consumers of cultural goods (CU). The welfare weight of the general public of nonusers (B_n) is assumed to be unity, meaning that EUR 1 to the median taxpayer is valued at EUR 1, while, for instance, the welfare weight of the poor benefiting from health research (included in AR) is > 1, and that of the shareholders of firms enjoying increased profits (included in TE) < 1. Moreover, in some cases, it would be worth considering a marginal cost of public funds ($mcpf$) > 1 because of the excess burden of distortionary taxation. In this context, the welfare-weighted NPV changes, and qualitatively, there may be winners and losers.

To be more concrete, let us consider the examples of the World Wide Web (WWW) and the HGP. In both cases, taxpayers, including the poor, have supported basic research that was then instrumental in producing knowledge as a public good, supplied for free to anybody. However, the benefits of such research have been distributed unevenly in the society. Investors in Internet companies and in the pharmaceutical industry, respectively, have been able to appropriate a considerable share of the value of economic innovations derived from this new knowledge. In principle, the profits of major Internet or pharmaceutical companies are taxed. Hence, part of this appropriation of the benefits goes back to the taxpayer, but the process is imperfect, to say the least. In fact, it is not entirely clear whether the current arrangements for basic research, where the costs are on the taxpayers and some of the benefits go to private investors downstream, tend to decrease or increase social inequality, which is a major concern of our times.

10.6 Caveats: What Can Go Wrong

Policymakers will never accept or reject an RI project based solely on its expected socioeconomic impact. Several other qualitative considerations are widely discussed in the science policy and management literature, and these have not been repeated here. In the past, for example, US scientific projects claiming to be relevant to national defense have received more money than other projects. Currently, some themes are very popular in the policy agendas of governments, such as health problems of the aging population, and one may expect a certain inclination by governments to fund RI projects related to disease correlated with age. However, CBA should not directly include such policy priorities in the project assessment model. They must be discussed separately.

Indeed, there are a number of caveats regarding the CBA approach advocated in this book. RI investments are sometimes financed within territorial development strategies (regional or national). Thus, an RI may become an element of the development path of a territory and of a smart specialization strategy. In such cases, relevant economic impacts in the form of employment or technological spillover effects on existing, or newly created, small medium enterprises (SMEs) in a region are included in the CBA, as described in the prior sections, but a regional breakdown of the effects may be helpful in dialogue with local stakeholders. Additional wider benefits—for example, in terms of contributions to regional gross domestic product (GDP)—are not included within the standard CBA model to avoid possible double-counting of benefits. The discussion of wider economic impacts is particularly developed in the transportation sector.[10]

Typically, local economic impacts include agglomeration economies, multiplier effects, labor supply impacts, the impact on competition, and changes in the values of land and housing. In an RI context, there also may be demonstration effects of outreach on the general public (particularly the young) about the role of science and technology. For instance, the proximity of universities to an RI project may convince a larger share of students to achieve degrees in one or more of the sciences, and this in turn could be correlated with the long-term regional growth rate (Valero and Van Reenen, 2016).

Further, an RI facility that attracts high-quality personnel, possibly from other regions of the country or abroad, may contribute to develop the local cultural environment. This, in turn, can contribute to an increase in social capital and, in some particularly beneficial cases, even to the improvement of the overall quality of institutions. *Social capital* is defined in the literature as being related to the network of informal relations that an individual is able to mobilize; for example, personal relations have been found instrumental in well-being and income (Knack and Keefer, 1997).

The quality of institutions may also be indirectly influenced by the location of an RI in a region, as far as the attraction of scientists and early career researchers from several countries have effects on the public discourse (e.g., in terms of a more cosmopolitan and open perspective on education, gender balance, religion, and other social institutions). There also may be displacement effects, as happens in transportation projects, which in some cases increase territorial polarization.

All these local impacts may be studied and reported on and may be interesting to some stakeholders. Nevertheless, it is important to avoid any double-counting.[11] Some of these effects are already accounted for within the direct effects heading; for example, wider economic impacts in terms of labor may have been already accounted for by using shadow wages. Second, in many cases, these local effects do not incorporate a proper counterfactual comparison (e.g., what would have happened without the project), and hence would not measure the relevant incremental benefits.

In most cases, it seems appropriate to supplement the CBA with well-designed, quali-quantitative impact analyses. In this regard, some of the indicators proposed in earlier literature may be informative, as they capture some angles or nuances that do not fit well in a welfare economics frame (see the "Further Reading" section).

Moreover, it should be further restated that a large part of the potential future benefits of science is unknown ex ante, and in some cases, even ex post in terms of causality from a specific discovery project to an economic innovation. Consequently, the actual meaning of an RI's CBA is to focus on what can be said quantitatively, with due caution, compared to what can be said only qualitatively. The final message is that even when all the benefits of science are unknown, the net social cost of investing in science is less than what financial accounts can show, even if this does not mean that all RI projects can pass the NPV test, or that they are risk free.

In fact, a number of things can go wrong in the assessment of an RI. To sum up, these include major mistakes in forecasting costs, as for any megaproject, as mentioned previously. Any evaluator of a major RI should possibly read *Tunnel Visions*, a very good book by Riordan, Hoddeson, and Kolb (2015), in order to understand the mistakes made in planning the Superconducting Super Collider (SSC). The initial cost forecast was in the range $1–3 billion in 1982, $3 billion in the conceptual design report (1986), and $5.9 billion at congressional approval (1989), but the cost had risen to more than $10 billion when the project was eventually discontinued by Congress in 1993. According to the authors, a major mistake was probably to designate the project as a "green field" one, rejecting the location at the Fermilab, a major national laboratory sponsored by the US Department of Energy (DoE).

Another example of delays and cost overruns is the nuclear fusion International Thermonuclear Experimental Reactor (ITER) project. According to Clery (2016b), who cite an independent review report, the project could be operative in 2025, with a deuterium-tritium "burning plasma" target shifted to 2032–2035, with a five-year delay, and only if it receives an extra \$4.6 billion. The total final costs could be three times the initial planned figure of \$11 billion.

Forecasting the benefits can go wrong for a number of other reasons, perhaps the most important one being failure to achieve the relevant scientific results that were expected. This may have consequences for the attractiveness of the RI for students and for early career researchers (ECRs), as well as on the impact of outreach on the general public—both critical components of the overall balance of the social benefits.

More generally, the main issue with making a socioeconomic impact assessment of an RI project is optimism bias. This is a particularly serious concern when the assessment is not delegated to independent experts, possibly on behalf of the funders or of so-called neutral institutions, but rather is performed by consultants directly hired by the RI managing body without a clear mandate. While the managers may have a serious interest in knowing the benefits and costs of an RI project, in some cases, they may see the assessment exercise as part of their external communication strategies. In fact many reports in the "gray" literature in this area are methodologically weak and tend to exaggerate some project benefits and to hide some costs, or simply mention the former but not the latter.

I hope that this book will instead encourage further scholarly empirical research on the socioeconomic impact of RIs. They are an increasingly important way to structure the production of knowledge at the frontier of science, and their socioeconomic impact deserves high-quality analysis. Some uncertainty is unavoidable, but impartiality and rigor are necessary.

10.7   Case Studies: The Net Benefits of Two Synchrotrons

To illustrate the bottom line of the project evaluation framework, in this section, I present the estimated NPV of two RIs for which specific types of benefits were discussed in previous chapters: the LHC and the CNAO. Clearly, the LHC and CNAO are just two examples of RIs. The LHC is an extreme case because of its long period of construction and operation; the high number of scientists, students, and postdoctoral researchers involved; the large number of firms in the supply chain; the externalities from the open access to software; the wide coverage in the media and attraction of onsite visitors; and the nature of a frontier basic research facility. CNAO is smaller and is similar to other medical research centers, but it shares with the LHC the contribution that particle physics research makes to its design and operation.

10.7.1 The Net Present Value of the Large Hadron Collider and of Its Upgrade (High-Luminosity Large Hadron Collider)

Florio, Forte, and Sirtori (2016) determine the net socioeconomic impact of the LHC as follows. The LHC's total present value to 2025 of operating and capital expenditure is estimated at EUR 13.5 billion (net of the cost of CERN and experimental collaborations' scientific personnel). The details of the total cost estimation were given in chapter 2 (section 2.8). Against these costs, there are the benefits arising from the impact in the scientific literature (chapter 3, section 3.8); the human capital effects on ECRs (chapter 4, section 4.5); technological spillovers to firms (chapter 5, section 5.6); the benefits of free software to users (chapter 6, section 6.3); the benefits of cultural goods (chapter 8, section 8.6); and the public good value (chapter 9, section 9.6). There are no direct benefits of the LHC for services provided outside scientific research (such as those mentioned in chapter 7 for synchrotron light sources), while indirect benefits from some innovations to a certain extent related to research at CERN (such as hadrontherapy, discussed next) are not estimated.

In terms of contributions to the sum of the social benefits (estimated at EUR 16.4 billion), the present value of the human capital effects and technological spillovers are the most important and of similar size, each contributing approximately to one-third of the benefits. Adding in the tiny secondary effect of publications (net of the direct value of LHC research output), approximately 68 percent of the socioeconomic benefits is related to professional activities (within firms, academia, and other organizations), while the remaining benefits spill over to the general public, either as a direct cultural (private) good or as a public good (a nonuse benefit) (see table 10.2 and figure 10.2). Any other unpredictable social benefits of future applications of scientific discoveries at the LHC are excluded from this analysis; they remain as a bonus for future generations, donated to them by current taxpayers.

Given the uncertainty surrounding the forecasts to 2025 as well as some estimates of past effects, a probability distribution of the LHC's NPV was estimated by running a Monte Carlo simulation [10,000 draws conditional on

Table 10.2
Summary of social benefits

Costs	13.5±0.4
Use Benefits	
Scientific publications	0.3±0.1
Human capital formation	5.5±0.3
Technological spillovers	5.3±1.7
Cultural effects	2.1±0.5
Nonuse Benefits	
Public good value	3.2±1.0

Source: Author's elaboration of data from Florio, Forte, and Sirtori (2016).

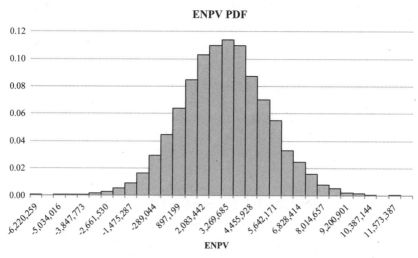

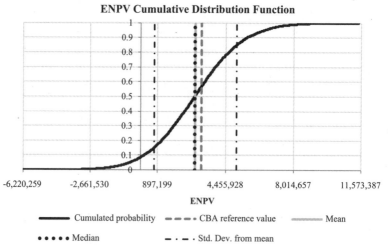

Figure 10.1
NPV probability density function and cumulative distribution function of the LHC. *Source*: Florio, Forte, and Sirtori (2016) and Florio, Bastianin, and Castelnovo (2017).

the probability density function (PDF) of nineteen stochastic variables (see Florio, Forte, and Sirtori, 2016, for the technical details)]. Each draw generates an economic NPV estimate in a state of the world supported by a random set of the possible values taken by the model's stochastic variables.

The final PDF and cumulative probability distribution for the NPV are shown in figure 10.1. The LHC's $\mathbb{E}$(NPV) is approximately EUR 2.9 billion, with a conditional probability of NPV < 0 smaller than 9 percent and with a 3σ Monte Carlo error below 2 percent (this is the estimated error arising from the fact that the number of draws is not infinite). The expected BCR is approximately 1.2 and the $\mathbb{E}$(IRR) is 4.7 percent. The NPV would be lower if an opportunity cost of public funds is considered because of distortionary taxation; however, it would still be positive for the typical current range of the excess burden of taxation in developed countries (Dahlby, 2008). Indeed, the European Commission (2014) does not recommend introducing a correction for the opportunity cost of public funds for projects funded by grants supported by international transfers because it would not be clear which is the relevant source of funding. For example, if marginal funding is from sovereign debt, for most of CERN's core member-states, the real interest rate on such debt for thirty-year bonds would be largely below both the SDR that is used here (3 percent, as recommended by European Commission, 2014) and the long-term rate of GDP growth.

More recently, an analysis was performed on the High-Luminosity LHC, an important upgrade of the LHC for which construction started in 2017 and operation is expected to start in 2025 and conclude in 2035. From a methodological

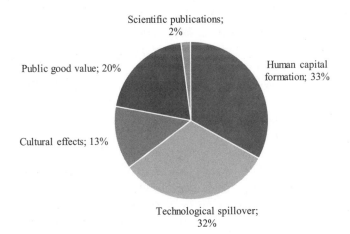

Figure 10.2

LHC: Share of total benefits by category. *Source*: Author's elaboration of data from Florio, Forte, and Sirtori (2016).

perspective, this CBA is interesting because, in many RI cases, there may be possible upgrades of existing facilities with different options to be evaluated.

I briefly report here the results of the CBA of the High-Luminosity LHC project, with a construction undiscounted cost of around 950 million Swiss francs (CHF). Details are available in Bastianin and Florio (2018). The counterfactual scenario is the continuation of the LHC without such an upgrade, while the High-Luminosity LHC has the objective of increasing the focus of the two intersecting particle beams, and consequently, the number of collisions and amount of data generated. This is also an example of the CBA of a "brownfield project," an investment incremental of the effectiveness of an existing RI.

For the CBA assessment, the High-Luminosity LHC scenario and the counterfactual scenario are identical during the 1993–2014 period (benefits and costs are the same as in Florio, Forte, and Sirtori, 2016); the first run of the LHC with the new High-Luminosity upgrade is expected around 2026–2027. The counterfactual scenario entails continuing to operate the LHC with ordinary consolidation activities; after 2031, data collection ends and CERN staff shift their attention to other scientific activities. The horizon of the analysis spans the 1993–2038 period.

The analysis was carried out in Swiss francs. The base year for social discounting and inflation adjustments is 2016, and the SDR[12] is 3 percent, as recommended by the European Commission's *CBA Guide* (European Commission, 2014). Values for the 1993–2016 period have been adjusted for inflation using the Consumer Price Index for Switzerland, sourced from the 2017 release of the World Economic Outlook Database of the International Monetary Fund (IMF). After 2016, prices are assumed to remain constant. No attempt has been made to model or forecast exchange rate variations. Bastianin and Florio (2018) conclude that the NPV of the High-Luminosity LHC project is positive at the end of the observation period. The ratio between incremental benefits and incremental costs of the High-Luminosity LHC with respect to continue operating the LHC under normal consolidation (i.e., without the High-Luminosity upgrade) is over 1.7, meaning that each Swiss franc invested in the High-Luminosity LHC upgrade project pays back approximately CHF 1.7 in societal benefits.

In terms of probabilities, simulations based on 50,000 Monte Carlo rounds show that there is a 94 percent chance of observing a positive NPV (i.e., a quantifiable economic benefit for the society). The attraction of CERN for ESRs is key for a positive CBA result. Technological spillovers are another very important ingredient in the CBA evaluation of CERN RIs.

10.7.2   The Net Present Value of CNAO

Figure 10.3, by comparison, reports the CBA results for the CNAO, which is very different from the LHC in terms of distribution of benefits. What follows draws heavily from Pancotti et al. (2015) and Battiston et al. (2016).

As already discussed in chapter 7, CNAO supplies hadrontherapy to patients affected by certain solid tumors. The CBA of health projects, such as a new therapy or a new hospital project, points to the decrease of mortality rates and the increase in quality-adjusted life as the core benefits. On these two variables (defined quantitatively in different ways), there is wide empirical literature. In the case of a health RI, it is the combination of clinical and research activities that generate such benefits through the experimentation on patients of innovative approaches to specific pathologies. The ingredients of the valuation, therefore, include the total frequency of patients by the various clinical treatments, the health benefit levels for each category of patients, and finally the value of statistical lives. This formula summarizes the estimation:

$$AR = \sum_{t=0}^{T} \frac{\sum_{cp=1}^{CP} \sum_{i=1}^{I} (N_{cp,i} * E_{cp}) * (YEAR_{cp,i} * VOLY_i) * QUAL_{cp}}{(1+SDR)^t}, \qquad (10.2)$$

where N is the number of patients; E is the share of patients who gain additional years of life compared to the identified counterfactual; $YEAR$ is the number of life-years gained, while $VOLY$ is the value of a life-year and $QUAL$ is the coefficient capturing the increased quality of life. $cp = (1, 2..23)$ is the clinical protocol; i $(1, ..6)$ is age class; t $(1, ... 30)$ years in the time horizon; and SDR is the social discount rate.

Benefits to patients differ according to the therapeutic protocol and the age of patients, because the comparative advantage relative to conventional therapies depends upon the probability of success in terms of life expectancy and quality. Thus, equation (10.2) should be interpreted as the cumulative effect of such different expectations.

Following OECD (2010), the core empirical variables are the value of statistical life (VOSL) and the value of a life-year (VOLY). VOLY and the VOSL are related as follows:

$$VOSL = \sum_{t=0}^{T-a} VOLY * (1+SDR)^{-t}, \qquad (10.3)$$

where a is the age of the individual or group considered, T is the life expectancy at birth, and SDR is an appropriate discount rate. VOLY is often estimated from VOSL as the average value of a life year, adjusted by the survival probabilities.

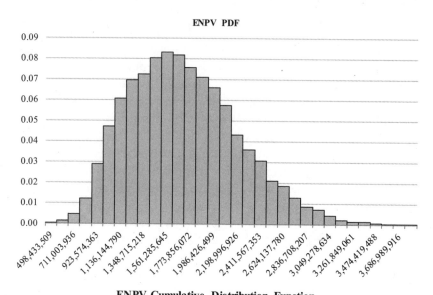

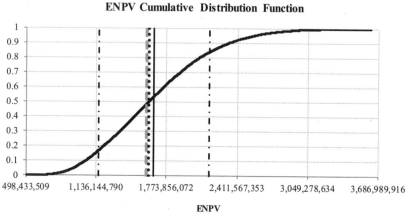

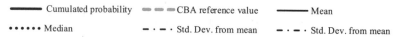

Figure 10.3

NPV probability density function and cumulative distribution function of CNAO. *Source*: Battiston et al. (2016) and Florio, Bastianin, and Castelnovo (2018).

In the case of CNAO, most of the benefits are related to the health effects of carbon ion therapy (as compared to the traditional treatment), which is the result of specific research and innovation. A minor but still relevant share of benefits comes from proton therapy, which is a more established approach but is refined further at CNAO through new applications. Other benefits are related to revenues from selling the beamline to other research teams, from the direct value of scientific publications, from technological spillovers, and from human capital formation, all estimated in the ways described in previous chapters.

After considering these benefits and the construction and operation costs over a time horizon of thirty years, the CNAO-expected NPV is conservatively estimated at EUR 1.6 billion. In addition, risk analysis shows that the project is affected by a low level of risk. The baseline BCR is 4.42, which is higher than for the LHC because of the more easily quantifiable direct social benefits in terms of the value of years of life saved.

10.8   Further Reading

This section briefly mentions some references on the following topics: NPV, other project economic performance indicators, and risk analysis; applications of CBAs to RIs; and other approaches.

10.8.1   Economic Performance Indicators and Risk Analysis

Many CBA manuals include good discussions on NPV, IRR, and BCR. Any of the following may be helpful for a wider presentation of these topics: Boardman et al. (2018), Brent (2017), De Rus (2010), Florio (2014), and, among the government guidelines, European Commission (2014) and HM Treasury (2018).

Among the abovementioned manuals, some have a wider discussion on how to deal with uncertainty. These include Boardman et al. (2018), Florio (2014), and Johansson and Kriström (2015). The classical studies are Reutlinger (1970) and Pouliquen (1970), but since then, what really makes the difference is the availability of computing resources and software to run the Monte Carlo analysis. Boardman et al. (2018), in their chapter 11, "Dealing with Uncertainty: Expected Values, Sensitivity Analysis, and the Value of Information," provide a good discussion and a simple case study. Johansson and Kriström (2015) provide a more advanced treatment of the subject. Florio and Lombardo (2014), deal with the conceptual foundations of Monte Carlo analysis and also discusses the treatment of mutually correlated stochastic variables. On applications of Monte Carlo methods, see, for instance, Robert and Casella (2004) and Balcombe and Smith (1999).

On Delphi methods, see, for example, Chang et al. (2002), Da Silveira et al. (2018), de Loë et al. (2016), Dalkey and Helmer (1963), and Rowe (2007).

10.8.2   CBA of Research Projects and Programs

The application of CBA methods to RIs is in its infancy, but not entirely new. Johnson and Kokus (1978) is the first example a CBA approach in the field of science that I am aware of. It studies four major activities of the NASA Technology Utilization program (a technology-transfer initiative started in 1963) for which it computes the BCR. A cost-benefit study was requested by Congress (Johnson and Kokus, 1978), and more than 700 interviews were conducted. The study uses a 10 percent discount rate, as required by the Office of Management and Budget, and on the benefit side, either a consumer surplus or a financial investment model. The objective of the interviews is reported as follows (Johnson and Kokus, 1978, 5):

The three primary data points for all interviews were: (a) the estimated costs and gross benefits, distributed over time, that the user attributed directly to receiving a specific information package (i.e., costs and benefits that would not have occurred, in the interviewee's opinion, without the information package); (b) the type of application achieved or expected for the technical information received; and (c) the estimated chance of success for expected applications.

In contemporary wording, this approach points to estimating a willingness to pay (WTP) for the value of information, and on average found a very high BCR of 7.5 for information projects. For engineering collaborative projects, the BCR was 4, also based on a simple estimation of success probabilities of innovations.

Garcia Montalvo et al. (2004) offer a combination of financial analysis, CBA, and input-output (I-O) analysis of the ALBA synchrotron light source, near Barcelona, Spain. In terms of CBA, they were able to forecast a positive NPV, a BCR of around 1.2, and an IRR of around 9 percent. To reach these conclusions, they consider as benefits the time saved by Spanish researchers who would have to travel to synchrotrons abroad, some human capital effects for skilled employees who would otherwise be unemployed or employed in less-qualified jobs, and some generic effects on economic activities based on an I-O model. Moreover, they consider some standard corrections such as the use of shadow wages, accounting for carbon dioxide emissions because of the electricity consumption, fiscal corrections, and other elements.

A special issue of the journal *Technological Forecasting and Social Change*, edited by Del Bo, Florio, and Forte (2016), discusses CBA and socioeconomic impact analysis, both in the field of particle accelerators and high-energy physics in general (I have already mentioned several articles from this source).

Following the European Commission (2014), JASPERS (an initiative for project assistance and capacity-building in the EU member-states by the European Commission, the European Investment Bank, and the European Bank for Reconstruction and Development) have prepared their guidelines in this area (see Teichmann and Swerdlow, 2016; Swerdlow, Teichmann, and Young, 2016). The initiative's staff has started to experiment with the use of CBA on the projects with which they assist. For a presentation of the CBA approach by JASPERS, see also JASPERS Networks,[13] and particularly the example by Teichman and Swerdlow (2016) about the European Supercomputer Research Center.

Some benefit-cost reports are in fact not standard BCAs. For example, Lee (2018) reports a BCR for the South Korea component of the ITER nuclear fusion project, but this is based on a very simplified approach, looking at some costs and some revenues of firms involved in procurement. EvaRIO (2013) discusses, mostly in quali-quantitative terms, the direct and indirect benefits of the EMBL-Bioinformatics Institute (see also chapter 6). The report is interesting, but it is not a standard BCA. Link and Scott (2004), while not on an RI but a small-scale research and development (R&D) project on wavelength measurement for optical fiber research, establish a link between CBA and the Griliches-Mansfield approach to the measurement of social returns (see chapter 5).

An interesting case study concerns the funding of medical imaging R&D supported by Canada Foundation for Innovation (CFI) and Canadian Institutes for Health Research (CIHR), two federal funding agencies. Their portfolio in this area was worth around 1 billion Canadian dollars (CAD) of grants (1999–2012).[14] After selecting the computed perfused tomography (CPT) projects (an imaging technique for the diagnosis of the stroke-affected brain), a report finds that R&D at four universities and associated hospitals leads to an acceleration of five years in the introduction of CPT. As the quality-adjusted life-years (QALYs) saved per patient is 0.12, given the estimates of a QALY and the number of treated patients in Canada, the report finds a BCR in the range of 1.5–2.3 or an IRR in the range 28%–46%. The cumulative additional QALYs are in the range 2,845–4,720 for the population of treated patients, depending on the uncertainty of some data and assumptions.

10.8.3 Other Economic Analysis Approaches

I-O models and impact multipliers have occasionally been applied to the study of some RIs. The I-O concept goes back to the work of Wassily Leontief, a former Soviet economist and later a professor at Harvard and a Nobel laureate. Leontief (1941) introduced for the first time the concept of a mathematical representation of the economy as a matrix (a double-entry table); also see

Leontief (1966). The objective was to study empirically, in a general equilibrium frame, changes of prices, outputs, investments, and incomes. I-O models quantify mutual interrelationships between various industries in a country or region in the form of a system of linear equations. Multipliers are computed as arising from the balances between the total input and the aggregate output of each commodity and service produced and consumed year by year.

Intuitively, an output multiplier of 3 of an investment in one sector would mean that the total increase of output could be three times the initial stimulus (if there is enough capacity in the economy to accommodate the quantity change). According to a review of Big Science project by Simmonds et al. (2013, 6): "The evaluations arrive at economic multipliers that typically range between 2 and 3, which is to say that every £1M in public expenditure is generating an additional £2M or £3M in wider economic activity through onward purchases within supply chains and the personal consumption of employees using their wages."

Examples of studies that use impact multipliers based on the I-O framework are given in box 10.1. For example, KPMG (2016) is an assessment report of SNOLAB, a neutrino detector facility 2,000 meters underground in Ontario, Canada. The report provides evidence of past and future capital and operational expenditures and aims at comparing them with economic benefits estimated with I-O techniques (with models developed by Statistics Canada). This allows direct, indirect, and induced impacts on added value to be computed. Over eight years, this is estimated at CAD 176.5 million nationwide. The report concludes that CAD 1.49 of GDP will be created for each dollar of public funding. Atkinson, Wolpe, and Kotze (2017) carry out a detailed impact study of the Square Kilometre Array (SKA) for Phase 1 of the South African component, including 197 dishes for radio-astronomy to be built between 2018 and 2025, at the cost of EUR 352 million (the total cost of Phase 2 is estimated at EUR 3–4 billion).

Other projects considered in these I-O studies include studies of the space economy in Europe, the impact of Fermilab (one of the US national laboratories funded by the DoE), the Daresbury synchrotron light source in the United Kingdom, the TRIUMF particle accelerator in Canada, and the ITER nuclear fusion project in France. While this research simply uses existing multiplier values, others use some software and plug into some RI-specific data, such as PricewaterhouseCoopers (2016), on the Earth observation program Copernicus, and Tripp and Grueber (2011), on the HGP. For instance, the study for Berkeley Lab calculates the indirect and induced impacts of the laboratory by running data on capital expenditure, payroll, and procurement through the IMPLAN (IMpact Analysis for PLANning) model, an I-O model developed by

Box 10.1
The Socioeconomic Impact of RIs with I-O Multipliers

Reference	Title of the Study	Contracting Authority/ Institution (Country)
Atkinson, Wolpe, and Kotze (2017)	"Socio-economic Assessment of SKA Phase 1 in South Africa"	Council for Scientific and Industrial Research (South Africa)
KPMG (2016)	"Assessment of SNOLAB— Final Report"	SNOLAB (Canada)
IDEA Consulting (2015)	"Economic Footprint of 9 European RTOs"	European Association of Research and Technology Organisations (EARTO) (Belgium)
Booz and Company (2014)	"Evaluation of Socio-Economic Impacts from Space Activities in the EU"	European Commission—DG GROW (European Union)
NRCC (2013)	"Return on Investment in Large Scale Research Infrastructure"	National Research Council Canada (Canada)
AEG (2011)	"The Economic Impact of Fermi National Accelerator Laboratory, New Light on Science"	University of Chicago (United States)
STFC (2010b)	"The Social and Economic Impact of the Daresbury Synchrotron Radiation Source, (1981–2008)"	Science and Technology Facilities Council (United Kingdom)
MMK Consulting (2009)	"Economic and Social Impacts of TRIUMF"	TRIUMF (Canada's particle accelerator center) (Canada)
EISS Group (2002)	"ITER at Cadarache: Socio-economic Environment Task SE1 Deliverable June 1, 2001"	ITER (International Thermonuclear Experimental Reactor) (France)
PricewaterhouseCoopers (2016)	"Study to Examine the Socioeconomic Impact of Copernicus in the EU Report on the Socio-economic Impact of the Copernicus Programme"	European Commission—DG GROW (European Union)
Tripp and Grueber (2011)	"Economic Impact of the Human Genome Project"	Life Technologies Foundation (United States)
CBRE CONSULTING (2010)	"Berkeley Lab—Economic Impact Study"	Lawrence Berkeley National Laboratory (United States)
Webb and White (2009)	"Economic impact of the John Innes Centre"	The John Innes Centre (United Kingdom)

Source: Selected references from Giffoni et al. (2018).

the US Department of Agriculture. The IMPLAN model was also used to calculate the return of investment of the HGP to the US economy. Similarly, and very recently, the I-O approach was applied to the European space program Copernicus by using the Cambridge Econometrics' E3ME model for modeling the wider effects on the European economy. Webb and White (2009) use an ad hoc I-O approach for the study of the impact of the John Innes Centre in Norwich, UK, a plant science, genetics, and microbiology research institute; see also Anderson Economic Group (2011) for information on Fermilab.

While the logic of an I-O-based impact multiplier may look similar to estimating a BCR, the main limitation of this accounting approach is that it applies nationwide average I-O coefficients to a specific project. This is, hence, not the empirical estimation of a project-specific impact, but, in fact, it is a simulation of the "what if" type. Moreover, economic output offers poor statistics for economic welfare effects when, as typical of scientific projects, many impacts are outside market transactions.

Epilogue

Research infrastructures (RIs) are a new form of public enterprise. They are *enterprises* because they need to efficiently combine capital, labor, and knowledge with budgetary autonomy and managerial discretion in order to produce a special good: new knowledge in the form of discoveries and inventions, supported by a sustained flow of experimental data, new methods of inquiry, and interpretations of evidence. They are *public* in two ways: because they are funded by governments, and because most of the science they produce is a public good.

The arena where these entities compete is not a market in the conventional sense. While in some cases, they can sell their services for a cost-reflective price, they often set that price to zero or to a modest level, given their public mission. Hence, the way they compete among them is ultimately based on the recognition they achieve within the global scientific communities. There are indeed successes and failures in the business of science, but a comparative metric of achievements is not readable in financial accounts.

Competition for funding is fierce, though. Only a fraction of all possible projects proposed by scientists will be financed by public-sector grants or private donations. Some existing facilities may starve for lack of funds or even be shut down because they have become obsolete. These key decisions lie in the hands of funders.

Managers of RIs face difficult tasks. They must steer, in a cost-effective way, an organization where hierarchies are often loose and always challenged by a special type of workers, with strong opinions and influences—namely, the scientists themselves. Nothing will happen if the scientific community is unconvinced that a project has value. At the same time, managers need to convince funders to ensure adequate resources for scientific work, at the right time, to fulfill the overarching goal of producing science on a large scale and for several direct and indirect users. The resources are often in the hands of a plurality of stakeholders. Thus, different "clients" need to be convinced by good arguments that an RI is valuable.

Mainstream economics is not at ease with the study of such organizations as these. Why would scientists spend so much effort at what they do if they were not motivated by higher pay? How is it possible to be efficient if market prices do not convey critical economic information? Why should taxpayers sacrifice some of their income to support research into questions that they cannot actually understand?

The difficulty of textbook economics to answer these three questions arises from an overly narrow view of what motivates people. From this perspective, RIs are similar to some not-for-profit education and health organizations, where dedicated, expert personnel work hard, keeping in mind a reward that has nothing to do with their income, but only the perception that they have a mission to accomplish. In fact, some career progress and monetary rewards may be attached to a scientific paper published in a blue-ribbon journal, but compared to the time and effort to achieve this, and compared to the pay at alternative jobs elsewhere, it seems that the main reward for scientists is heightened reputation (including the desire to show they are better than their peers), and some psychological satisfaction for the work they do.

Economists are increasingly realizing that intrinsic motivation is crucial to the understanding of the behavior and performance of certain individuals and teams. Hence, we can answer to a certain extent the first question: intrinsic motivation, competition with peers, and reputational mechanisms are the keys.

This book contributes to answering the second question. If the efficiency metric is a comparison of social benefit and costs, within the scientific communities that are managing RIs, there are mechanisms in place to minimize costs and to maximize certain types of benefits. Usually, budget constraints are binding. Governments are not ready to disburse any sum at any time to fulfill any investment project. Hence, managers of an RI know that each cent they save has value. At the same time, they know that the more new knowledge, human capital, technological innovations, outreach, and cultural goods that they produce, the more likely they are to survive and flourish in the competition for funds. In the words of the *Economist* (2013):

Big Science projects differ from companies in important ways. They are publicly financed and do not seek profits.... Yet, like companies, they must innovate furiously, make the most of limited resources and beat rivals to breakthroughs... Their aims are often clear-cut—find the Higgs, sequence the genome, potter around Mars—but the means to attaining them are anything but.

Hence, says the *Economist*, rival teams, or individuals within teams, must be free to make their own proposal or to criticize those of others, to discuss ideas openly, and to challenge authority. Hierarchy must be soft and based on merit and consensus to the greatest possible extent. This is per se a powerful efficiency mechanism.

The third question, however, is much more difficult to answer. In Europe alone, there are perhaps around 300 major RIs and a large number of smaller ones, each of which is funded by citizens who are largely unware of what scientists are looking for with these projects.

What is really surprising is not that scientists do science, but that the remaining 99.9 percent of society is implicitly or explicitly willing to pay for it, even when it takes the form of large-scale, capital-intensive, and in a sense high-risk projects, because by definition, research does not know what it wants to know.

Governments and funding agencies are imperfect mediators of social preferences. Hence, the true question is not why governments support science and RIs, albeit often inadequately, but why citizens apparently support research on neutrinos, black holes, gravitational waves, genes of marine species, and thousands of other objects unknown to laypeople.

From this perspective, the answer implicitly given in this book is twofold: Citizens support the work of scientists, first, because we—laypeople—suspect that something useful will be generated by all this supposedly useless research; and second, because we simply like that we, as humans, know more about the universe, and we are happy that there are experts finding all this out for us.

In the words of Abraham Flexner, the founding director of the Institute for Advanced Study in Princeton, and author of *The Usefulness of Useless Knowledge*, originally published in 1939 (Flexner, 2017, 51–52):

> From a practical point of view, intellectual and spiritual life is, on the surface, a useless form of activity, in which men indulge because they procure for themselves greater satisfaction than are otherwise obtainable. ... I shall concern myself with the question of the extent to which the pursuit of these useless satisfactions proves unexpectedly the source from which undreamed-of utility is derived.

Hence, scientists work for us, even when they think they are working for themselves, to satisfy their own curiosity. What I have tried to show in this book is that to a certain extent, it is possible to measure both the societal use and nonuse value of science when it is produced and organized in the form of large scale RIs. Because RIs are a form of public enterprises, and in a precise sense are the infrastructures of knowledge production, I have tried to analyze their workings through the lenses of social cost-benefit analysis (CBA) of public investment. This admittedly is a partial perspective. The surprising finding is that a machine as costly as the Large Hadron Collider (LHC), entirely devoted to discover things for which we currently have no use, passes a cost-benefit test. This result is obtained under conservative assumptions, and without considering the unknown future use value (if any) of discoveries. At least in this case, as in several others, taxpayers are well served by investing in science.

However, what I also have tried to show is that not all RIs would pass a social cost-benefit test. The difference between rhetorical arguments and empirical analysis is that in the former, you always want to predict a positive outcome, and in the latter, you are not sure until the end. And even then, there is uncertainty. The scientific case may be more important than the socioeconomic impact argument. But I see some progress in a public discourse on science that carefully considers measurable social benefits against costs, albeit with such uncertainty, along with the scientific case.

There is a more general lesson for government in this inquiry. While science is the business of scientists, analyzing its socioeconomic impact usually is not. Scientists should do their job in as unconstrained a fashion as possible. Governments and their policy advisors should do their own jobs, which unavoidably have to do with setting missions and priorities, and these will constrain what scientists can do, at least when they need capital-intensive projects. This tension is unavoidable and should be understood and accepted as such.

Governments have the ultimate responsibility for determining the final distribution of the social benefits of scientific knowledge. Access to data from the Human Genome Project (HGP) is available for free; the data cannot be patented, but new cures for genetic disorders cannot be had for free. The European Organization for Nuclear Research (CERN) has donated the World Wide Web to everybody, but Internet companies have earned huge profits building on it and are now at the top of corporate global rankings by profits and market value. Global positioning system (GPS) services, supported by a publicly funded constellation of satellites, is available for free, but apps in any smartphone use information of the geographic positioning of their users for commercial purposes.

There is perhaps a tension, if not a contradiction, between the public-enterprise nature of RIs and the private appropriation of their ultimate economic benefits by private companies. It is the tension between the selfishness embedded in our current economic arrangement and the generosity and openness of the organization of contemporary science.

Our societies would be better places to live if we could learn lessons from large-scale scientific enterprises. For example, providers of health services and of education, directly or indirectly funded by governments, are often huge organizations with a rigid governance. They attract, however, dedicated staff with intrinsic motivation and the sense of a public mission. Reform of such organizations is often proposed in terms of mimicking business and markets, including with monetary rewards for managers and privatization of assets.

Providers of health, education, and some other knowledge-based public services instead should learn from the emergence of RIs and their characteristics: clear missions, budgetary autonomy, managerial discretion, competition of ideas for solutions among teams and individuals, governance based on merit and bottom-up recognition of leaders, reputational mechanisms, transparency, and openness. Perhaps the RI paradigm points to a new avenue in the governance of knowledge-based public organizations beyond science.

Acknowledgments

The origins of this book are, on the one hand, my long-standing interest in social cost-benefit analysis (CBA) of infrastructure projects and, on the other, a lifelong fascination with science as a public good. Combining both perspectives was challenging, and I needed the help of many colleagues and friends. I am especially grateful to Stefano Forte, Department of Physics, University of Milan, for having co-led the initial CBA of the Large Hadron Collider (LHC), bringing into the project the perspective of a theoretical physicist and his personal interest in the socioeconomic impact of science. He patiently explained to me several aspects of the particle physics research environment. Emanuela Sirtori, economist at the Centre for Industrial Studies (CSIL), was the key coauthor (with Stefano Forte) of our first paper on the socioeconomic impact of the LHC. Chiara Pancotti (CSIL) was the lead author for the National Centre of Oncological Hadrontherapy (CNAO) case study.

Over the years, I have had the opportunity to involve several other colleagues in my research agenda. They particularly include, at the University of Milan, Andrea Bastianin, Paolo Castelnovo, Stefano Clò, Chiara Del Bo, Alfio Ferrara, Carlo Fiorio, Marta Marsilio, Francesco Rentocchini, and Silvia Salini; at CSIL, Gianni Carbonaro, Ugo Finzi, Mario Genco, Francesco Giffoni, Julie Pellegrin, and Silvia Vignetti; at the University of Rome Tre, Anna Giunta; and at Durham University, Riccardo Scarpa.

The scientists and staff at European Organization for Nuclear Research (CERN) were extremely collaborative and helpful. I enjoyed their warm interest in the project, particularly from Fabiola Gianotti (CERN Director General), Michael Benedikt and Johannes Gutleber (Future Circular Collider Study), Frédérick Bordry (Director for Accelerators and Technology), Tiziano Camporesi (Compact Muon Solenoid [CMS]), Peter Jenni (ATLAS), Lucio Rossi (High-Luminosity LHC project leader), Herwig Schopper (Director General Emeritus), Florian Sonneman (Head of Finance), Anders Unnervik

(Procurement and Industrial Services), and Rüdiger Voss (President, European Physical Society). Among the many other CERN and Collaborations experts who offered their advice, I particularly want to thank Giovanni Anelli, Andrzej Charkiewicz, Alfred Daljievec, Gijs De Rijk, Albert De Roeck, Alberto Di Meglio, Fido Dittus, Laure Esteveny, Irene del Rosario Crespo Garrido, Paolo Giacomelli, James Gilles, Anna Godinho, Frederic Hemmer, Efrat Tal Hod, Kate Kahle, Rolf Landua, Olivier Martin, Pere Mato, Axel Naumann, Alberto Pace, Fons Rademakers, Lucia Silvestris, Frank Zimmermann, and several others; with sincere apologies for not citing all of them.

For the other case studies considered in this book, I am most grateful to several experts for their input. For CNAO, Sandro Rossi (General Director), Ugo Amaldi, Giuseppe Battistoni, Erminio Borloni, Alexandra Donika Chiaramonte, Maria Rosaria Fiore, Maria Vittoria Livraga, Paola Mella, Silvia Meneghello, Marco Pullia, and Francesca Valvo. For ALBA, Caterina Biscari (General Director), Miguel Angel Aranda, David Fernández, Gaston Garcia, A. B. Martínez, Ramon Pascual, Francis Pérez, Immaculada Ramos, Alejandro Sánchez, Jose G. Montalvo (Pompeu Fabra University), and Franscesc Subirada (Director DG Research Generalitat de Catalunya). For European Molecular Biology Laboratory–European Bioinformatics Institute (EMBL-EBI), Mary Barlow, Alex Bateman, Cath Brooksbank, Ian Dunham, Adrian Legar, Sarah Morgan, Abel Ureta-Vidal (Eagle Genomics), and Jessica Vamathevan. For ELIXIR, Niklas Blomberg, Corinne Martin, Susanna Repo, and Andrew Smith. For the European Space Agency (ESA), Elisabeth Ackerler, Luca Del Monte, and Alessandra Tassa. For the Agenzia Spaziale Italiana, Simonetta Di Ciaccio, and Danilo Rubini. For the US Department of Energy (DoE), Rick Borchelt, Benjamin Brown, Eric Colby, Howard Gruespecht, Barbara Helland, Stephen Meador, and Jim Siegrist. For the National Institutes of Health (NIH), Diana Finzi and Michael Tartakovsky (CIO NIAID, Office of Cyber Infrastructure and Computational Biology). For the Broad Institute, Giulio Genovese and Marta Florio, my daughter. For the Square Kilometre Array (SKA), Simon Berry.

I have also benefited from conversations on research infrastructure (RI) and CBA with several people in different international organizations. At the European Commission (DG Research, DG Regional and Urban Policy, DG Ecfin, DG Connect), particularly Philippe Froissard, Antonio Di Giulio, Mikel Landabaso, Andrea Mairate, Katja Reppel, Reka Rozsavolgyi, and Witold Willak. At the European Investment Bank (including EIB Institute and JASPERS), José Doramas Jorge-Calderon, Antonella Calvia-Götz, Luisa Ferreira, Mark Mawhinney, Tanja Tanayama, Davide Sartori, Robert Swerdlow, and Dorothea Teichman. At the OECD Global Science Forum, Lucilla Alagna, Jean Moulin, and Frédéric Sgard.

Other experts from whom the research project has benefited in various ways include Scott Farrow (University of Maryland, Baltimore County), Massimo Ferrario (INFN), Andrew Harrison and Isabelle Boscaro-Clarke (Diamond Light Source), Diana Hicks (Georgia Institute of Technology), Per-Olov Johansson (Stockholm School of Economics), Henning Kroll (Fraunhofer Institute for Systems and Innovation Research), Yusuke Kuwayama (Resources for the Future), Sylvie Niessen (French Ministry of Education and Research—Department of Research Infrastructures), Emile Quinet (Paris School of Economics), and Giorgio Rossi (European Strategy Forum on Research Infrastructures [ESFRI] and University of Milan).

I have also benefited from questions and comments by participants in several events where some results of the research and related topics were presented, in Barcelona, Brussels, Durbach, Geneva, Luxembourg, Madrid, Milan, Paris, Rome, Washington, D.C., and elsewhere.

In the last several years, I have been involved in various projects, which have been instrumental in financially supporting my research. These include 2013–2015, the research project "Cost/Benefit Analysis in the Research, Development, and Innovation Sector," sponsored by the European Investment Bank University Research Sponsorship programme (EIBURS); 2014, European Commission "Guide to Cost-Benefit Analysis of Major Projects in the Context of EU Regional Policies"; 2014–2017, Jean Monnet Network "Services of General Interest in the EU: a Citizens' Perspective on Public Versus Private Provision (EUsers)"; 2016–2019, a study on the socioeconomic impact assessments of post-LHC scenarios in the framework of the FCC study, a collaboration agreement between CERN and the University of Milan (ref. KE3044/ATS); a special contribution from the University of Milan; 2018–2020, Horizon 2020 project "Charting Impact Pathways of Investment in Research Infrastructures"; 2018–2020, a study of socioeconomic impacts of public interventions in the spatial sector, a collaboration agreement between the University of Milan and the Italian Space Agency.

This book partly draws from several previously published preprints and articles. I am grateful to the publishers, coauthors, and research team members for their permission to use this material, particularly articles published in the following journals: *CERN Courier, European Journal of Physics, Industrial and Corporate Change, Journal of Economic Policy Reform, Nuclear Instruments and Method in Physics Research-A, Research Policy*, and *Technological Forecasting and Social Change*. These articles include Battistoni et al. (2016); Camporesi et al. (2017); Carrazza et al. (2016); Castelnovo et al. (2018); Catalano et al. (2018); Del Bo (2016); Florio, Forte, and Sirtori (2016); Florio et al. (2016); Florio and Sirtori (2016); Bastianin and Florio (2018); Bastianin,

Florio, and Giffoni (2018); Florio, Bastianin, and Castelnovo (2018); Florio, Giffoni, and Catalano (2018); and Florio et al. (2018).

Several people have agreed to read and comment on earlier drafts, and their constructive criticism was very helpful: Three anonymous reviewers appointed by the publisher for the review of the initial project; Peter Jenni and Rüdiger Voss, appointed by CERN DG; and several others on specific chapters: S. Carrazza, A. Di Meglio, A. Godinho, J. Gutleber, F. Hemmer, L. Silvestris, G. Catalano, F. Giffoni, C. Pancotti, S. Vignetti, E. Sirtori, S. Forte, F. Rentocchini, S. Salini, C. Martin, A. Smith, M. Barlow, P. Markova, A. L. Wegener, S. Berry, A. Tassa, C. Biscari, G. Garcia, D. Finzi, and Marta Florio. Emily Taber, editor at MIT Press, offered her very helpful advice at several stages of the project and smoothly and competently managed the submission and review process.

Jane Pennington has revised the English of the initial draft, and Marinella Manghina and Martina Gazzo contributed to the text editing. I am grateful to Susan McClung, copyeditor, and Helen Wheeler, Westchester Publishing Services. Finally, this book has enormously benefited from the research assistance and personal contribution of Valentina Morretta, postdoc fellow of the University of Milan, who has helped me chapter by chapter, including statistical analysis and extensive bibliographical research. At the same time, Gelsomina Catalano has effectively managed the coordination of several team activities, including data collection and surveys. More than 3,000 people kindly accepted to participate in personal interviews or online surveys, including firm managers, early career researchers, students, team leaders, scientists, and taxpayers.

Notes

Introduction

1. See Galison and Hevly (1992), Gosling (1999), Hughes (2002).
2. See Martin and Irvine (1984), Hallonsten (2016).
3. https://www.nasa.gov/vision/space/features/jfk_speech_text.html, accessed on January 5, 2018.
4. https://history.fnal.gov/testimony.html, accessed on January 5, 2018.
5. Environmental cleanup refers particularly to radioactive waste. See DoE (2017b, 1).
6. Beyond nuclear energy.
7. The DoE manages its own specific security clearance process; see Berry (2015) at http://www.securityclearanceblog.com/2015/07/the-department-of-energy-security-clearance-process.html, accessed on January 31, 2018.
8. https://www.nih.gov/about-nih/who-we-are/history, accessed on January 31, 2018.
9. https://www.mpg.de/en, accessed on April 4, 2018.
10. https://www.helmholtz.de/en/, accessed on April 4, 2018.
11. https://www.nature.com/scitable/topicpage/dna-sequencing-technologies-key-to-the-human-828, accessed on January 31, 2018.
12. https://www.genome.gov/sequencingcosts/, accessed on January 31, 2018.
13. See Chalmers (2013) for the origin of the idea.
14. https://clinicaltrials.gov/, accessed on January 31, 2018.
15. http://vaccineenterprise.org/.
16. See Malone and Bernstein (2015), for definitions and examples of collective intelligence.
17. https://www.popsci.com/science/article/2011-07/supersized-10-most-awe-inspiring-projects-universe#page-6, accessed on February 2, 2018.
18. https://www.scientificamerican.com/article/how-big-is-science/, accessed on June 14, 2018.
19. An update list has been released in 2018: http://roadmap2018.esfri.eu/, accessed on September 25, 2018.
20. For the RI road map in Germany, see BMBF (2013).
21. http://cache.media.enseignementsup-recherche.gouv.fr/file/Infrastructures_de_recherche/74/5/feuille_route_infrastructures_recherche_2016_555745.pdf, accessed on June 14, 2018.
22. https://www.education.gov.au/2016-national-research-infrastructure-roadmap, accessed on February 1, 2018.
23. See Giudice (2010), Baggott (2012), and Maiani and Bassoli (2013). The definitive study on the SSC failure is the impressive book by Riordan, Hoddeson, and Kolb (2015).

24. *RHIC* refers to the Relativistic Heavy Ion Collider.

25. See, for example, the *Journal of Benefit Cost Analysis* and the activities of the Society for Benefit Cost Analysis at https://benefitcostanalysis.org/, accessed on April 9, 2018.

Chapter 1

1. See, for example, Griliches (1980), Adams (1990), Romer (1990), and Barro and Sala-i-Martin (2004).

2. These figures and those given later in this chapter are at purchasing power parity (PPP).

3. See, e.g., Baum and Tolbert (1985).

4. See, e.g., Viscusi and Aldy (2003) and other references in the "Further Reading" section later in this chapter.

5. I'm grateful to Diana Hicks for suggesting this example. According to the *Maritime Executive 2019* (https://www.maritime-executive.com/article/u-s-navy-opts-for-block-buy-for-two-ford -class-carriers, accessed on March 19, 2019), "at $13 billion in construction costs, not including $5 billion for R&D, the USS *Ford* is the most expensive self-propelled ship ever built."

6. The European Social Survey is a network established to develop, store, and study long time series of data used to monitor and interpret changes in European social attitudes and behavior patterns.

7. https://fcc-cdr.web.cern.ch/.

8. See National Human Genome Research Institute (https://www.genome.gov/12011239/a-brief -history-of-the-human-genome-project/, accessed on April 10, 2018).

9. Projects aimed at tackling climate change are an extreme example. See the *Stern Review* (HM Treasury, 2006).

10. An externality can also be a cost, for example: pollution.

11. A comparative assessment of the advantages and disadvantages of the CERN Large Electron-Positron (LEP) collider was conducted by Martin and Irvine (1984). This exercise shows that even very large and cutting-edge accelerators might have some rival projects.

12. Other examples include seismographic stations.

13. See Chapter 8 in Florio (2014).

14. See Torrance (1986).

15. See Florio (2014) for a discussion of this topic in the CBA context.

16. See Boadway (2006).

17. See De Rus (2010).

18. In a partial equilibrium frame, this is equivalent to the producer surplus (the difference between price and marginal costs).

19. This is represented by the consumer surplus, the difference between WTP and price, when looking at a specific market.

20. Florio (2014) discusses in detail the empirical issues involved.

21. See https://www.atomicheritage.org/history/german-atomic-bomb-project, accessed March 19, 2019.

22. See A. Einstein, "On My Participation in the Atom Bomb Project" (http://www.atomicarchive .com/Docs/Hiroshima/EinsteinResponse.shtml, accessed on February 6, 2018).

23. For example, scientists sometimes pay submission fees or charges to be published Open Access, e.g., $5200 at Cell, see https://www.cell.com/cell-systems/authors, accessed March 19, 2019.

24. See "System of National Accounts 2008" (http://unstats.un.org/unsd/nationalaccount/sna2008 .asp, accessed on April 11, 218).

25. There were 7.8 million researchers in 2013, according to UNESCO (https://en.unesco.org /unesco_science_report/figures).

26. See Krutilla (1967).

27. This includes reports, documents, preprints, working papers, and other materials published outside scholarly publication outlets.

28. See BBMRI-ERIC (2016): http://www.bbmri-eric.eu/ (accessed on June 14, 2018). "Performance indicators for BBMRI-ERIC v 1.2."

Chapter 2

1. http://www.esa.int/Our_Activities/Human_Spaceflight/International_Space_Station/How _much_does_it_cost, accessed on February 8, 2018.

2. See the "Further Reading" section about the details of ESFRI accounting conventions, under revision in 2018.

3. https://www.accountingtools.com/articles/what-is-a-sunk-cost.html, accessed on June 3, 2018.

4. For a further discussion of the issue, see Belli et al. (2001), European Investment Bank (2013), European Commission (2014), and Florio (2014).

5. See OECD (2014a) to explore further the topic of internationally distributed RIs.

6. https://www.ligo.caltech.edu/page/about, accessed on February 8, 2018.

7. https://www.skatelescope.org/project/, accessed on April 16, 2018.

8. https://www.theguardian.com/science/across-the-universe/2017/apr/21/space-debris-must-be -removed-from-orbit-says-european-space-agency, accessed on June 3, 2018.

9. https://www.eiroforum.org, accessed on April 16, 2018.

10. https://press.cern/facts-and-figures/budget-overview, accessed on April 16, 2018.

11. RAMIRI online handbook, http://www.ramiri-blog.eu/, accessed on April 18, 2018.

12. As mentioned, the Future Circular Collider study at CERN started in 2014 (see https://fcc .web.cern.ch/Pages/default.aspx) for an infrastructure proposed to follow the LHC around 2040.

13. The condition is that the net benefits do not change sign during the project's life (e.g., because of high decommissioning costs). Otherwise, more than one FIRR value may make the NPV equal zero. Additionally, FIRR cannot be used when time-varying discount rates are used.

14. In contrast, economic performance should be positive; see chapter 10.

15. See Wright and Robbie (1996); Manigart et al. (1997).

16. See www.capital-investment.co.uk, accessed on April 17, 2018.

17. For example, Brigham, Gapenski, and Daves (1999).

18. For a review, see Manigart et al. (2000), which examined the valuation methods used by venture capitalists in the United States, Great Britain, France, Belgium, and the Netherlands.

19. Only upgrade costs related to the so-called Phase 1 have been considered, these being sustained to optimize the physics potential of LHC experiments. The High-Luminosity LHC upgrade is not considered.

20. Current CHF values have first been accounted in constant CHF 2013 by considering the yearly change of average consumption prices from the IMF World Economic Outlook (October 2013); then expressed in euros at the exchange rate 1 CHF = 0.812 EUR (European Central Bank average of daily rates for the year 2013).

21. A sensitivity analysis of the impact of apportioning a higher share of overhead costs to LHC shows that the economic NPV (see chapter 10) remains positive up to a 75 percent share attributed to LHC, without changing any other hypothesis.

22. These mainly take the form of equipment made available for free to CERN by third parties and for which, in Annual Accounts (Financial Statements) 2008 (CERN/2840 CERN/FC/5337),

a cumulative asset value of CHF 1.47 billion is recorded, combining in-kind contribution to the LHC machine and the detectors. The attribution year by year of this cumulative figure has been done, assuming the same trend as for CERN procurement expenditures.

23. Based on the Draft Medium Term Plan 2014 (personal communication, April 2, 2014). Again, I have implemented an apportionment to LHC of each expenditure item. As all values were given in constant CHF 2014, these were first converted to CHF 2013 and then future values discounted to 2013 levels by the 0.03 rate.

24. For details on the written sources of the data, see Florio, Forte, and Sirtori (2016), where the CERN and Collaborations documents are listed.

25. As mentioned in the previous section, the team has not included decommissioning costs as no reliable information was available. For the same reason, there are no forecasts of accidents or negative externalities.

26. https://cds.cern.ch/record/2255762/files/CERN-Brochure-2017–002-Eng.pdf, accessed on June 3, 2018.

27. https://indico.cern.ch/event/417101/contributions/1878771/attachments/859917/1202641 /Rossi.pdf.

28. An update list has been released in 2018: http://roadmap2018.esfri.eu/, accessed on September 25, 2018.

29. www.riportal.eu, accessed on April 17, 2018. See also the MERIL portal https://portal.meril .eu/meril/.

30. http://www.era-instruments.eu/downloads/recommendations_3.pdf, accessed on June 14, 2018.

Chapter 3

1. See Strachey (1948) for a famous biographical sketch.

2. http://www.lbl.gov/laboratory-organization-chart/, accessed on February 16, 2018.

3. AT&T Inc. is an American multinational conglomerate holding company, headquartered in Dallas. For the history of the Bell Labs, see https://www.bell-labs.com/about/history-bell-labs/ and Gerther (2013).

4. At the LHC, something like a million gigabytes per second of information is produced, "sufficient to saturate every hard disk of the planet in about a day" (Giudice 2010, 135).

5. For example, Pinski and Narin (1976) and Martin (1996).

6. Ashby 2003.

7. In the case of perfect competition, marginal WTP and MPC converge to the market price equilibrium, the point where, in a standard undergraduate microeconomics textbook, the supply curve crosses the demand curve.

8. European Commission (2014), and Johansson and Kriström (2015).

9. https://unstats.un.org/unsd/nationalaccount/docs/sna2008.pdf, accessed on June 28, 2018.

10. Skilled labor in CBA is sometimes given a high shadow wage relative to observed wage, if labor markets are distorted. I discuss in chapter 4 the issue of scientists' wage.

11. These techniques are discussed by Carrazza, Ferrara, and Salini (2016).

12. These issues are discussed by (Carrazza, Ferrara, and Salini 2014).

13. https://clarivate.com.

14. http://esi.incites.thomsonreuters.com/IndicatorsAction.action?Init=Yes&SrcApp=IC2LS&SID =H5-K0r3yJdKUIycW9a9hYMuBOmUkEgVPBdf-18x2dQIznv4FLYI7soix2BefrHXeAx3Dx3DO YJAx2FPaxxcYVFvKcHUBQ9QAx3Dx3D-9vvmzcndpRgQCGPd1c2qPQx3Dx3D-wx2BJQh9G KVmtdJw3700KssQx3Dx3D, accessed on June 13, 2018.

15. See chapter 5, particularly section 5.4.

16. Hagström 1965; de Solla Price 1970.

17. These techniques are discussed in Carrazza, Ferrara, and Salini (2014, 2016).

18. See Hicks 2004; Nederhof 2006.

19. For a review of the literature on scientists' research productivity, see IVA (2012).

20. The *h-index* (Hirsch 2005) aims at providing a robust single-number metric of an academic's impact by combining influence with quantity. A scientist has an *h-index* equal to N if his or her papers have at least N citations each and the other papers have no more than N citations each. Hence, a scientist with an *h-index* of 20 has produced twenty articles with at least twenty citations each.

21. See Simkin and Roychowdhury (2007); Broadus (1983).

22. See, for instance, Batista, Campiteli, and Kinouchi (2006) or the alternative provided by the Publish or Perish software program.

23. See, for instance, Iglesias and Pecharromán (2007), Radicchi, Fortunato, and Castellano (2008), and Malesios and Psarakis (2012).

24. See, for instance, Sidiropoulos, Katsaros, and Manolopoulos (2007).

25. Citations' skewness was identified early on by de Solla Price (1965).

26. See Moed et al. (1985).

27. See King 1987.

28. For simplicity, the median time needed to download, read someone else's paper, and decide to cite it can be set equal to one hour.

29. http://inspirehep.net, accessed on April 19, 2018.

30. http://science.thomsonreuters.com/products/esi, accessed on February 19, 2018.

31. https://bibliotek.usn.no/publication-points/category26628.html, accessed on February 20, 2018.

Chapter 4

1. https://www.uef.fi/documents/10184/39908/UEF+Research+Infrastructure+Programme.pdf /4a38104f-f12e-4feb-a085-6f382a04531c, accessed on June 5, 2018.

2. https://www.lhc-closer.es/taking_a_closer_look_at_lhc/0.lhc_trigger, accessed on April 24, 2018.

3. See Aad et al. (2012).

4. "CERN Impact on People's Careers," presented at CERN Council Open Session on December 15, 2018. https://indico.cern.ch/event/681590/contributions/2792974/attachments/1576283 /2489237/careers_cern_council.pdf, accessed on June 13, 2018.

5. https://alumni.web.cern.ch accessed on March 1, 2018.

6. Fig. 4.1 draws from an extended more recent survey; see Giacomelli (2019).

7. https://www.embl.de/training/eipp/mission/mission-full/index.html, accessed on June 5, 2018.

8. EMBL (2016) and https://www.embl.org/, accessed on June 20, 2018.

9. See www.ramiri-blog.eu., accessed on March 1, 2018.

10. http://www.ramiri.eu/, accessed on June 20, 2019.

11. A very selective sample of such papers includes Schultz (1961), Mincer (1974), Psacharopoulos and Patrinos (2004), and Blaug (1987).

12. See DoE (2017a, 31).

13. See Kolb and Fry (1974).

14. See also Kolb (2015) for a review and update of the approach.

15. https://cos.northeastern.edu/biology/experiential-learning/, accessed on April 23, 2018.

16. See Rodrigues et al. (2007); Sharma et al. (2007); O'Byrne et al. (2008); Institute of Physics (2012); and Nielsen (2014).

17. See Choi et al. (2011).

18. See Islam et al. (2015).

19. See Shelley (1994).

20. See Keaveny and Inderrieden (2000) and Fernandez-Mateo (2009).

21. See, for reviews, Psacharopoulos (1994), Psacharopoulos (1995), Psacharopoulos and Patrinos (2004), and Heckman, Lance, and Todd (2006).

22. See Blöndal, Field, and Girouard (2002), Boarini and Strauss (2007).

23. See Wooldridge (2010).

24. This means the European Union member-states except Malta, plus Iceland, Norway, Switzerland, and Turkey.

25. https://cds.cern.ch/collection/CERN%20Annual%20Personnel%20Statistics?ln=en, accessed on April 23, 2018.

26. See the "Further Reading" section.

27. See, for example, http://www.payscale.com/research/US/Job=Electronics Engineer/Salary, accessed on March 7, 2018.

28. With regard to CERN technical students, it was assumed that only 10 percent will go either to research centers or academia and 45 percent to industry and other areas. With regard to the other students, it was assumed destination in research and academia that 60 percent will go in research and academia and 20 percent will go in industry and other areas. Interviews with experts (including headhunters who regularly monitor CERN students) have confirmed this distribution.

29. See http://www.oecd.org/pensions.

30. The survey was designed and managed by the University of Milan research group in the framework of the Future Circular Collider (FCC) Study and was implemented online between March and May 2018. See Catalano et al. (2018).

31. https://www.ons.gov.uk/employmentandlabourmarket/peopleinwork/earningsandworkinghours /bulletins/annualsurveyofhoursandearnings/2014-11-19#tab-Key-points, accessed March 21, 2019.

32. http://www.aps.org/careers/statistics/index.cfm; see also https://www.aip.org/statistics accessed on March 7, 2018.

33. https://www.payscale.com/, accessed on June 18, 2018.

34. https://www.academics.com accessed on March 7, 2018.

35. https://www.payscale.com/research/US/Job=Physicist/Skill, accessed on March 7, 2018.

36. This was part of the Snowmass Young Physicists project.

37. © European Physical Society. Reproduced by permission of IOP Publishing. All rights reserved.

38. See Schweitzer et al. (2014); Hogue, Dubois, and Fox-Cardamone (2010); and Ng and Wiesner (2007).

39. See Hazari et al. (2010); Institute of Physics (2012); and Lissoni et al. (2011).

40. See Islam et al. (2015); Jusoh, Simun, and Choy Chong (2011); and Shelley (1994).

41. See Schweitzer et al. (2014).

42. See Catalano et al. (2018).

Chapter 5

1. See Zuniga and Correa (2013).

2. See Mazzucato (2018).

3. Some sections of the chapter draw from Florio et al. (2017); Castelnovo et al. (2018); Florio, Forte, and Sirtori (2016); Florio et al. (2016); and Florio et al. (2018).

4. See Crépon, Duguet, and Mairesse (1998).

5. See, among others, Scherer (1965), Schmookler (1966), and Hall, Jaffe, and Trajtenberg (2005).

6. See SQW (2008).

7. See Belussi and Caldari (2009) for the origin and development of the Marshallian district concept.

8. See Griliches (1979), Romer (1990), and Foray (2004).

9. This means that every additional investment of capital and labor yields more than proportionate returns.

10. See also Nemet (2012) on wind power projects.

11. See Stokey (1988), for one.

12. See Edquist and Zabala-Iturriagagoitia (2012), Håkansson et al. (2009), and Phillips et al. (2006).

13. See Edquist and Zabala-Iturriagagoitia (2012).

14. See Cano-Kollmann, Hamilton, and Mudambi (2017); Edquist (2011); Rothwell (1994); and Lundvall (1993, 1985).

15. See Cohen and Levinthal (1990).

16. See also Bacchiocchi and Montobbio (2009).

17. See Acs, Audretsch, and Feldman (1992) and Feldman and Florida (1994).

18. https://ec.europa.eu/digital-single-market/en/public-procurement-innovative-solutions, accessed on May 2, 2018.

19. See OECD (2017a).

20. See Newcombe (1999) and Mowery and Rosenberg (1979).

21. See Unnervik (2009).

22. http://science.howstuffworks.com/innovation/inventions/top-5-nasa-inventions.htm#page=1, accessed on March 16, 2018.

23. See "The Economic Impact of the US Space Program," https://er.jsc.nasa.gov/seh/economics .html.

24. See, for instance, Pearce, Atkinson, and Mourato (2006); the Asian Development Bank (2013); and Florio (2014).

25. See Liyanage and Boisot (2011).

26. For example. Mansfield et al. (1977); Hall, Mairesse, and Mohnen (2009).

27. Difference-in-difference, discontinuity design, matching approach, etc.

28. See, for example, Mouqué (2012).

29. See Narin and Olivastro (1992).

30. See Jaffe, Trajtenberg, and Henderson (1993); Foray (2004); and Deng (2008).

31. See, for instance, Caballero and Jaffe (1993); Trajtenberg, Henderson, and Jaffe (1997); Jaffe and Trajtenberg (1999); and Squicciarini, Dernis, and Criscuolo (2013).

32. See Gupta et al. (2010) and Pearl (2017).

33. See European Commission (2006).

34. See Squicciarini, Dernis, and Criscuolo (2013).

35. See Pakes (1986), Schankerman and Pakes (1986), Schankerman (1998), and Lanjouw (1998).

36. See Serrano (2008), Sneed and Johnson (2009), Leone and Oriani (2008), and Sakakibara (2010).

37. See Harhoff et al. (1999); Harhoff, Scherer, and Vopel (2003a, 2003b); and Gambardella, Harhoff, and Verspagen (2008)

38. See Griliches (1981); Cockburn and Griliches (1988); Hall, Jaffe, and Trajtenberg (2005); and Hall, Thoma, and Torrisi (2007).

39. See Lerner (1994) and Hsu and Ziedonis (2008).

40. See European Commission (2006).

41. The considered countries are Denmark, France, Germany, Hungary, Italy, the Netherlands, Spain, and the United Kingdom.

42. See the NASA online spinoff database, available at http://spinoff.nasa.gov/spinoff/database, accessed on March 16, 2018.

43. https://dictionary.cambridge.org/dictionary/english/spin-off, accessed on May 2, 2018.

44. https://definitions.uslegal.com/s/spin-off/, accessed on May 2, 2018.

45. Some possible sources of information include Eurostat Business Demography Statistics (2012), European Commission (2011, 2013), and European Investment Bank (2013).

46. See Schmied (1977); Bianchi-Streit et al. (1984); Autio, Bianchi-Streit, and Hameri (2003); Autio, Hameri, and Vuola (2004); and Autio (2014).

47. See Unnervik (2009) and Autio, Bianchi-Streit, and Hameri (2003)

48. https://orbis.bvdinfo.com, accessed on May 2, 2018.

49. Data on procurement commitment by country provided by CERN staff, October 2013.

50. Source of data: World Bank (http://databank.worldbank.org/data/home.aspx, accessed on March 16, 2018).

51. See Autio, Hameri, and Vuola (2004) and Nordberg, Campbell, and Verbeke (2003).

52. See, among others, Terleckyj (1974, 1980).

53. See, for example, Bernstein (1988) and Bernstein and Nadiri (1991).

54. See also Andersen and Åberg (2017).

55. This includes Diamond Synchrotron [Diamond Light Source Ltd., funded by the Science and Technology Facilities Council (STFC) at 86 percent and by the Wellcome Trust at 14 percent]; Royal Research Ship James Cook (funded by the Natural Environment Research Council); ISIS Neutron Source, Second Target Station (STFC); Accelerators and Lasers in Combined Experiments (ALICE), formerly known as Energy Recovery Linac Prototype (funded by STFC); Halley VI Antarctic Research Station (funded by Natural Environment Research Council); High-End Computing Terascale Resource, HECToR (funded by Engineering and Physical Sciences Research Council and operated by STFC Daresbury Laboratory); Muon Ionisation Cooling Experiment (MICE), hosted by STFC.

56. Taking log-transformations of patent count is a common practice in the literature investigating the determinants of patenting: see, for example, Rong, Wu, and Boeing (2017), Li X. (2012), and Aghion et al. (2013).

Chapter 6

1. https://www.facebook.com/broadinstitute/videos/2083299018376884/, accessed on May 7, 2018.

2. https://www.broadinstitute.org/news/broad-institute-sequences-its-100000th-whole-human-genome-national-dna-day, accessed on May 7, 2018.

3. https://webfoundation.org/about/vision/history-of-the-web/, accessed on May 7, 2018.

4. See Hoffmann, Nordberg, and Boisot (2011).

5. See Giudice (2010).

6. http://highscalability.com/blog/2012/9/11/how-big-is-a-petabyte-exabyte-zettabyte-or-a-yottabyte.html, accessed on May 7, 2018.

7. https://blogs.cisco.com/sp/the-zettabyte-era-officially-begins-how-much-is-that, accessed on May 7, 2018.

8. http://www.dtc.umn.edu/mints/home.php, accessed on May 7, 2018.

9. https://code.facebook.com/posts/229861827208629/scaling-the-facebook-data-warehouse-to -300-pb/, accessed on May 7, 2018.

10. https://home.cern/about/computing, accessed on May 7, 2018.

11. https://root.cern.ch/drupal/content/download-statistics.

12. http://geant4.web.cern.ch/geant4/license/, accessed on March 23, 2018.

13. This is my main source for the rest of this section.

14. http://cerncourier.com/cws/article/cern/69615, accessed on May 7, 2018.

15. This important point was suggested by Johannes Gutleber, CERN, along with many other comments on a previous version of this chapter.

16. https://webfoundation.org/about/vision/history-of-the-web/, accessed on May 7, 2018.

17. A primer about genome sequencing is available in the National Human Genome Research Institute website, which is part of the NIH (https://www.genome.gov/27565109/the-cost-of -sequencing-a-human-genome/), accessed on May 7, 2018.

18. http://web.ornl.gov/sci/techresources/Human_Genome/research/bermuda.shtml#1, accessed on May 7, 2018.

19. Subscribed by the Wellcome Trust, the UK Medical Research Council, the NIH National Center for Human Genome Research, the US Department of Energy (DoE), the German Human Genome Programme, the European Commission, Human Genome Organisation (HUGO), and the Human Genome Project of Japan.

20. https://www.embl.de/aboutus/general_information/organisation/member_states/, accessed on June 23, 2018.

21. https://www.ebi.ac.uk, accessed on March 23, 2018.

22. https://code.facebook.com/posts/229861827208629/scaling-the-facebook-data-warehouse-to -300-pb/, accessed on 23, 2018.

23. https://www.elixir-europe.org/, accessed on May 25, 2018.

24. In 2014, EMBL Council established ELIXIR's legal framework (EMBL/2013/16/Rev1), See the organization's annual implementation report (ELIXIR, 2016). See also https://cordis.europa .eu/result/rcn/144033_en.html, accessed on May 25, 2018.

25. See https://www.elixir-europe.org/about-us/why-needed, accessed on May 25, 2018.

26. EMBL is the ELIXIR member, while EMBL-EBI is an ELIXIR node.

27. To explore ELIXIR's nodes, see https://www.elixir-europe.org/about-us/who-we-are/nodes, accessed on May 25, 2018.

28. https://www.elixir-europe.org/platforms, accessed on May 25, 2018.

29. https://www.elixir-europe.org/use-cases, accessed on May 25, 2018.

30. https://drive.google.com/file/d/1NZJHfQA70YWXqLf7yc9f66iIPCedu21b/view

31. This is the European Genome-Phenome Archive.

32. For the full list, see https://www.elixir-europe.org/platforms/data/core-data-resources, accessed on May 25, 2018.

33. www.proteinatlas.org, accessed on May 25, 2018.

34. See the ELIXIR report "Public Data Resources as a Business Model for SMEs," available at https://drive.google.com/file/d/1Xhi7X-XeJqi0JDfYM_50VhlXgXPgAX0q/view, accessed on May 25, 2018.

35. Patterns of database citation in articles and patents indicate long-term scientific and indus- try value of biological data resources (https://f1000research.com/articles/5–160/v1, accessed on May 25, 2018).

36. See also https://drive.google.com/file/d/0B8in1NtGRloYcFNLSFpZWWthYzg/view, accessed on May 25, 2018.

37. https://opensource.org/osd, accessed on March 19, 2019.

38. https://report.nih.gov/NIHfactsheets/ViewFactSheet.aspx?csid=45, accessed on March 23, 2018.

39. https://www.ebi.ac.uk/training, accessed on March 23, 2018.

40. For other examples of open-source hardware, see https://opensource.com/resources/what -open-hardware, accessed on March 19, 2019.

41. https://www.lsst.org, accessed on March 23, 2018.

42. https://www.jwst.nasa.gov, accessed on March 23, 2018.

43. http://worldwidetelescope.org, accessed on March 23, 2018.

44. www.ivoa.net, accessed on March 23, 2018.

45. http://www.unoosa.org/oosa/en/ourwork/psa/schedule/2017/workshop_italy_openuniverse .html, accessed on March 23, 2018.

46. https://www.genome.gov/10001688/international-hapmap-project/, accessed on March 19, 2019.

47. https://www.broadinstitute.org/genomics, accessed on March 23, 2018.

48. https://software.broadinstitute.org/gatk/, accessed on May 7, 2018.

49. https://www.broadinstitute.org/data-sciences-platform, accessed on March 23, 2018.

50. https://www.humanbrainproject.eu/en/about/governance/framework-partnership-agreement/, accessed on March 23, 2018.

51. https://www.nih.gov/about-nih/what-we-do/nih-almanac/national-institute-allergy-infectious -diseases-niaid, accessed on May 7, 2018.

52. https://www.niaid.nih.gov/research/bioinformatics-resource-centers, accessed on May 7, 2018.

53. NASA Technology Transfer Program: Hierarchical Image Segmentation; available at https:// technology.nasa.gov//t2media/tops/pdf/GSC-TOPS-14.pdf, accessed on May 25, 2018.

54. In particular, see the chapters "The Knowledge Commons Framework" in Strandburg, Frischmann, and Madison (2017); "Leviathan in the Commons: Biomedical Data and the State," in Jorge L. Contreras; "Centralization, Fragmentation, and Replication in the Genomic Data Commons," by Peter Lee; and "Genomic Data Commons," by Barbara J. Evans.

Chapter 7

1. For C-ion, see https://www.ptcog.ch/index.php/facilities-in-operation, accessed on May 6, 2019.

2. http://www.esrf.eu/files/live/sites/www/files/UsersAndScience/UserGuide/Applying/Advice _on_writing_a_good_proposal.pdf, accessed on May 3, 2018.

3. http://www.esrf.eu/, accessed on May 11, 2018.

4. https://www.cells.es/en/users/applying-for-beam-time, accessed on May 29, 2018.

5. http://www.esrf.eu/home/UsersAndScience/Publications/Highlights/highlights-2015/industrial -research.html, accessed on May 3, 2018.

6. https://als.lbl.gov, accessed on May 11, 2018.

7. https://www.cells.es/en/about/2018-rates-of-utilization, accessed on May 29, 2018.

8. https://lightsources.org/lightsources-of-the-world/, accessed March 19, 2019.

9. It was recently announced that the High-Luminosity LHC will use a superconducting cable (120,000 amperes, based on MgB2) studied by ASG Superconductors; see http://www .asgsuperconductors.com, accessed July 2018.

10. www.copernicus.eu, accessed on June 23, 2018.

11. www.emergency.copernicus.eu, accessed on June 23, 2018.

12. See Brent (2006).

13. The database is available at www.bvdinfo.com/en-gb/home.

14. This includes manufacture of basic metals (24); manufacture of computer, electronic, and optical products (26); manufacture of electrical equipment (27); manufacture of machinery and other equipment (28); telecommunications (61); and computer programming, consultancy, and related activities (62).

15. See Fokas et al. (2009), Loeffler and Durante (2013), and Tsujii et al. (2014).

16. See Karnofsky and Burchenal (1949).

17. See Landefeld and Seskin (1982); Viscusi and Aldy (2003); Viscusi (2018); and Abelson (2003).

18. See Landefeld and Seskin (1982).

19. http://www.copernicus.eu/sites/default/files/documents/Copernicus_Programme_v2.pdf.

20. This is based on on-site interviews (April 2018), personal communications, and other materials available at www.cells.es.

21. See NAO (2016), figure 10.

22. http://www.graphene.manchester.ac.uk, accessed on May 3, 2018.

23. This list is not intended to be exhaustive.

24. This does not hold true for projects within the EU Emissions Trading System (ETS).

25. Examples of pollutants are nitrogen oxide (NOx), sulfur dioxide (SO_2), particulate matter (PM10 and PM2.5), nonmethane volatile organic compounds (NMVOC) as a precursor of ozone (O_3), and ammonia (NH_3).

26. See European Commission (2004).

27. See the Clean Air for Europe (CAFE) Programme, at https://www.eea.europa.eu/themes/air /links/research-projects/clean-air-for-europe-programme-cafe, accessed on June 25, 2018.

28. See New Energy Externalities Development for Sustainability (http://www.needs-project.org/, accessed on June 25, 2018).

29. www.externe.info, http://www.externe.info/externe_d7/.

30. The estimation of this benefit involves two main sources of difficulties. First, predicting when an actual disaster will occur and at what intensity is not always possible. Second, the effectiveness of the investments is estimated through vulnerability assessments that include a degree of uncertainty. Therefore, the avoided damages are probabilistic, at best.

31. Some examples of existence values retrieved from the literature are provided in chapter 9.

32. Alternatively, a project is carried out within the RI.

33. This section draws from Clò, Florio, and Morretta (2019).

34. This section draws from Florio, Bastianin, and Castelnovo (2017).

35. See Panofsky (1997).

36. See Chernyaev and Varzar (2014).

Chapter 8

1. https://www.kennedyspacecenter.com/explore-attractions/behind-the-gates/fly-with-an -astronaut, accessed on May 21, 2018.

2. https://www.nasa.gov/sites/default/files/files/Taylor-Communication-Strategy-NAC-ATaylor -072015_v8-TAGGED2.pdf, accessed on May 21, 2018.

3. See CERN (2017a, 6).

4. See CERN (2011, 2).

5. https://en.wikipedia.org/wiki/Kim_Kardashian, accessed on March 19, 2019.

6. For example, the EXPO 2015 in Milan, with an initial forecast of 20 million visitors and an estimated number of around 21 million. Ansa (www.ansa.it), retrieved on October 29, 2015.

7. For example, see Schopper (2009).

8. See, for instance, Clawson and Knetsch (1966); Caulkins, Bishop, and Bouwes (1986); Cuccia and Signorello (2000); and Bedate, Herrero, and Ángel Sanz (2004).

9. With regard to WTP for recreational sites/activities, see Sorg and Loomis (1984), Pearce (1993), Loomis and Walsh (1997), and Mendes and Proença (2005).

10. For reviews, see Garrod and Willis (1991), Tietenberg and Lewis (2009), and Hanley and Barbier (2009), for instance.

11. See, for instance, Bedate, Herrero, and Sanz (2004); Poor and Smith (2004); Alberini and Longo (2005)' and Ruijgrok (2006).

12. This is what the authors define as "entitativity."

13. See Snowball (2007).

14. https://earthsky.org/space/apollo-and-the-moon-landing-hoax, accessed on March 19, 2019.

15. https://eyewire.org/explore, accessed on May 21, 2018.

16. See Serjeant (2017).

17. https://www.zooniverse.org/, accessed on May 21, 2018.

18. https://daily.zooniverse.org/2017/02/28/new-project-muon-hunter/, accessed on March 23, 2018.

19. https://veritas.sao.arizona.edu/outreach-aamp-education-mainmenu-80, accessed on March 23, 2018.

20. These include Galaxy Zoo Supernovae, Galaxy Zoo Hubble, Moon Zoo, Old Weather, The Milky Way Project, Planet Hunters, Solar Stormwatch.

21. The median value of leisure time for the European countries in figure 8.1 is around EUR 7.8 per hour.

22. CERN internal reports and personal communications, 2018.

23. https://edition.cnn.com/2016/04/29/tech/cern-weasel-irpt/index.html, accessed on May 21, 2018.

24. https://www.higgshunters.org, accessed on May 21, 2018.

25. Data for on-site visitors since 2004 to 2013 are provided by the Communication Groups of CERN and by each Collaboration.

26. The three zones and the share of visitors for each zone are based on data provided by the CERN Communication Group (personal communication, October 2013); additional costs are estimated, including for accommodation and meals (data from the CERN website).

27. Source: CERN and Communication Groups in the main Collaborations.

28. For social media usage, data were provided by CERN and Collaborations, attributing to the LHC 80 percent of the hits to CERN-related social media and 100 percent of those related to Collaborations.

29. Personal communication.

30. Each regular maintenance period lasts 4 days. However, only 2 days can be used for organizing tours.

31. Following a triangular probability distribution with modal value of EUR 10. The range of variation is based on the authors' own experience.

32. Following a triangular probability distribution with modal value of EUR 70. The range of variation is based on data available on www.booking.com.

33. Following a triangular probability distribution with modal value of EUR 50.

34. HEATCO values also are taken as reference by the European Commission (2014).

35. Using the average GDP growth in Europe, the average HEATCO values of time in the European Union have been adjusted to the year 2013.

36. Commenting on Bakhshi, Freeman, and Hitchen (2009).

37. https://www.statista.com/statistics/550501/journalistic-online-content-willingness-to-pay-in -germany/, accessed on March 17, 2018.

Chapter 9

1. http://www.gtc.iac.es, accessed on May 22, 2018.

2. The project was cofunded by the European Regional Development Fund and by other partners; thus, taxpayers of other countries contributed as well. https://research.ufl.edu/publications/explore /past/fall2009/story_4/index.html, accessed March 19, 2019.

3. See, for example, Boardman et al. (2018).

4. https://www.worldwildlife.org/about/financials, accessed on May 22, 2018.

5. See De Roeck (2016).

6. See Daily (1997), OECD (1999), Turner (1999), and Pearce, Atkinson, and Mourato (2006).

7. With regard to "bequest value," see, for instance, Walsh, Loomis, and Gillman (1984) and Schuster, Cordell, and Phillips (2005).

8. With regard to "existence value," see, for instance, Boyle and Bishop (1987) and Blomquist and Whitehead (1995).

9. See Johansson and Kriström (2015).

10. See also Brookshire, Eubanks, and Randall (1983) and Boardman et al. (2018).

11. See Wang (2018).

12. https://home.cern/about/accelerators/future-circular-collider, accessed on June 18, 2018.

13. On these initial concerns about the HGP, see Wilkie (1993).

14. https://www.genome.gov/11006943/.

15. http://nautil.us/issue/48/chaos/if-et-calls-think-twice-about-answering, accessed on May 22, 2018

16. https://www.un.org/disarmament/, accessed on May 22, 2018.

17. See Weikard (2005) and Boudreaux et al. (1999).

18. See Dana (2004).

19. Empirical studies that use CV include, for example, Greenley, Walsh, and Young (1981); Walsh, Loomis, and Gillman (1984); Carson (1998); Togridou, Hovardas, and Pantis (2006); Sattout, Talhouk, and Caligari (2007); and Vesely (2007).

20. See Carson and Groves (2007) and Carson (2012).

21. See, for example, Roson (2000) and Laird and Mackie (2009).

22. See Weimer (2017) for a further discussion of this topic.

23. https://www.evri.ca/, accessed on May 22, 2015.

24. See Tohmo (2004).

25. See Willis (1994) and Tuan and Navrud (2008).

26. See Barlow (2008)

27. https://www.agenziaentrate.gov.it/, accessed on June 21, 2018.

Chapter 10

1. See also the literature reviewed on RI evaluation by Pancotti, Pellegrin, and Vignetti (2014) and Manganote, Schulz, and de Brito Cruz (2016).

2. See, for example, Martin (1996), Salter and Martin (2001), Martin and Tang (2007), and OECD (2019).

3. See Steuer (1986). For a more general discussion of the methodological issues affecting composite indices, see the recent review of the literature by Greco et al. (2017), which focuses particularly on problems arising from weighting, aggregation, and robustness.

4. https://www.sciencealert.com/this-is-the-watery-grave-the-iss-will-dive-into-in-2024-or -beyond-2, accessed on April 4, 2018.

5. See, for example, Weitzman (2001) and Broughel (2017).

6. See Jones, Domingos, and Moura (2014).

7. See, inter alia, Zerbe and Dively (1994), De Rus (2010), and Florio (2014).

8. To explore this topic further, see chapter 9 in De Rus (2010) and chapter 8 in Florio (2014).

9. See Florio (2014).

10. See, inter alia, Venables (2007) and Betancor, Hernández, and Socorro (2013).

11. See Jorge-Calderón (2014) and Tresch (2008).

12. For the motivation of using of a 3 percent SDR for Switzerland as well, see, for example, https://trimis.ec.europa.eu, accessed on May 23, 2018.

13. http://www.jaspersnetwork.org/display/EVE/Cost-Benefit+Analysis+Forum+meeting+on+R DI+infrastructures, accessed on April 4, 2018.

14. "Pilot Socioeconomic Impact Analysis of CFI and CIHR: Medical Imaging R&D," released in 2013, https://www.innovation.ca/results-impacts, accessed on March 21, 2019.

References

Aad, G., et al. 2012. "Observation of a new particle in the search for the Standard Model Higgs boson with the ATLAS detector at the LHC." *Physics Letters B, 716*(1), 1–29.

Abelson, Peter. 2003. "The value of life and health for public policy." *Economic Record, 79*(Special Issue).

Abelson, P. 2008. "Establishing a Monetary Value for Lives Saved: Issues and Controversies," WP 2008-02. In *Cost-Benefit Analysis*. Office of Best Practice Regulation, Department of Finance and Deregulation, Sydney University.

Åberg, Susanne, and Anna Bengtson. 2015. "Does CERN procurement result in innovation?" *Innovation: The European Journal of Social Science Research, 28*(3), 360–383.

Abt, Helmut. 2007. "The publication rate of scientific papers depends only on the number of scientists." *Scientometrics, 73*(3), 281–288.

Abt, Helmut, and Eugene Garfield. 2002. "Is the relationship between numbers of references and paper lengths the same for all sciences?" *Journal of the Association for Information Science and Technology, 53*(13), 1106–1112.

Acs, Zoltan J., David B. Audretsch, and Maryann P. Feldman. 1992. "Real effects of academic research: Comment." *American Economic Review, 82*(1), 363–367.

Acs, Zoltan, Saul Estrin, Tomasz Mickiewicz, and László Szerb. 2014. "The Continued Search for the Solow Residual: The Role of National Entrepreneurial Ecosystem." IZA DP No. 8652.

AdAge. 2015. *What It Costs: Ad Prices from TV's Biggest Buys to the Smallest Screens.* http://adage.com/article/news/costs-ad-prices-tv-mobile-billboards/297928/. Access date: March 18, 2019.

Adams, James D. 1990. "Fundamental stocks of knowledge and productivity growth." *Journal of Political Economy, 98*(4), 673–702.

Aghion, Philippe, John Van Reenen, and Luigi Zingales. 2013. "Innovation and institutional ownership." *American Economic Review, 103*(1), 277–304.

Aghion, Philippe, Ufuk Akcigit, and Peter Howitt. 2015. "Lessons from Schumpeterian growth theory." *American Economic Review, 105*(5), 94–99.

Alberini, Anna, and Alberto Longo. 2005. "The value of cultural heritage sites in Armenia: Evidence from a travel cost method study." Fondazione Eni Enrico Mattei, No. 112, Milano.

Almirall Pilar Garcia (ed.), Montse Moix Bergadà, and Pau Queraltó Ros. 2008. *The Socio-Economic Impact of the Spatial Data Infrastructure of Catalonia.* European Commission JRC, EUR 23300 EN–2008, Luxembourg.

American Economic Group (AEG). 2011. Anderson Economic Group LLC. Economic Impact of Fermi National Accelerator Laboratory, University of Chicago.

Anderson, Jacob, et al. 2013a. *Benefits to the US from Physicists Working at Accelerators Overseas.* Snowmass Electronic Proceedings Community Summer Study, Minneapolis, arXiv:1312.4884.

Anderson, Jacob et al. 2013b. *Snowmass 2013 Young Physicists Science and Career Survey Report.* arXiv:1307.8080v3.

Andersen, Paul H., and Susanne Åberg. 2017. "Big-science organizations as lead users: A case study of CERN." *Competition & Change, 21*(5), 345–363.

Andrés, Ana. 2009. *Measuring Academic Research: How to Undertake a Bibliometric Study.* Oxford: Chandos Publishing.

Anex, Robert P. 1995. "A travel-cost method of evaluating household hazardous waste disposal services." *Journal of Environmental Management, 45*(2), 189–198.

Anselin, Luc, Attila Varga, and Zoltan Acs. 2000. "Geographical spillovers and university research: A spatial econometric perspective." *Growth and Change, 31*(4), 501–515.

April. 2007. *White Paper: The Economic Models of Free Software,* Collection April, Paris. https://www.april.org/files/economic-models_en.pdf. Access date: March 18, 2019.

Arenius, Marko, and Max Boisot. 2011. "A tale of four suppliers." In Max Boisot, Marcus Nordberg, Said Yami, and Bertrand Nicquevert (eds.), *Collisions and Collaboration: The Organization of Learning in the ATLAS Experiment at the LHC* (pp. 135–159). Oxford: Oxford University Press.

Arrow, Kenneth J. 1962. "The economic implications of learning by doing." *Review of Economic Studies, 29*(3), 155–173.

Arrow, Kenneth J., and Anthony C. Fisher. 1974. "Environmental preservation, uncertainty, and irreversibility." *Quarterly Journal of Economics, 88*(2), 312–319.

Arrow, K., R. Solow, P. R. Portney, E. E. Leamer, R. Radner, and H. Schuman. 1993. "Report of the NOAA panel on contingent valuation." *Federal Register, 58*(10), 4601–4614.

Aschhoff, Birgit, and Wolfgang Sofka. 2009. "Innovation on demand—Can public procurement drive market success of innovations?" *Research Policy, 38*(8), 1235–1247.

Ashby, Neil. 2003. "Relativity in the global positioning system." *Living Reviews in Relativity, 6*(1), 1.

Asian Development Bank. 2013. *Cost-Benefit Analysis for Development: A Practical Guide.* Mandaluyong City, Philippines: Asian Development Bank, 2013.

Ates, Gülay, and Angelika Brechelmacher. 2013. "Academic career paths." In *The Work Situation of the Academic Profession in Europe: Findings of a Survey in Twelve Countries* (pp. 13–35). Dordrecht, the Netherlands: Springer.

Atkinson, Doreen, Rae Wolpe, and Hendrik Kotze 2017. *Socio-economic Assessment of SKA Phase 1 in South Africa.* Prepared as part of the Strategic Environmental Assessment (SEA) for the Phase 1 of the Square Kilometre Array (SKA) radio telescope, South Africa.

Atkinson, Giles, Susana Mourato, and David P. Pearce. 2006. *Cost-Benefit Analysis and the Environment: Recent Developments.* Paris: OECD Publishing.

Atkinson, G., N. A. Braathen, S. Mourato, and B. Groom. 2018. *Cost Benefits Analysis and the Environment: Further Developments and Policy Use.* Paris: OECD Publishing.

Australian Academy of Sciences. 2016. *Discovery Machines—Accelerators for Science, Technology, Health, and Innovation.* https://www.science.org.au/support/analysis/reports/future-science-discovery-machines-accelerators-science-technology-health. Access date: March 18, 2019.

Autio, Erkko. 2014. "Innovation from Big Science: Enhancing Big Science Impact Agenda." Department for Business Innovation and Skills, Imperial College Business School, London.

Autio, Erkko, Marilena Bianchi-Streit, and Ari P. Hameri. 2003. *Technology Transfer and Technological Learning through CERN's Procurement Activity,* CERN—Education and Technology Transfer Division.

Autio, Erkko, Ari-Pekka Hameri, and Olli Vuola. 2004. "A framework of industrial knowledge spillovers in big-science centers." *Research Policy, 33*(1), 107–126.

Bacchiocchi, Emanuele, and Fabio Montobbio. 2009. "Knowledge diffusion from university and public research: A comparison between US, Japan and Europe using patent citations." *Journal of Technology Transfer, 34*(2), 169–181.

Báez, Andrea, and Luis César Herrero. 2012. "Using contingent valuation and cost-benefit analysis to design a policy for restoring cultural heritage." *Journal of Cultural Heritage, 13*(3), 235–245.

Baggott, Jim. 2012. *Higgs: The Invention and Discovery of the "God Particle."* Oxford: Oxford University Press.

Bajari, Patrick, and Steven Tadelis. 2001. "Incentives versus transaction costs: A theory of procurement contracts." *Rand Journal of Economics*, 387–407.

Bakhshi, Hasan, Alan Freeman, and Graham Hitchen. 2009. Measuring intrinsic value–how to stop worrying and love economics." https://mpra.ub.uni-muenchen.de/14902/1/MPRA_paper _14902.pdf. Access date: March 19, 2019.

Balakrishnan, Narayanaswamy. 1992. *Handbook of the Logistic Distribution*. New York: Marcel Dekker.

Balcombe, K. G., and Laurence E. D. Smith. 1999. "Refining the use of Monte Carlo techniques for risk analysis in project planning." *Journal of Development Studies*, *36*(2), 113–135.

Bania, Neil, Randall W. Eberts, and Michael S. Fogarty. 1993. "Universities and the startup of new companies: Can we generalize from Route 128 and Silicon Valley?" *Review of Economics and Statistics*, 761–766.

Barro, Robert J., and Xavier Sala-I-Martin. 2004. *Economic Growth*. Cambridge, MA: MIT Press.

Basili, Marcello, and Fulvio Fontini. 2005. "Quasi-option value under ambiguity." *Economics Bulletin*, *4*(3), 1–10.

Bastianin, Andrea, and Massimo Florio. 2018. *Social Cost Benefit Analysis of HL-LHC*. No. CERN-ACC–2018-0014. FCC-DRAFT-MGMT–2018-001. http://cds.cern.ch/record/2319300. Access date: March 19, 2019.

Bastianin, Andrea, Massimo Florio, and Francesco Giffoni. 2018. "LHC upgrade brings benefits beyond physics." *CERN Courier.*

Batista, Pablo D., Mônica G. Campiteli, and Osame Kinouchi. 2006. "Is it possible to compare researchers with different scientific interests?"*Scientometrics*, *68*(1), 179–189.

Battistoni, Giuseppe, Mario Genco, Marta Marsilio, Chiara Pancotti, Stefano Rossi, and Silvia Vignetti (2016). "Cost–benefit analysis of applied research infrastructure: Evidence from health care." *Technological Forecasting and Social Change*, *112*, 79–91.

Baum, Warren, and Stokes Tolbert. 1985. "Investing in development: Lessons of World Bank experience." *Finance and Development*, *22*(4), 25.

BBC. 2013. *The Economic Value of the BBC: 2011/12 A Report by the BBC*. http://downloads.bbc .co.uk/aboutthebbc/insidethebbc/howwework/reports/pdf/bbc_economic_impact_2013.pdf. Access date: March 18, 2019.

BBMRI-ERIC. 2016. "Performance indicators for BBMRI-ERIC v 1.2," http://www.bbmri-eric .eu/. Access date: March 21, 2019.

Beagrie, Neil, and John Houghton. 2016. "The value and impact of the European bioinformatics institute." EMBL-EBI. http://vuir.vu.edu.au/33707/1/EBI-impact-report.pdf. Access date: March 18, 2019.

Beaumont, Christopher N., Alyssa Goodman, Sarah Kendrew, Jonathan Williams, and Robert Simpson. 2014. "The Milky Way Project: Leveraging citizen science and machine learning to detect interstellar bubbles." *Astrophysical Journal Supplement Series*, *214*(1), 3.

Bedate, Ana, Luis César Herrero, and José Ángel Sanz. 2004. "Economic valuation of the cultural heritage: Application to four case studies in Spain." *Journal of Cultural Heritage*, *5*(1), 101–111.

Behrens, Werner, and Peter M. Hawranek. 1991. *Manual for the Preparation of Industrial Feasibility Studies*. Vienna: United Nations Industrial Development Organization.

Belgian Science Policy Office. 2013. *Annual Report on Science and Technology Indicators for Belgium.* https://biblio.ugent.be/publication/4223722/file/6807091.pdf. Access date: March 18, 2019.

Belli, Pedro, Jock R. Anderson, Howard N. Barnum, John A. Dixon, and Jee Peng Tan. 2001. *Economic Analysis of Investment Operations—Analytical Tools and Practical Applications*. Washington, DC: World Bank Institute (WBI).

Belussi, Fiorenza, and Katia Caldari. 2009. "At the origin of the industrial district: Alfred Marshall and the Cambridge school." *Cambridge Journal of Economics*, *33*(2), 335–355.

Bengtson, Anna, and Susanne Åberg. 2016. "Found in translation? On the transfer of technological knowledge from science to industry." In *Extending the Business Network Approach* (pp. 227–245). London: Palgrave Macmillan.

Benjamin, Marina. 2003. *Rocket Dreams: How the Space Age Shaped Our Vision of a World Beyond*. New York: Simon and Schuster.

Benson, Charlotte, and John Twigg. 2004. *Measuring Mitigation: Methodologies for Assessing Natural Hazard Risks and the Net Benefits of Mitigation—A Scoping Study.* International Federation of Red Cross and Red Crescent Societies, ProVention Consortium, Geneva, Switzerland.

Bernstein, Jeffrey I. 1988. "Costs of production, intra- and interindustry R&D spillovers: Canadian evidence." *Canadian Journal of Economics*, 324–347.

Bernstein, Jeffrey I., and M. Ishaq Nadiri. 1991. *Product Demand, Cost of Production, Spillovers, and the Social Rate of Return to R&D*. No. 3625. Cambridge, MA: National Bureau of Economic Research.

Berry, John V. 2015. "The Department of Energy Security Clearance Process." http://www.securityclearanceblog.com/2015/07/the-department-of-energy-security-clearance-process.html. Access date: March 18, 2019.

Betancor, O., A. Hernández, and Maria P. Socorro. 2013. *Revision of Infrastructure Project Assessment Practice in Europe Regarding Impacts on Competitiveness*. I-C-EU Project, European Commission, Bruxelles.

Bezdek, Roger H., and Robert M. Wendling. 1992. "Sharing out NASA's spoils." *Nature*, *355*(6356), 105.

Bianchi-Streit, Marilena, R. Buude, H. Helwig Schmied, B. Schorr, N. F. Blackburne, H. Reitz, and B. Sagnell. 1984. *Economic Utility Resulting from CERN Contracts (Second Study)* (No. CERN–84–14). CERN, Geneva.

Biotechnology and Biological Sciences Research Counci (BBSRC). 2015. *Building the Bioeconomy. Impact Report 2015*. https://bbsrc.ukri.org/documents/impact-report–2015-pdf/. Access date: March 18, 2019.

Birch, Stephen, and Cam Donaldson. 1987. "Applications of cost-benefit analysis to health care: Departures from welfare economic theory." *Journal of Health Economics*, *6*(3), 211–225.

BIS. 2010. *Economic Impacts of the UK Research Council System: An Overview*. https://assets.publishing.service.gov.uk/government/uploads/system/uploads/attachment_data/file/32477/10-917-economic-impacts-uk-research-council-system.pdf.

Bitzer, Jürgen, and Philipp JH Schröder. 2006. *The Economics of Open Source Software Development*. New York: Elsevier Science Inc.

Blaug, Mark. 1987. *The Economics of Education and the Education of an Economist.* Cheltenham, UK: Edward Elgar Publishing.

Blomquist, G. C., P. A. Coomes, C. Jepsen, B. C. Koford, and K. R. Troske. 2014. "Estimating the social value of higher education: willingness to pay for community and technical colleges." *Journal of Benefit-Cost Analysis*, *5*(1), 3–41.

Blomquist, Glenn C., and John C. Whitehead. 1995. "Existence value, contingent valuation, and natural resources damages assessment." *Growth and Change*, *26*(4), 573–589.

Blöndal, Sveinbjörn, Simon Field, and Nathalie Girouard. 2002. *Investment in Human Capital through Upper-Secondary and Tertiary Education*. OECD Economic Studies, *1*, 41–89.

BMBF. 2013. *Roadmap for Infrastructure*. A pilot project of the Federal Ministry of Education and Research. https://www.bmbf.de/pub/Roadmap_Research_Infrastructures.pdf. Access date: March 18, 2019.

Boadway, Robin. 2006. "Principles of cost-benefit analysis." *Public Policy Review*, *2*(1), 1–44.

Boardman, Anthony E., David Greenberg, Aidan R. Vining, and David L. Weimer. 2006. *Cost Benefit Analysis—Concepts and Practice*, 3rd ed. London: Pearson.

Boardman, Anthony E., David Greenberg, Aidan R. Vining, and David L. Weimer. 2018. *Cost Benefit Analysis—Concepts and Practice*, 4th ed., reissued. New York: Cambridge University Press.

Boarini, R. and H. Strauss (2007), "The Private Internal Rates of Return to Tertiary Education: New Estimates for 21 OECD Countries." *OECD Economics Department Working Papers*, No. 591. Paris: OECD Publishing.

Boisot, Max, Markus Nordberg, M., Said Yami, and Bertrand Nicquevert (eds.). 2011. *Collisions and Collaboration: The Organization of Learning in the ATLAS Experiment at the LHC*. Oxford: Oxford University Press.

Booz and Company. 2014. "Evaluation of socio-economic impacts from space activities in the EU." European Commission—DG Grow, Brussels.

Boston Consulting Group (BCG). 2009. *Willingness to Pay for News Online: Key Findings from an International Survey*. http://la-rem.eu/wp-content/uploads/2013/12/BCG-Monde.pdf. Access date: 3/18/2019.

Boudreaux, Donald J., Roger E. Meiners, and Todd J. Zywicki. 1999. "Talk is cheap: The existence value fallacy." *Environmental Law, 29*(1999), 765.

Boulding, K. E. (1945). "The concept of economic surplus." *American Economic Review, 35*(5):851–869.

Bousfield, D., McEntyre, J., Velankar, S., Papadatos, G., Bateman, A., Cochrane, G., … and Blomberg, N. 2016. *Patterns of Database Citation in Articles and Patents Indicate Long-Term Scientific and Industry Value of Biological Data Resources*. F1000Research, 5: ELIXIR-160. https://f1000research.com/articles/5-160/v1. Access date: March 19, 2019.

Boyer, Marcel, and Jacques Robert. 2006. *The Economics of Free and Open Source Software: Contributions to a Government Policy on Open Source Software*. CIRANO Project Reports 2006rp-03, CIRANO, Montreal.

Boyle, Kevin J., and Richard C. Bishop. 1987. "Valuing wildlife in benefit-cost analyses: A case study involving endangered species." *Water Resources Research, 23*(5), 943–950.

Brendle, P., P. Cohendet, J. A. Heraud, R. Larue de Tournemine, H. Schmied, D. Vitry, and E. Zuscovitch. 1980. "Les effets economiques induits de l' ESA." *ESA Contract Report, 3*, 17.

Brent, R. J. 2003. "Cost–benefit analysis and the human capital approach." In *Cost–Benefit Analysis and Health Care Evaluations* (pp. 271–292). Cheltenham, UK: Edward Elgar Publishing.

Brent, Robert J. 2006. *Applied Cost-Benefit Analysis*. Cheltenham, UK: Edward Elgar Publishing.

Brent, Robert J. 2017. *Advanced Introduction to Cost–Benefit Analysis*. Cheltenham, UK: Edward Elgar Publishing.

Bressan, Beatrice, and Max Boisot. 2011. "The individual in the ATLAS Collaboration: A learning perspective." In Max Boisot, Marcus Nordberg, Said Yami, and Bertrand Nicquevert (eds.), *Collisions and Collaboration: The Organization of Learning in the ATLAS Experiment at the LHC* (pp. 201–225). Oxford: Oxford University Press.

Brigham, Eugene F., Louis C. Gapenski, and Phillip Daves, P. R. 1999. *Intermediate Financial Management*, 6th ed. The Dryden Press.

Broadus, Robert N. 1983. "An investigation of the validity of bibliographic citations." *Journal of the Association for Information Science and Technology, 34*(2), 132–135.

Brookshire, David S.; Larry Eubanks, Alan Randall. 1983. "Estimating option prices and existence values for wildlife resources." *Land Economics, 59*(1), 1–15.

Broughel, James. 2017. *The Social Discount Rate: A Baseline Approach*. Mercatus Center, George Mason University Working paper, Arlington.

Broughton, Richard. 2016. *How Much Is an Hour of TV Viewing Worth?* https://www.ampereanalysis.com/blog/23ba0327-ff03-48bf-9b3a-c422409ccd1b. Access date: March 18, 2019.

Burton, Robert E., and R. W. Kebler. 1960. "The "half-life" of some scientific and technical literatures." *Journal of the Association for Information Science and Technology, 11*(1), 18–22.

Byford, Sarah, David J. Torgerson, and James Raftery. 2000. "Economic note: cost of illness studies." *BMJ (Clinical Research Ed.)*, *320*(7245), 1335.

Caballero, Ricardo J., and Adam B. Jaffe. 1993. "How high are the giants' shoulders: An empirical assessment of knowledge spillovers and creative destruction in a model of economic growth." *NBER Macroeconomics Annual*, *8*, 15–74.

Callaert, Julie, B. Van Looy, A. Verbeek, K. Debackere, and B. Thijs. 2006. "Traces of prior art: An analysis of non-patent references found in patent documents." *Scientometrics*, *69*(1), 3–20.

Cameron, A. Colin, and Pravin K. Trivedi. 2005. *Microeconometrics: Methods and Applications*. New York: Cambridge University Press.

Campbell, Harry F., and Richard P. C. Brown. 2003. *Benefit-Cost Analysis: Financial and Economic Appraisal Using Spreadsheets*. Cambridge: Cambridge University Press.

Camporesi, Tiziano. 2001. "High-energy physics as a career springboard." *European Journal of Physics*, *22*(2), 159–148.

Camporesi, Tiziano, Gelsomina Catalano, Massimo Florio, and Francesco Giffoni. 2017. "Experiential learning in high energy physics: A survey of students at the LHC." *European Journal of Physics*, *38*(2), 025703.

Cano-Kollmann, Marcelo, Robert D. Hamilton, and Ram Mudambi. 2017. "Public support for innovation and the openness of firms' innovation activities." *Industrial and Corporate Change*, *26*(3), 421–442.

Card, David. 1999. "The causal effect of education on earnings." In O. Ashenfelter and Card D. (eds.), *Handbook of Labor Economics* (Vol. 3a, 1801–1863). Amsterdam: Elsevier.

Carrazza, Stefano, Alfio Ferrara, and Silvia Salini. 2014. *Research Infrastructures in the LHC Era: A Scientometric Approach*. Working Paper no.12. Department of Economics, Management and Quantitative Methods, Università degli Studi di Milano.

Carrazza, Stefano, Alfio Ferrara, and Silvia Salini. 2016. "Research infrastructures in the LHC era: A scientometric approach." *Technological Forecasting and Social Change*, *112*, 121–133.

Carson, Richard T. 1998. "Valuation of tropical rainforests: philosophical and practical issues in the use of contingent valuation." *Ecological Economics*, *24*(1), 15–29.

Carson, Richard T. 2012. "Contingent valuation: A practical alternative when prices aren't available." *Journal of Economic Perspectives*, *26*(4): 27–42.

Carson, Richard T., N. E. Flores, K. M. Martin, and J. L. Wright. 1996. "Contingent valuation and revealed preference methodologies: Comparing the estimates for quasi-public goods." *Land Economics*, 80–99.

Carson, Richard T., Nicholas E. Flores, and Norman F. Meade. 2001. "Contingent valuation: Controversies and evidence." *Environmental and Resource Economics*, *19*(2), 173–210.

Carson, Richard T., and Theodore Groves. 2007. "Incentive and informational properties of preference questions." *Environmental and Resource Economics*, *37*(1), 181–210.

Cassis, Youssef, Giuseppe De Luca, and Massimo Florio. 2016. *Infrastructure Finance in Europe: Insights into the History of Water, Transport, and Telecommunications*. Oxford: Oxford University Press.

Castelnovo, Paolo, Massimo Florio, Stefano Forte, Lucio Rossi, and Emanuela Sirtori. 2018. "The economic impact of technological procurement from large-scale research infrastructure: evidence from the large hadron-collider at CERN," *Research Policy*, *47*(9), 1853–1867.

Catalano, Gelsomina, Massimo Florio, and Francesco Giffoni. 2018. "Should governments fund basic science? Evidence from a willingness-to-pay experiment in five universities." *Journal of Economic Policy Reform*, 1–18.

Catalano, Gelsomina, Florio Massimo, Morretta Valentina, and Portaluri Tommaso. 2018. "The Value of Human Capital Formation at CERN." CERN-ACC–2018-0025. http://cds.cern.ch/record /2635864/files/CERN-ACC-2018-0025.pdf?version=3. Access date: March 19, 2019.

Caulkins, Peter P., Richard C. Bishop, and Nicolaas W. Bouwes Sr. 1986. "The travel cost model for lake recreation: A comparison of two methods for incorporating site quality and substitution effects." *American Journal of Agricultural Economics*, *68*(2), 291–297.

CBRE Consulting. 2010. *Berkeley Lab Economic Impact Study*. https://www.lbl.gov/wp-content /uploads/2014/06/CBRE-LBNL-Economic-Impact-Study-FINAL.pdf. Access date: March 18, 2019.

CERN. 2011. *CERN Communications Strategy 2012–2016*. CERN/DG/CO/2011–1, October 12, 2011, Draft–5. https://communications.web.cern.ch/sites/communications.web.cern.ch/files/down loads/CERNComms_draft5_12oct2011.pdf. Access date: March 19, 2019.

CERN. 2017a. *CERN's Communications Strategy 2017–2020*. CH–1211. Geneva, Switzerland.

CERN. 2017b. *FAQ: LHC the Guide*. https://cds.cern.ch/record/2255762/files/CERN-Brochure–2017 -002-Eng.pdf. Access date: March 18, 2019.

Chalmers, Alan F. 2013. *What Is This Thing Called Science?* Indianapolis, IN: Hacket Publishing.

Chang, Pao Cheng, Chien P. Wang, Benjamin J. Yuan, and Kai-Ting Chuang. 2002. "Forecast of development trends in Taiwan's machinery industry." *Technological Forecasting and Social Change*, *69*(8), 781–802.

Chernyaev, A. P., and S. M. Varzar 2014. "Particle accelerators in modern world." *Physics of Atomic Nuclei*, *77*(10), 1203–1215.

Chesbrough, Henry. 2003. *Open Innovation: The New Imperative for Creating and Profiting from Technology*. Cambridge, MA: Harvard Business School Press.

Chial, Heidi. 2008. "DNA sequencing technologies key to the Human Genome Project." *Nature Education*, *1*(1), 219.

Choi, Serene, H. J. Maren Funk, Susan Roelofs, Martha Alvarez-Elizondo, and Timo Nieminen. 2011. *International Research Work Experience of Young Females in Physics*. arXiv:1101.1758.

Chrysostomou, Antonio. 2017. *The Square Kilometre Array: An Example of International Cooperation*. http://www.unoosa.org/documents/pdf/psa/activities/2017/OpenUniverse/slides /Presentation27.pdf. Access date: March 18, 2019.

Chyi, Hsiang Iris. 2005. "Willingness to pay for online news: An empirical study on the viability of the subscription model." *Journal of Media Economics*, *18*(2), 131–142.

Clarke, Steven, Mark Mawhinney, Robert Swerdlow, and Dorothee Teichmann. 2013. *Project Preparation and CBA of RDI Infrastructure Projects*. JASPERS Knowledge Economy and Energy Division Staff Working Paper.

Clawson, Marion, and J. L. Knetsch. 1966. *Economics of Outdoor Recreation*. Baltimore: Johns Hopkins Press.

Clery, Daniel. 2016a. "Forbidden planets." *Science*, *353*(6298), 438–441.

Clery, Daniel. 2016b. "UPDATED: Panel backs ITER fusion project's new schedule, but balks at cost." *Science*. https://www.sciencemag.org/news/2016/04/updated-panel-backs-iter-fusion-project -s-new-schedule-balks-cost. Access date: March 21, 2019.

Clinch, J. Peter, and John D. Healy. 2001. "Cost-benefit analysis of domestic energy efficiency." *Energy Policy*, *29*(2), 113–124.

Clò, S., Massimo Florio and Valentina Morretta. 2019. *The Socioeconomic Impact of Earth Observation: Preliminary Evidence about Cosmo Sky Med*. Working paper of Department of Economics, Management and Quantitative Methods.

Clò, Stefano, Massimo Florio, Valentina Morretta, and Davide Vurchio. 2019. *Earth Observation in a Cost-Benefit Analysis Perspective: Cosmo SkyMed Satellites of the Italian Space Agency*. Departmental Working Papers 2019-06, Department of Economics, Management and Quantitative Methods at Università degli Studi di Milano.

Cockburn, Iain M., and Zvi Griliches. 1988. "Industry effects and appropriability measures in the stock markets valuation of R&D and patents" *American Economic Review*, *78*(2), 419–23.

Cohen, Wesley M., and Daniel A. Levinthal. 1990. "Absorptive capacity: A new perspective on learning and innovation," *Administrative Science Quarterly*, *35*(1), 128–152.

Collins, Harry. 2017. *Gravity's Kiss: The Detection of Gravitational Waves*. Cambridge, MA: MIT Press.

Cook-Degan, Robert, Rachel Ankeny, and Kathryn Maxson Jones 2017. "Sharing data to build a medical information commons: From Bermuda to the Global Alliance." *Annu Rev Genomics Hum Genet.*, *18*, 389–415.

Conrad, Jon M. 1980. "Quasi-option value and the expected value of information." *The Quarterly Journal of Economics*, *94*(4), 813–820.

Contreras, J. L. 2017. "Leviathan in the commons: Biomedical data and the state." In Strandburg, K. J., B. M. Frischmann, and M. J. Madison, eds., *Governing Medical Knowledge Commons* (pp. 19–46). New York: Cambridge University Press, 19–46.

Cornes, Richard, and Todd Sandler. 1996. *The Theory of Externalities, Public Goods, and Club Goods*, 2nd ed. Cambridge: Cambridge University Press.

COST Office. 2010. *Benefits of Research Infrastructures beyond Science The Example of the Square Kilometre Array (SKA)*. European Cooperation in Science and Technology, Final Report, March 30 and 31, Rome.

Cottrell, Geoff. 2016. *Telescopes: A Very Short Introduction*. Oxford: Oxford University Press.

Cowan, Robin, and Natalia Zinovyeva 2013. "University effects on regional innovation." *Research Policy*, *42*(3), 788–800.

Crépon, Bruno, Emmanuel Duguet, and Jacques Mairesse. 1998. "Research, innovation and productivity: An econometric analysis at the firm level." *Economics of Innovation and New Technology*, *7*(2), 115–158.

CrowdFlower. 2016. Data Science Report. https://visit.figure-eight.com/rs/416-ZBE-142/images/CrowdFlower_DataScienceReport_2016.pdf. Access date: March 18, 2019.

Cuccia, Tiziana, and Giovanni Signorello. 2000. "A contingent valuation study of willingness to pay for visiting a city of art: The case study of Noto (Italy)." *Proceedings of the 11th International Conference on Cultural Economics*. Minneapolis, May 28–31.

Da Silveira Jr., Luiz A. B., Eduardo Vasconcellos, Luis Guedes, Luis Fernando Guedes., and Renato M. Costa, 2018. "Technology roadmapping: A methodological proposition to refine Delphi results." *Technological Forecasting and Social Change*, *126*, 194–206.

Daffara, Carlo. 2009. *Economic Free Software Perspectives*. Technology Innovation Management Review, August. https://timreview.ca/article/277. Access date: March 18, 2019.

Dahlby, Bev. 2008. *The Marginal Cost of Public Funds: Theory and Applications*. Cambridge, MA: MIT Press.

Daily, Gretchen (ed.). 1997. *Nature's Services: Societal Dependence on Natural Ecosystems*. Washington, DC: Island Press.

Dalgaard, Carl-Johan, Nikolai Kaarsen, Ola Olsson, and Pablo Selaya. 2018. "Roman Roads to Prosperity: Persistence and Non-Persistence of Public Goods Provision." CEPR discussion paper 12745.

Dalkey, Norman, and Olaf Helmer. 1963. "An experimental application of the Delphi method to the use of experts." *Management Science*, *9*(3), 458–467.

Dana, David A. 2004. "Existence value and federal preservation regulation." *Harvard Environmental Law Review*, *28*, 343.

Danish Agency for Science. 2008. *Evaluation of Danish Industrial Activities in the European Space Agency (ESA)*. Assessment of the economic impacts of the Danish ESA membership. Technical report, Danish Agency for Science.

Dasgupta, Partha. 2007. "Commentary: The Stern Review's economics of climate change," *National Institute Economic Review*, *199*(4), 4–7.

Dasgupta, Partha. 2008. "Discounting climate change," *Journal of Risk and Uncertainty*, *37*(2), 141–169.

de Bruijn, Hans, and Martijn Leijten. 2008. "Management Characteristics of Mega-Projects." In Hugo Priemus, Bent Flyvbjerg, Bert van Wee (eds.), *Decision-Making on Mega-Projects*.

Cost-Benefit Analysis, Planning and Innovation (pp. 23–40). Cheltenham, UK: Edward Elgar Publishing.

Del Bo, Chiara F. 2016. "The rate of return to investment in R&D: The case of research infrastructures." *Technological Forecasting and Social Change, 112,* 26–37.

Del Bo, Chiara, Carlo Fiorio, and Massimo Florio. 2011 "Shadow wages for the EU regions." *Fiscal Studies, 32*(1), 109–143.

Del Bo, Chiara, Massimo Florio, and Stefano Forte. 2016. "The social impact of research infrastructures at the frontier of science and technology: The case of particle accelerators: Editorial introduction." *Technological Forecasting & Social Change, 112,* 1–3

De Loë, Rob C., Natalya Melnychuk, Dan Murray, and Ryan Plummer. 2016. "Advancing the state of policy Delphi practice: A systematic review evaluating methodological evolution, innovation, and opportunities." *Technological Forecasting and Social Change, 104,* 78–88.

Deng, Yi. 2008. "The value of knowledge spillovers in the US semiconductor industry." *International Journal of Industrial Organization, 26*(4), 1044–1058.

De Roeck, Albert. 2016. "The probability of discovery," *Technological Forecasting and Social Change, 112,* 13–19.

De Rus, Ginés. 2010. *Introduction to Cost-Benefit Analysis: Looking for Reasonable Shortcuts.* Cheltenham, UK: Edward Elgar Publishing.

Desaigues, B., et al. 2011. "Economic valuation of air pollution mortality: A 9-country contingent valuation survey of value of a life year (VOLY)." *Ecological Indicators, 11*(3), 902–910.

de Solla Price, Derek J. 1963. *Little Science, Big Science*, 10th ed. New York: Columbia University Press, 12–19.

de Solla Price, Derek J. 1965 "Networks of scientific papers." *Science, 149* (3683), 510–515.

de Solla Price, Derek J. 1970. "Citation measures of hard science, soft science, technology, and nonscience." In C. E. Nelson, and D. K. Pollock (eds.), *Communication among Scientists and Engineers* (pp. 3–22). Lexington, MA: DC Heath and Company.

de Solla Price, Derek J. 1986. *Little Science, Big Science ... and Beyond.* New York: Columbia University Press.

Di Meglio Alberto, Maria Girone, Andrew Purcell, and Fons, Rademakers. 2017. *CERN Openlab White Paper on Future ICT Challenges in Scientific Research* (CERN). http://inspirehep.net /record/1650102.

Diamond, Peter A., and Jerry A. Hausman. 1994. "Contingent valuation: Is some number better than no number?" *Journal of Economic Perspectives, 8*(4), 45–64.

Dickson, Matt, and Colm Harmon. 2011. "Economic returns to education: What we know, what we don't know, and where we are going—some brief pointers." *Economics of Education Review, 30*(6), 1118–1122.

Dijkgraaf, Robbert. 2017. "The world of tomorrow." In Abraham Flexner, *The Usefulness of Useless Knowledge* (pp. 1–49). Princeton, NJ: Princeton University Press.

Dixit, Avinash K., and Robert S. Pindyck. 1994. *Investment under Uncertainty.* Princeton, NJ: Princeton University Press.

Donatello, Michael C. 2013. "Assessing audiences' willingness to pay and price response for news online." Dissertation, the University of North Carolina at Chapel Hill.

Dosi, G., C. Freeman, R. C. Nelson, G. Silverman, and L. Soete. 1988. *Technical Change and Economic Theory.* London: Pinter.

Dou, Wenyu. 2004. "Will Internet users pay for online content?" *Journal of Advertising Research, 44*(4), 349–359.

Drèze, Jean, and Nicholas Stern. 1987. "The theory of cost-benefit analysis." In A. J. Auerbach and M. Feldstein (eds.), *Handbook of Public Economics* (Vol. 2, pp. 909–989). Amsterdam: North Holland.

Drèze, Jean, and Nicholas Stern. 1990. "Policy reform, shadow prices, and market prices." *Journal of Public Economics, 42*(1), 1–45.

Druckman, James N., and Cindy D. Kam. 2011. "Students as experimental participants." In Druckman et al. (eds.), *Cambridge Handbook of Experimental Political Science* (pp. 41–57). New York: Cambridge University Press.

DTZ. 2009. *Economic Impact of the John Innes Centre*. Technical report, Edinburgh: DTZ.

Dupuit, Jules. 1844. *De la mesure de l'utilité des travaux publics*, Annales des ponts et chaussées, Second series, 8 (translated as "On the measurement of the utility of public works"). *International Economic Papers*, 2(1952), 83–110.

Economist. 2013. "Titan of innovation. What can business learn from Big Science?" April 27. https://www.economist.com/business/2013/04/27/titans-of-innovation. Access date: March 21, 2019.

Edler, Jakob, and Luke Georghiou. 2007. "Public procurement and innovation-Resurrecting the demand side." *Research Policy*, 36(7), 949–963.

Edquist, Charles. 2011. "Design of innovation policy through diagnostic analysis: Identification of systemic problems (or failures)." *Industrial and Corporate Change*, 20(6), 1725–1753.

Edquist, Charles, and Jon Mikel Zabala-Iturriagagoitia. 2012. "Public procurement for innovation as mission-oriented innovation policy."*Research Policy*, 41(10), 1757–1769.

Edquist, Charles, N. S. Vonortas, Jon Mikel Zabala-Iturriagagoitia, and J. Edler. 2015. *Public Procurement for Innovation*. Cheltenham, UK: Edward Elgar Publishing.

Egghe, Leo, and I. K. Ravichandra Rao. 1992. "Citation age data and the obsolescence function: Fits and explanations." *Information Processing & Management*, 28(2), 201–217.

EIROforum. 2015. *Long-Term Sustainability of Research Infrastructures*, EIRO forum discussion paper.

EISS Group. 2002. ITER at Cadarache: Socio-economic environment, Task SE1, Deliverable 1.

Eliasson, Gunnar. 2010. *Advanced Public Procurement as Industrial Policy: The Aircraft Industry as a Technical University*. Vol. 34. New York: Springer Science & Business Media.

ELIXIR. 2016. *Annual Report 2016*. https://www.elixir-europe.org/system/files/elixir_annual_report_layout_v16_digital_singles.pdf. Access date: March 18, 2019.

EMBL Alumni. 2017. *Alumni Impact Report 2017*. https://www.embl-hamburg.de/aboutus/alumni/about-us/our-alumni/impact_report/index.html. Access date: March 18, 2019.

Esparza, José, and Tadataka Yamada. 2007. "The discovery value of "Big Science." *Journal of Experimental Medicine*, Apr. 16; 204(4): 701–704.

European Commission. 2004. HEATCO: *Developing Harmonised European Approaches for Transport Costing and Project Assessment*, Deliverable 5, Brussels. https://trimis.ec.europa.eu/sites/default/files/project/documents/20090918_161442_29356_HEATCO%20-%20Final%20Report.pdf. Access date: March 21, 2019.

European Commission. 2006. *Study on Evaluating the Knowledge Economy. What Are Patents Actually Worth? The Value of Patents for Today's Economy and Society*, Final Report Tender No MARKT/2004/09/E, Lot 2, July 23.

European Commission. 2008a. *Guide to Cost-Benefit Analysis of Investment Projects (Structural Funds, Cohesion Fund and Instrument for Pre-Accession)*, European Commission.

European Commission. 2008b. IMPACT: *Internalisation Measures and Polices for All External Cost of Transport, Handbook on Estimation of External Costs in the Transport Sector*, Version 1.1, Brussels.

European Commission. 2011. *Innovation Union Competitiveness Report*. http://ec.europa.eu/research/innovation-union/pdf/competitiveness-report/2011/iuc2011-full-report.pdf. Access date: March 18, 2019.

European Commission. 2012. *Socio-Economic Impact Assessment for Research Projects. Sequoia Projects,* http://www.lse.ac.uk/media@lse/research/SEQUOIA/SEQUOIA_D3.3b_final_modif_md_v2.pdf. Access date: March 21, 2019.

European Commission. 2013. *Innovation Union Competitiveness Report*. https://ec.europa.eu/research/innovation-union/pdf/competitiveness_report_2013.pdf. Access date: March 18, 2019.

European Commission. 2014. *Guide to Cost-Benefit Analysis of Investment Projects Economic Appraisal Tool for Cohesion Policy 2014–2020*, Directorate-General for Regional and Urban Policy.

European Commission. 2015. *Horizon 2020 Indicators. Assessing the Results and Impact of Horizon.* https://ec.europa.eu/programmes/horizon2020/en/news/horizon-2020-indicators-assessing -results-and-impact-horizon. Access date: March 21, 2019.

European Commission. 2016a. *Long-Term Sustainability of Research Infrastructures.* Nonpaper Stakeholders Workshop of November 25, 2016.

European Commission. 2017a. *Horizon 2020, Work Programme 2016–2017.* http://ec.europa.eu /research/participants/data/ref/h2020/wp/2016_2017/main/h2020-wp1617-infrastructures_en.pdf. Access date: March 18, 2019.

European Investment Bank (EIB). 2013. *The Economic Appraisal of Investment Projects at the EIB.* http://www.eib.org/attachments/thematic/economic_appraisal_of_investment_projects_en .pdf. Access date: March 18, 2019.

European Molecular Biology Laboratory (EMBL). 2016. *2016 Annual Report.* https://www.embl .de/aboutus/communication_outreach/publications/annual_report/annual-report-2016.pdf. Access date: March 1, 2019.

European Molecular Biology Laboratory–European Bioinformatics Institute (EMBL-EBI). 2016. *Impact Report 2016.* https://www.ebi.ac.uk/sites/ebi.ac.uk/files/groups/external_relations /Documents/EMBL-EBI_Impact_Report_2016_web.pdf. Access date: March 18, 2019.

European Molecular Biology Laboratory–European Bioinformatics Institute (EMBL-EBI). 2017. *The European Bioinformatics Institute Impact Report 2016.* European Molecular Biology Laboratory (EMBL). www.ebi.ac.uk/about/our-impact. Access date: March 18, 2019.

European Network of Transmission System Operators for Electricity (ENTSOE). 2015. *Guideline for Cost Benefit Analysis of Grid Development Projects.* https://docstore.entsoe.eu/Documents /SDC%20documents/TYNDP/ENTSO-E%20cost%20benefit%20analysis%20approved%20 by%20the%20European%20Commission%20on%204%20February%202015.pdf. Access date: March 18, 2019.

European Space Agency (ESA). 2012. *Design of a Methodology to Evaluate the Direct and Indirect Economic and Social Benefits of Public Investments in Space*, Technical Note 3, Amsterdam.

European Space Agency (ESA). 2013. *How Many Space Debris Objects Are Currently in Orbit?* European Space Agency. http://www.esa.int/Our_Activities/Space_Engineering_Technology/Clean _Space/How_many_space_debris_objects_are_currently_in_orbit. Access date: March 18, 2019.

European Strategy Forum on Research Infrastructures (ESFRI). 2013. *Indicators of Pan-European Relevance of Research Infrastructures.* https://ec.europa.eu/research/infrastructures/pdf/FI13_46 _14_ESFRI%20Indicators_report_4.pdf. Access date: March 18, 2019.

European Strategy Forum on Research Infrastructures (ESFRI). 2016a. *Public Roadmap 2018 Guide.* December 9, 2016. http://www.esfri.eu/sites/default/files/docs/ESFRI_Roadmap_2018 _Public_Guide_f.pdf. Access date: March 18, 2019.

European Strategy Forum on Research Infrastructures (ESFRI). 2016b. *Strategy Report on Research Infrastructure, Roadmap 2016.* http://www.esfri.eu/sites/default/files/20160308 _ROADMAP_single_page_LIGHT.pdf. Access date: March 18, 2019.

European Strategy Forum on Research Infrastructures (ESFRI). 2016c. *Working Group on Innovation, Report to ESFRI*—March 2016, FI16–56-05.

European Strategy Forum on Research Infrastructures (ESFRI). 2017. *Long-Term Sustainability of Research Infrastructures,* Long-Term Sustainability Working Group. https://ec.europa.eu/research /infrastructures/pdf/esfri/publications/esfri_scripta_vol2.pdf. Access date: March 18, 2019.

European Strategy Forum on Research Infrastructures (ESFRI). 2018. *Strategy Report on Research Infrastructures, Roadmap 2018.* http://roadmap2018.esfri.eu/. Access date: March 18, 2019.

Eurostat Business Demography Statistics. 2012. http://ec.europa.eu/eurostat/web/structural-business -statistics/entrepreneurship/business-demography. Access date: March 18, 2019.

Evans, B. J. 2017. "Genomic data commons." In K. J. Strandburg, B. M. Frischmann, and M. J. Madison (eds.), *Governing Medical Knowledge Commons* (pp. 46–74). New York: Cambridge University Press.

EvaRIO. 2013. *EvaRIO Case Study: EMBL-EBI, Part of the Intermediate Report: Preliminary Report on Results—Deliverable D5.1.* Guittard, Moritz, Wolff, Matt.

Fahlander, Claes. 2016. "Engaging Local Industry in the Development of Basic Research Infrastructure and Instrumentation–The Case of HIE-ISOLDE and ESS Scandinavia." *AIP Conference Proceedings, 1753*(1).

Farrow, Scott, and Adam Rose. 2018. "Welfare analysis: Bridging the partial and general equilibrium divide for policy analysis."*Journal of Benefit-Cost Analysis, 9*(1), 67–83.

Feather, Peter, and W. Douglass Shaw. 1999. "Estimating the cost of leisure time for recreation demand models."*Journal of Environmental Economics and Management, 38(1), 49–65.*

Feldman, Maryann P., and Richard Florida. 1994. "The geographic sources of innovation: Technological infrastructure and product innovation in the United States." *Annals of the Association of American Geographers, 84*(2), 210–229.

Fernandez-Mateo, Isabel. 2009. "Cumulative gender disadvantage in contract employment." *American Journal of Sociology, 114*(4), 871–923.

Ferrara, Alfio, and Silvia Salini. 2012. "Ten challenges in modeling bibliographic data for bibliometric analysis." *Scientometrics, 93*(3), 765–785.

Fisher, Anthony C., and W. Michael Hanemann. 1987. "Quasi-option value: Some misconceptions dispelled." *Journal of Environmental Economics and Management, 14*(2), 183–190.

Flexner, A. (1944) 2017. *The Usefulness of Useless Knowledge.* Princeton, NJ: Princeton University Press.

Florio, Massimo. 2006. "Cost–benefit analysis and the European Union cohesion fund: On the social cost of capital and labour." *Regional Studies, 40*(2), 211–224.

Florio, Massimo. 2014. *Applied Welfare Economics: Cost-Benefit Analysis of Projects and Policies.* London: Routledge.

Florio, M., and F. Giffoni. 2018. *Scientific Research at CERN as a Public Good: A Survey to French Citizens.* CERN-ACC–2018-0024, August 23, 2018.

Florio, Massimo, and Emanuela Sirtori. 2016. "Social benefits and costs of large scale research infrastructures." *Technological Forecasting and Social Change, 112*, 65–78.

Florio, Massimo, Andrea Bastianin, and Paolo Castelnovo. 2018. "The socio–economic impact of a breakthrough in the particle accelerators' technology: A research agenda." *Nuclear Instruments and Method in Physics Research—A, 909*, 21–26.

Florio, Massimo, Stefano Forte, and Emanuela Sirtori. 2016. "Forecasting the socio-economic impact of the Large Hadron Collider: A cost–benefit analysis to 2025 and beyond." *Technological Forecasting and Social Change, 112*, 38–53.

Florio, Massimo, Valentina Morretta, and Witold Willak. 2018. "Cost-benefit analysis and European Union cohesion policy: Economic versus financial returns in investment project appraisal." *Journal of Benefit-Cost Analysis, 9*(1), 147–180.

Florio, Massimo, Stefano Forte, Chiara Pancotti, Emanuela Sirtori, and Silvia Vignetti. 2016. *Exploring Cost-Benefit Analysis of Research, Development and Innovation Infrastructures: An Evaluation Framework.* arXiv:1603.03654.

Florio, Massimo, Francesco Giffoni, Anna Giunta, and Emanuela Sirtori. 2018. "Big science, learning, and innovation: Evidence from CERN procurement." *Industrial and Corporate Change, 27*(5), 1–22.

Flyvbjerg, Bent. 2006. "Five misunderstandings about case-study research." *Qualitative Inquiry, 12*(2), 219–245.

Flyvbjerg, Bent, Nils Bruzelius, and Werner Rothengatter. 2003. *Megaprojects and Risk: An Anatomy of Ambition.* Cambridge: Cambridge University Press.

Fokas, E., G. Kraft, H. An, and R. Engenhart-Cabillic. 2009. "Ion beam radiobiology and cancer: Time to update ourselves." *Biochimica et Biophysica Acta (BBA), 1796*(2), 216–229.

Foray, Dominique. 2004. *Economics of Knowledge*. Cambridge, MA: MIT Press.

Freeman, A. Myrick. 1984. "The sign and size of option value." *Land Economics, 60*(1), 1–13.

Frick, Bernd, and Michael Maihaus. 2016. "The structure and determinants of expected and actual starting salaries of higher education students in Germany: Identical or different?" *Education Economics*, 374–392.

Friedman, Milton. 1962. *Capitalism and Freedom*. Chicago: University of Chicago Press.

Fujiwara, Daniel. 2013. *Museums and Happiness: The Value of Participating in Museums and the Arts*. http://happymuseumproject.org/wp-content/uploads/2013/04/Museums_and_happiness _DFujiwara_April2013.pdf. Access date: March 18, 2019.

Galiani, Sebastian, and Ramiro H. Gálvez. 2017. *The Life Cycle of Scholarly Articles across Fields of Research*. No. 23447. National Bureau of Economic Research.

Galison, Peter, and W. Bruce Hevly. 1992. *Big Science: The Growth of Large-Scale Research*. Stanford, CA: Stanford University Press.

Gambardella, Alfonso, Dietmar Harhoff, and Bart Verspagen. 2008. "The value of European patents." *European Management Review, 5*(2), 69–84.

Garcia Montalvo, José, Josep Raya, Ferran Sancho, and Julia Bosch. 2004. "*Analisis coste beneficio y estudio de impacto economico de una fuente de luz de sincrotron en el valles occidental.*" *Coneixement i Societat*, 09, Barcelona, http://www.econ.upf.edu/~montalvo/wp/cis09_montalvo .pdf. Access date: March 21, 2019.

Garrett, M. A., J. M. Cordes, D. R. Deboer, J. L. Jonas, S. Rawlings, and R. T. Schilizzi. 2010. *Square Kilometre Array: A Concept Design for Phase 1*. arXiv:1008.2871.

Garrod, Guy, and Ken Willis. 1991. *Economic Valuation of the Environment: Methods and Case Studies*. Northampton, UK: Edgar Elgar.

Gertner Jon. 2013. *The Idea Factory: Bell Labs and the Great Age of American Innovation*. New York: Penguin.

Giacomelli, Paolo. 2019. "Assessing CERN impact on careers." *CERN Courier*, March–April, 55–56.

Giffoni, F., Silvia Vignetti, Henning Kroll, Andrea Zenker, Torben Schubert, and Elina Griniece. 2018. *State of Play—Literature Review*, Ri-Paths Project, Horizon 2020 Programme Open Innovation and Open Science Research Infrastructures.

Gittelman, Michelle, and Bruce Kogut. 2003. "Does good science lead to valuable knowledge? Biotechnology firms and the evolutionary logic of citation patterns." *Management Science, 49*(4), 366–382.

Giudice, Gian Francesco. 2010. *A Zeptospace Odyssey: A Journey into the Physics of the LHC*. New York: Oxford University Press.

Gosling, Francis George. 1999. *The Manhattan Project: Making the Atomic Bomb*. Collingdale, PA: Diane Publishing.

Gramlich, Edward M. 1994. "Infrastructure investment: A review essay." *Journal of Economic Literature, 32*(3), 1176–1196.

Greco, Salvatore, Alessio Ishizaka, Menelaos Tasiou, and Gianpiero Torrisi. 2017. "On the methodological framework of composite indices: A review of the issues of weighting, aggregation, and robustness." *Social Indicators Research*: 1–34. https://doi.org/10.1007/s11205 -017-1832-9.

Green Land and EARSC. 2016. *Copernicus Sentinels' Products Economic Value: A Case Study of Forest Management in Sweden,* http://earsc.org/news/copernicus-sentinels-products-economic -value-study. Access date: March 18, 2019.

Greenley, Douglas A., Richard G. Walsh, and Robert A. Young. 1981. "Option value: Empirical evidence from a case study of recreation and water quality." *Quarterly Journal of Economics, 96*(4), 657–673.

Griliches, Zvi. 1958. "Research costs and social returns: Hybrid corn and related innovations." *Journal of Political Economy*, *66*(5): 419–431.

Griliches, Zvi. 1979. "Issues in assessing the contribution of research and development to productivity growth." *Bell Journal of Economics*, 92–116.

Griliches, Zvi. 1980. *R&D and the Productivity Slowdown*. National Bureau of Economic Research (NBER), Working Paper No. 434, Cambridge MA.

Griliches, Zvi. 1981."Market value, R&D, and patents." *Economics Letters*, *7*(2), 183–187.

Griliches, Zvi. 1998. "Issues in assessing the contribution of research and development to productivity growth." In Griliches (ed.), *R&D and Productivity: The Econometric Evidence* (pp. 17–45). Chicago: University of Chicago Press.

Guerzoni, Marco, and Emilio Raiteri. 2015. "Demand-side vs. supply-side technology policies: Hidden treatment and new empirical evidence on the policy mix." *Research Policy*, *44*(3), 726–747.

Guha-Sapir, Debarati, Indhira Santos, and Alexandre Borde. 2013. *The Economic Impacts of Natural Disasters*. New York: Oxford University Press.

Gupta, H., S. Kumar, S. K. Roy, and R. S. Gaud. 2010. "Patent protection strategies." *Journal of Pharmacy and Bioallied Sciences*, *2*(1), 2.

Hagström, Warren O. 1965. *The Scientific Community*. New York: Basic Books.

Håkansson, H., D. Ford, L.-E. Gadde, I. Snehota, and A. Waluszewski. 2009. *Business in Networks*. Chichester, UK: John Wiley & Sons.

Hall, Bronwyn. H. 1996. "The Private and Social Returns to Research and Development." In Bruce L. R. Smith and Claude E. Barfield (eds.), *Technology, R&D, and the Economy*. Washington, DC: Brookings Institution.

Hall, Bronwyn H. 2007. *Measuring the Returns to R&D: The Depreciation Problem*. No. 13473. National Bureau of Economic Research (NBER), Cambridge MA. http://www.nber.org/papers/w13473. Access date: March 18, 2019.

Hall, Bronwyn H., Adam Jaffe, and Manuel Trajtenberg. 2005. "Market value and patent citations." *RAND Journal of Economics*, 16–38.

Hall, Bronwyn H., Grid Thoma, and Salvatore Torrisi. 2007. *The Market Value of Patents and R&D: Evidence from European Firms*. NBER Working Paper 13426. Cambridge, MA.

Hall, Bronwyn H., Jacques Mairesse, and Pierre Mohnen. 2009. "Measuring the returns to R&D." In Bronwyn H. Hall and Nathan Rosenberg (eds.), *Handbook of the Economics of Innovation* (Vol. 2, pp. 1033–1082). North-Holland.

Hallin, Emil. 2012. *Economic, Societal, and Technological Impact of Large-Scale Research Facilities: A View from Canada*. https://indico.desy.de/indico/event/5340/session/1/material/slides/0?contribId=6. Access date: 3/18/2019.

Hallonsten, Olof. 2016. *Big Science Transformed*. Palgrave Macmillan.

Hallonsten, Olof, Mats Benner, and Gustav Holmberg. 2004. *Impacts of Large-Scale Research Facilities: A Socio-economic Analysis*. School of Economics and Management, Lund University, Lund, Sweden.

Han, B. O., and John Windsor. 2011. "User's willingness to pay on social network sites." *Journal of Computer Information Systems*, *51*(4), 31–40.

Hanley, Nick, and Edward Barbier. 2009. *Pricing Nature: Cost-Benefit Analysis and Environmental Policy*. Cheltenham, UK: Edward Elgar Publishing.

Hanley, Nick, and Clive L Spash, 1993. *Preferences, Information and Biodiversity Preservation*. Working Paper, Series 93/12, University of Stirling, Division of Economics.

Hargreaves, Ian. 2011. "Digital Opportunity: A Review of Intellectual Property and Growth." Independent report.

Harhoff, Dietmar, Francis Narin, Frederic M. Scherer, and Katrin Vopel. 1999. "Citation frequency and the value of patented inventions." *Review of Economics and Statistics*, *81*(3), 511–515.

Harhoff, D., Frederic M. Scherer, and Katrin Vopel. 2003a. "Citations, family size, opposition and the value of patent rights," *Research Policy*, 32, pp. 1343–1363.

Harhoff, D., Frederic M. Scherer, and Katrin Vopel. 2003b. "Exploring the tail of patented invention value distribution." In O. Granstrand (ed.), *Economics, Law, and Intellectual Property: Seeking Strategies for Research and Teaching in a Developing Field* (pp. 279–308). Boston: Kluwer Academic Publishers.

Harmon, Colm, Hessel Oosterbeek, and Ian Walker. 2003. "The returns to education: Micro-economics." *Journal of Economic Surveys*, *17*(2), 115–156.

Harzing, Anne-Wil. 2010. *Citation Analysis across Disciplines: The Impact of Different Data Sources and Citation Metrics*. University of Melbourne.

Hazari, Z., G. Sonnert, P. M. Sadler, and M. C. Shanahan. 2010. "Connecting high school physics experiences, outcome expectations, physics identity, and physics career choice: A gender study." *Journal of Research in Science Teaching*, *47*(8), 978–1003.

Heckman, James J., Lance J. Lochner, and Petra E. Todd. 2006. "Earnings functions, rates of return, and treatment effects: The Mincer equation and beyond." *Handbook of the Economics of Education*, *1*, 307–458.

Helmers, Christian, and Henry Overman. 2017a. *The Impact of Big Research Infrastructure Facilities on Local Scientific Output*. CEPR Policy portal. https://voxeu.org/article/impact-big-research-infrastructure-facilities-local-scientific-output.

Helmers, Christian, and Henry G. Overman. 2017b. "My precious! The location and diffusion of scientific research: Evidence from the Synchrotron Diamond Light Source." *Economic Journal*, *127*(604), 2006–2040.

Hermansson, Gabriele. 2013. "One dollar news: User needs and willingness to pay for digital news content." Dissertation thesis, Graduate School of Business and Law, RMIT University, Melbourne City, Australia.

Heuer, Chris, 2012. "Measuring and capturing the value of social media." *Wall Street Journal*. http://deloitte.wsj.com/cio/2012/10/25/measuring-and-capturing-the-value-of-social-media-investments/. Access date: March 18, 2019.

Hicks, Diana. 2004. "The four literatures of social science." In H. F. Moed, W. Glänzel, and U. Schmoch (eds.), *Handbook of Quantitative Science and Technology Research* (pp. 473–496). Dordrecht, Netherlands: Springer.

Hicks, D., P. Wouters, L. Waltman, S. De Rijcke, and I. Rafols. 2015. "The Leiden Manifesto for research metrics." *Nature*, *520*(7548), 429.

Higgs, Peter W. 1964. "Broken symmetries and the masses of gauge bosons." *Physical Review Letters*, *13*(16), 508.

Hirsch, Jorge E. 2005. "An index to quantify an individual's scientific research output." *Proceedings of the National Academy of Sciences of the United States of America*, *102*(46), 16569–16572.

HM Treasury. 2006. *Stern Review: The Economics of Climate Change*. London. http://mudancasclimaticas.cptec.inpe.br/~rmclima/pdfs/destaques/sternreview_report_complete.pdf. Access date: March 18, 2019.

HM Treasury. 2018. *Green Book, Central Government Guidance on Appraisal and Evaluation*. https://assets.publishing.service.gov.uk/government/uploads/system/uploads/attachment_data/file/685903/The_Green_Book.pdf. Access date: March 18, 2019.

Hoffmann, Hans F., Markus Nordberg, and Max Boisot. 2011. "ATLAS and e-Science." In Max Boisot, Marcus Nordberg, Said Yami, and Bertrand Nicquevert (eds.), *Collisions and Collaboration: The Organization of Learning in the ATLAS Experiment at the LHC* (pp. 247–267). Oxford: Oxford University Press.

Hogue, Mary, Cathy LZ Dubois, and Lee Fox-Cardamone. 2010. "Gender differences in pay expectations: The roles of job intention and self-view." *Psychology of Women Quarterly*, *34*(2), 215–227.

Hood, Leroy, and Lee Rowen. 2013. "The Human Genome Project: Big Science transforms biology and medicine." *Genome Medicine*, *5*(9), 79.

Hotelling, Harold. 1947. "An economic study of the monetary evaluation of recreation in national parks." Letter of June 18, 1947, to Newton B. Dury. Included in the report *The Economics of Public Recreation*: 1949. Land and Recreational Planning Division, National Park Service, Washington, D.C.

Hough, J. R. 1994. "Educational cost-benefit analysis—education." *Education Economics*, *2*(2), 93–128.

Hsu, David H., and Rosemarie H. Ziedonis. 2008. "Patents as quality signals for entrepreneurial ventures." *Academy of Management Proceedings, 2008*(1).

Hughes, Jeff A. 2002. *The Manhattan Project: Big Science and the Atom Bomb*. New York: Columbia University Press.

IDEA Consulting. 2015. Economic footprint of 9 European RTOs in 2015–2016. IDEA REPORT for EARTO. https://www.earto.eu/earto-publishes-new-economic-footprint-study-the-impact-of-9 -rtos-in-2015-2016/. Access date: March 18, 2019.

Iglesias, Juan, and Carlos Pecharromán. 2007. "Scaling the h-index for different scientific ISI fields." *Scientometrics*, *73*(3), 303–320.

Institute of Physics (IOP). 2011. *Physics in Ireland: The Brightest Minds Go Further*. http://www .iopireland.org/publications/iopi/file_51212.pdf.

Institute of Physics (IOP). 2012. *The Career Paths of Physics Graduates. A Longitudinal Study 2006–2010*. Institute of Physics Report, London.

Irwin, Douglas A., and Peter J. Klenow. 1994. "Learning-by-doing spillovers in the semiconductor industry." *Journal of Political Economy*, *102*(6), 1200–1227.

Irvine, John, and Ben R. Martin. 1984. "CERN: Past performance and future prospects: II. The scientific performance of the CERN accelerators." *Research Policy*, *13*(5), 247–284.

Islam, Talat, Ishfaq Ahmed, Zainab Khalifah, Misbah Sadiq, and Muhammad Faheem. 2015. "Graduates' expectation gap: the role of employers and Higher Learning Institutes." *Journal of Applied Research in Higher Education*, *7*(2), 372–384.

IVA (Kungl. Ingenjörsvetenskapsakademien). 2012. *Review of Literature on Scientists' Research Productivity*. https://www.iva.se/globalassets/rapporter/agenda-for-forskning/review20of20literat ure20on20scientistse2809920research20productivity1.pdf. Access date: March 18, 2019.

Jaffe, Adam B. 1989. "Real effects of academic research." *American Economic Review*, *79*(5), 957–970.

Jaffe, Adam B., and Manuel Trajtenberg. 1999. "International knowledge flows: Evidence from patent citations." *Economics of Innovation and New Technology*, *8*(1–2), 105–136.

Jaffe, Adam B., Manuel Trajtenberg, and M. S. Fogarty. 2000. "Knowledge spillovers and patent citations: Evidence from a survey of inventors." *American Economic Review*, *90*(2000) 215–218.

Jaffe, Adam B., Manuel Trajtenberg, and Rebecca Henderson. 1993. "Geographic localization of knowledge spillovers as evidenced by patent citations." *Quarterly Journal of Economics*, *108*(3), 577–598.

Johannesson, M., and P. O. Johansson 1996. "To be or not to be, that is the question: An empirical study of the WTP for an increased life expectancy at an advanced age." *Journal of Risk and Uncertainty*, *13*, 163–174.

Johansson, Per-Olov. 1987. *The Economic Theory and Measurement of Environmental Benefits*. Cambridge University Press.

Johansson, Per-Olov, and Bengt Kriström. 2015. *Cost-Benefit Analysis for Project Appraisal*. Cambridge: Cambridge University Press.

Johansson, Per-Olov, and Bengt Kriström. 2018. *Cost-Benefit Analysis*. Cambridge: Cambridge University Press.

Johnson, F. Douglas, and Martin Kokus. 1978. *NASA Technology Utilization Program: A Cost Benefit Evaluation*. Space Congress Proceedings.

Johnston, R. J., et al. 2017. "Contemporary guidance for stated preference studies." *Journal of the Association of Environmental and Resource Economists*, *4*(2), 319–405.

Jones, Heather, Filipe Moura, and Tiago Domingos. 2014. "Transport infrastructure project evaluation using cost-benefit analysis." *Procedia-Social and Behavioral Sciences*, *111*, 400–409.

Jorge-Calderón, Doramas. 2014. *Aviation Investment: Economic Appraisal for Airports, Air Traffic Management, Airlines and Aeronautics*. Ashgate Publishing Limited.

Josephson, Paul R. 2010. *Lenin's Laureate: Zhores Alferov's Life in Communist Science*. Cambridge, MA: MIT Press.

Jura Consultant. 2005. *Bolton's Museums, Library and Archive Services an Economic Valuation*. Bolton Metropolitan Borough Council and MLA North West. http://webarchive.nationalarchives.gov.uk/20120215214224/http://research.mla.gov.uk/evidence/documents/bolton_main.pdf. Access date: March 19, 2019.

Jusoh, M., M. Simun, and S. Choy Chong. 2011. "Expectation gaps, job satisfaction, and organizational commitment of fresh graduates: Roles of graduates, higher learning institutions and employers." *Education + Training*, *53*(6), 515–530.

Kahle, Logan Q., Maureen E. Flannery, and John P. Dumbacher. 2016. "Bird-window collisions at a west-coast urban park museum: Analyses of bird biology and window attributes from Golden Gate Park, San Francisco." *PLoS ONE*, *11*(1), e0144600.

Kamer, P. M., 2005. *Brookhaven National Lab Economic Impact Report*. Brookhaven National Laboratory. https://www.bnl.gov/NYCCS/impact.php.

Karnofsky, David A., and Joseph H. Burchenal. 1949. *Evaluation of Chemotherapeutic Agents*. New York: Columbia University Press.

Keaveny, Timothy J., and Edward J. Inderrieden. 2000. "Gender differences in pay satisfaction and pay expectations." *Journal of Managerial Issues*, 363–379.

Keef, Stephen P., and Melvin L. Roush. 2001. "Discounted cash flow methods and the fallacious reinvestment assumption: A review of recent texts." *Accounting Education*, *10*(1), 105–116.

Kerr, Clark. 1963. *The Uses of the University*. London: Harvard University Press.

King, Jean. 1987. "A review of bibliometric and other science indicators and their role in research evaluation. *Journal of Information Science*, *13*(5), 261–276.

Knack Stephen, and Philip Keefer. 1997. "Does social capital have an economic payoff? A cross-country investigation." *Quarterly Journal of Economics*, *112*(4), 1251–1288.

Knapp, Alex, 2012. "How much does it cost to find a Higgs boson?" *Forbes*. https://www.forbes.com/sites/alexknapp/2012/07/05/how-much-does-it-cost-to-find-a-higgs-boson/#b740d1a39480. Access date: March 19, 2019.

Kogut, Bruce, and Anca Metiu. 2001. "Open-source software development and distributed innovation." *Oxford Review of Economic Policy*, *17*(2), 248–264.

Kolb, David. 1984. *Experiential Learning as the Science of Learning and Development*, 1st ed. Englewood Cliffs, NJ: Prentice Hall.

Kolb, David. 2015. *Experiential Learning: Experience as the Source of Learning and Development*, 2nd ed. Pearson Education.

Kolb, David, and Ronald Eugene Fry. 1974. "Toward an applied theory of experiential learning." In C. Cooper (ed.), *Theories of Group Process* (pp. 33–57). London: John Wiley.

KPMG. 2014. *Australia: National Collaborative Research Infrastructure Strategy Project Review—Overarching Report,* https://docs.education.gov.au/node/38305. Access date: 3/19/2019.

KPMG. 2016. *Assessment of Snolab Final Report*. KPMG LLP.

Krutilla, John V. 1967. "Conservation reconsidered." *American Economic Review*, *57*(4), 777–786.

Kunreuther, H., and Michel-Kerjan, E. 2014. "Economics of natural catastrophe risk insurance." In M. J. Machina and W. K. Viscusi (eds.), *Handbook of the Economics of Risk and Uncertainty* (Vol. 1, pp. 651–695). Elsevier.

Lafleur, Claude. 2010. "Costs of US piloted programs" *The Space Review*, 8, http://www .thespacereview.com/article/1579/1. Access date: March 19, 2019.

Laibson, David. 1997. "Golden eggs and hyperbolic discounting." *Quarterly Journal of Economics*, *112*(2), 443–477.

Laird, James, and Peter Mackie. 2009. *Review of Economic Assessment in Rural Transport Appraisal,* Scottish government. https://trid.trb.org/view/909111. Access date: March 19, 2019.

Landefeld, J. Steven, and Eugene P. Seskin. 1982 "The economic value of life: Linking theory to practice." *American Journal of Public Health*, *72*(6), 555–566.

Lanjouw, Jean Olson. 1998. "Patent protection in the shadow of infringement: Simulation estimations of patent value." *Review of Economic Studies*, 65, 671–712.

Larivière, Vincent, Éric Archambault, and Yves Gingras. 2008. "Long-term variations in the aging of scientific literature: From exponential growth to steady-state science (1900–2004)." *Journal of the Association for Information Science and Technology*, *59*(2), 288–296.

Lawrence Berkeley National Laboratory (LBNL). 2001. *Ernest Orlando Lawrence Berkeley National Laboratory, Economic Impact 2001*. Technology Transfer Department, University of California, Berkeley, CA. http://www.osti.gov/scitech/servlets/purl/813372-mR0sHR/native/. Access date: March 19, 2019.

Lawton Smith, H., and K. Ho. 2006. "Measuring the performance of Oxford University, Oxford Brookes University and the government laboratories' spin-off companies." *Research Policy*, *35*(10), 1554–1568.

Lebrun, Philippe, and Thomas Taylor. 2017. "Managing the laboratory and large projects." In C. Fabjan, T. Taylor, D. Treille, and H. Wenninger (eds.), *Technology Meets Research. 60 Years of CERN Technology: Selected Highlights* (pp. 393–417). Singapore: World Scientific Publishing.

Le Goff, Jean-Marie, Rolf Heuer, Jean-Pierre Koutchouk, Steinar Stapnes, and Svetlomir Stavrev. 2011. *CERN Studies on the Socio-economic Impacts of Particle Physics.* https://cds.cern.ch/record /1431474?ln=en. Access date: March 19, 2019.

Lee, Hyeon Gon. 2018. *Case Study of Socio-economic Impact of Research Infrastructures: ITER Korean Project*. International Workshop on Establishing a Reference Framework for Assessing the SEIRI, GSF, OECD 19–20, Paris.

Lee, P. 2017. "Centralization, fragmentation, and replication in the genomic data commons." In K. J. Strandburg, B. M. Frischmann, and M. J. Madison (eds.), *Governing Medical Knowledge Commons* (pp. 46–74). New York: Cambridge University Press.

Lemieux, Thomas. 2006. "The 'Mincer equation' thirty years after schooling, experience, and earnings." In Shoshanna Grossbard (ed.), *Jacob Mincer: A Pioneer of Modern Labor Economics* (pp. 127–145). New York: Springer.

Lemire, Daniel. 2014. *The Financial Value of Open Source Software*. https://lemire.me/blog/2014 /04/14/the-financial-value-of-open-source-software/. Access date: March 19, 2019.

Leone, Maria Isabella, and Raffaele Oriani. 2008. "Explaining the Remuneration Structure of Patent Licenses." In *2008 Annual Meeting of the Academy of Management, 8–13 August*. http:// www.realoptions.org/papers2010/219.pdf. Access date: March 19, 2019.

Leontief, W. 1941. *The Structure of the American Economy 1919–1929*. New York: Oxford University Press.

Leontief, W. 1966. *Input-Output Economics*. New York: Oxford University Press.

Lerner, J. 1994. "The importance of patent scope: an empirical analysis." *RAND Journal of Economics*, 319–333.

Lerner, Josh, and Jean Tirole. 2002. "Some simple economics of open source." *Journal of Industrial Economics,* *50*(2), 197–234.

Lerner, Josh, and Jean Tirole. 2005a. "The economics of technology sharing: Open source and beyond." *Journal of Economic Perspectives*, *19*(2), 99–120.

Lerner, Josh, and Jean Tirole. 2005b. "The scope of open source licensing." *Journal of Law, Economics, and Organization*, *21*(1), 20–56.

Li, Jiang, and Y. Fred Ye. 2014. "A probe into the citation patterns of high-quality and high-impact publications." *Malaysian Journal of Library & Information Science, 19*(2), 17–33

Li, Wendy C. Y. 2012. *Depreciation of Business R&D Capital.* No. 22473. National Bureau of Economic Research. U.S. Bureau of Economic Analysis.

Li, Xibao. 2012 "Behind the recent surge of Chinese patenting: An institutional view." *Research Policy, 41*(1), 236–249.

Lichtenberg, Frank R. 1988. "The private R and D investment response to federal design and technical competitions." *American Economic Review, 78*(3), 550–559.

Lindhjem, H., S. Navrud, N. A. Braathen, and V. Biausque. 2011. "Valuing mortality risk reductions from environmental, transport, and health policies: A global meta-analysis of stated preference studies." *Risk Analysis, 31*(9), 1381–1407.

Link, Albert N., and John T. Scott. 2004. "Evaluating public sector R&D programs: The advanced technology program's investment in wavelength references for optical fiber communications." *Journal of Technology Transfer, 30*(1–2), 241–251.

Linstone, Harold A., and Murray Turoff (eds.). 1975. *The Delphi Method.* Reading, MA: Addison-Wesley.

Lissoni, Francesco, Jacques Mairesse, Fabio Montobbio, and Michele Pezzoni. 2011. "Scientific productivity and academic promotion: A study on French and Italian physicists." *Industrial and Corporate Change, 20*(1), 253–294.

Little, Ian M., and James A. Mirrlees. 1974. *Project Appraisal and Planning for Developing Countries.* New York: Heinemann Educational Books.

Liyanage, Shantha, and Max Boisot. 2011. "Leadership in the ATLAS collaboration." In Max Boisot, Marcus Nordberg, Said Yami, and Bertrand Nicquevert (eds.), *Collisions and Collaboration: The Organization of Learning in the ATLAS Experiment at the LHC* (pp. 226–247). Oxford: Oxford University Press.

Loeffler, Jay S., and Marco Durante. 2013. "Charged particle therapy—optimization, challenges, and future directions." *Nature Reviews Clinical Oncology, 10*(7), 411.

Loomis, John B., and Richard G. Walsh. 1997. *Recreation Economic Decisions: Comparing Benefits and Costs.* Edmonton, Canada: Venture Publishing Inc.

Lundvall, Bengt-Ake. 1985. "Product innovation and user-producer interaction." In *The Learning Economy and the Economics of Hope* (pp. 19–58). London: Anthem Press.

Lundvall, Bengt-Åke. 1993. "User–producer relationships, national systems of innovation and internationalization." In D. Forey and C. Freeman (eds.), *Technology and the Wealth of the Nations—The Dynamics of Constructed Advantage.* London: Pinter.

Macilwain, Colin. 2010. "Science economics: What science is really worth." *Nature, 465*(7299), 682–685.

Maessen, Kas, Algis Krupavičius, and Ricardo Migueis. 2016. *Strategic Priorities, Funding and Pan-European Co-operation for Research Infrastructures in Europe.* Science Europe Working Group on Research Infrastructures, Survey Report.

Maihaus, Michael. 2014. *The Economics of Higher Education in Germany: Salary Expectations, Signaling, and Social Mobility.* Paderborn, Germany: Tectum Verlag.

Malesios, C. C., and Stelios Psarakis. 2012. "Comparison of the h-index for different fields of research using bootstrap methodology." *Quality & Quantity, 48*(1), 521–545.

Malone, Thomas W., and Michael S. Bernstein. 2015. *Handbook of Collective Intelligence.* Cambridge, MA: MIT Press.

Manganote, Edmilson J. T., Peter A. Schulz, and Carlos Henrique de Brito Cruz. 2016. "Effect of high energy physics large collaborations on higher education institutions citations and rankings." *Scientometrics, 109*(2), 813–826.

Manigart, Sophie, Koen De Waele, Mike Wright, Ken Robbie, Philippe Desbrières, H. Sapienza, and A. Beekman. 2000. "Venture capital, investment appraisal and accounting information: A comparative study of the US, UK, France, Belgium and Holland." *European Financial Management, 6,* 380–404.

Manigart, Sophie, Mike Wright, Ken Robbie, Philippe Desbrières, and Koen De Waele. 1997. "Venture capitalists' appraisal of investment projects: An empirical European study." *Entrepreneurship Theory and Practice, 21*(4), 29–43.

Mansfield, Edwin. 1991. "Academic research and industrial innovation." *Research Policy, 20*(1), 1–12.

Mansfield, Edwin. 1998. "Academic research and industrial innovation: An update of empirical findings 1." *Research Policy, 26*(7–8), 773–776.

Mansfield, Edwin, and Jeong-Yeon Lee. 1996. "The modern university: Contributor to industrial innovation and recipient of industrial R&D support." *Research Policy, 25*(7), 1047–1058.

Mansfield, Edwin, John Rapoport, Anthony Romeo, Samuel Wagner, and George Beardsley. 1977. "Social and private rates of return from industrial innovations." *Quarterly Journal of Economics, 91*(2), 221–240.

Marks, N. 1998. "Synchrotron Radiation Projects of Industrial Interest." *EPAC.* Vol. 98. CLRC Daresbury Laboratory, Warrington, UK. http://accelconf.web.cern.ch/AccelConf/e98/papers/wei05b .pdf. Access date: March 20, 2019.

Marshall, Alfred, 1890. *Principles of Economics.* London: Macmillan.

Martin, Ben R. 1996. "The use of multiple indicators in the assessment of basic research." *Scientometrics, 36*(3), 343–362.

Martin, Ben R., and John Irvine. 1984a. "CERN: Past performance and future prospects: CERN's position in world high-energy physics." *Research Policy 13*(4), 183–210.

Martin, Ben R., and John Irvine. 1984b. "CERN: Past performance and future prospects: III. CERN and the future of world high-energy physics." *Research Policy, 13*(6), 311–342.

Martin, Ben R., and Puay Tang. 2007. *The Benefits from Publicly Funded Research.* Science Policy Research Unit, University of Sussex, Brighton, UK.

Mazzucato, Mariana. 2015. *The Entrepreneurial State: Debunking Public vs. Private Sector Myths.* Anthem Press.

Mazzucato, Mariana. 2016. "From market fixing to market-creating: A new framework for innovation policy." *Industry and Innovation, 23*(2), 140–156.

Mazzucato, Mariana. 2018. *Mission-Oriented Research and Innovation in the European Union: A Problem-Solving Approach to Fuel Innovation-Led Growth.* European Commission, Bruxelles. https://ec.europa.eu/info/sites/info/files/mazzucato_report_2018.pdf. Access date: March 20, 2019.

McIntosh, Emma, Philip M. Clarke, Emma J. Frew, and Jordan J. Louviere. 2010. *Applied Methods of Cost–Benefit Analysis in Health Care.* New York: Oxford University Press.

McMillan, G. Steven, Francis Narin, and David L. Deeds. 2000. "An analysis of the critical role of public science in innovation: The case of biotechnology. *Research Policy, 29*(1), 1–8.

McNeil, Laurie, and Paula Heron. 2017. "Preparing physics students for 21st-century careers." *Physics Today, 70*(11), 38–43.

McPherson, Amanda, Brian Proffitt, and Ron Hale-Evans. 2008. *Estimating the Total Cost of a Linux Distribution.* The Linux Foundation. https://www.linux.com/publications/estimating-total -cost-linux-distribution. Access date: March 20, 2019.

Mendes, Isabel, and Isabel Proenca. 2005. *Estimating the Recreation Value of Ecosystems by Using a Travel Cost Method Approach.* Working Paper 2005/08, Department of Economics at the School of Economics and Management (ISEG), Technical University of Lisbon, Portugal.

Messina, Vincenzina, and Valentina Bosetti. 2003. "Uncertainty and option value in land allocation problems." *Annals of Operations Research, 124*(1–4), 165–181.

Mincer, Jacob. 1958. "Investment in human capital and personal income distribution." *Journal of Political Economy, 66*(4), 281–302.

Mincer, Jacob. 1974. *Schooling, Experience, and Earnings.* New York: Columbia University Press.

Ministère de l'Éducation Nationale, de l'Enseignemen Superieur et de la Recherche. 2016. "French National Strategy on Research Infrastructures." Paris. https://cache.media.enseignementsup -recherche.gouv.fr/file/Infrastructures_de_recherche/16/4/infrastructures_UK_web_615164.pdf. Access date: March 20, 2019.

Mitchell, Robert Cameron, and Richard T. Carson. 1989. *Using Surveys to Value Public Goods: The Contingent Valuation Method.* Baltimore: Johns Hopkins University Press.

MMK Consulting. 2009. *Economic and Social Impacts of TRIUMF.* http://www.triumf.ca/sites /default/files/MMK-EconomicImpact-Study-vFINAL.pdf. Access date: March 20, 2019.

Moed, H. F., W. J. M. Burger, J. G. Frankfort, and A. F. Van Raan. 1985. "The use of bibliometric data for the measurement of university research performance." *Research Policy, 14*(3), 131–149.

Monjon, Stephanie, and Patrick Waelbroeck. 2003. "Assessing spillovers from universities to firms: Evidence from French firm-level data." *International Journal of Industrial Organization, 21*(9), 1255–1270.

Montenegro, Claudio E., and Harry A. Patrinos. 2014. *Comparable Estimates of Returns to Schooling around the World.* Policy Research Working Paper 7020, World Bank Group.

Mouqué, Daniel. 2012. "What are counterfactual impact evaluations teaching us about enterprise and innovation support." *Regional Focus* 2.

Mowery, David, and Nathan Rosenberg. 1979. "The influence of market demand upon innovation: a critical review of some recent empirical studies." *Research Policy, 8*(2), 102–153.

Mrozek, Janusz R., and Laura O. Taylor. 2002. "What determines the value of life? A meta-analysis." *Journal of Policy Analysis and Management, 21*(2), 253–270.

Multihazard Mitigation Council (MMC). 2005. *Natural Hazard Mitigation Saves: An Independent Study to Assess the Future Savings from Mitigation Activities.* Vol. 2—Study Documentation. Washington, DC: Multihazard Mitigation Council.

Mulvey, Patrick, and Jack Pold. 2015. *Physics Masters One Year after Degree Results from the Follow-Up Survey of Master's Recipients, Classes of 2012, 2013, and 2014 Combined.* American Institute of Physics, https://www.aps.org/careers/statistics/upload/masters-oneyear1215.pdf. Access date: March 20, 2019.

Mulvey, Patrick, and Jack Pold. 2016. *Physics Doctorates Initial Employment Data from the Degree Recipient Follow-Up Survey for the Classes of 2013 and 2014.* American Institute of Physics. www.aip.org/statistics. Access date: March 20, 2019.

Mulvey, Patrick, and Jack Pold. 2017. *Physics Bachelors: Initial Employment.* American Institute of Physics. https://www.aip.org/sites/default/files/statistics/employment/bachinitemp-p–14.1.pdf. Access date: March 20, 2019.

Mustar, Philippe. 1997. "How French academics create hi-tech companies: the conditions for success or failure." *Science and Public Policy, 24*(1), 37–43.

Nadiri, M. Ishaq. 1993. *Innovations and Technological Spillovers.* No. 4423. National Bureau of Economic Research (NBER).

Nadiri, M. Ishaq, and Ingmar R. Prucha. 1993. *Estimation of the Depreciation Rate of Physical and R&D Capital in the US Total Manufacturing Sector.* NBER Working Paper No. 4591.

Narin, Francis, and Dominic Olivastro. 1992. "Status report: linkage between technology and science." *Research Policy, 21*(3), 237–249.

National Aeronautics and Space Administration (NASA). 2015. "Final memorandum, audit of NASA's management of International Space Station operations and maintenance contracts" (IG-15-021; A-14-023-00). NASA Office of Inspector of General Office of Audits, Washington, DC. https://oig.nasa.gov/audits/reports/FY15/IG-15-021.pdf. Access date: March 20, 2019.

National Aeronautics and Space Administration (NASA). 2018a. *FY 2019 Budget Estimate.* https://www.nasa.gov/sites/default/files/atoms/files/fy19_nasa_budget_estimates.pdf. Access date: March 20, 2019.

National Aeronautics and Space Administration (NASA). 2018b. *Spinoff*, Technology Transfer Program. https://spinoff.nasa.gov/Spinoff2018/pdf/Spinoff2018.pdf. Access date: March 20, 2019.

National Audit Office (NAO). 2016. *BIS's Capital Investment in Science Projects.* Report by the Comptroller and Auditor General. https://www.nao.org.uk/wp-content/uploads/2016/03/Capital-investment-in-science-projects.pdf. Access date: March 20, 2019.

National Institutes of Health (NIH). 2018. *Strategic Plan for Data Science.* Draft version. https://grants.nih.gov/grants/rfi/NIH-Strategic-Plan-for-Data-Science.pdf. Access date: March 20, 2019.

National Research Council Canada (NRCC). 2013. *Return on Investment in Large-Scale Research Infrastructure.* http://www.triumf.ca/sites/default/files/HAL-ReturnOnInvestmentStudy-May-2013.pdf. Access date: March 20, 2019.

Nederhof, Anton J. 2006. "Bibliometric monitoring of research performance in the social sciences and the humanities: A review." *Scientometrics, 66*(1), 81–100.

Nelson, Robert H. 1997. "Does existence value exist? Environmental economics encroaches on religion." *The Independent Review, 1*(4), 499–521.

Nemet, Gregory F. 2012. "Inter-technology knowledge spillovers for energy technologies." *Energy Economics, 34*(5), 1259–1270.

Newcombe, Robert. 1999. "Procurement as a learning process." In *Profitable Partnering in Construction Procurement*, 285–94.

Newman, John Henry Cardinal.1992 (orig. 1854). *The Idea of a University.* Notre Dame, IN: University of Notre Dame Press.

Nielsen, Trine Louise Brøndt. 2014. "From master's programme to labour market: A study on physics graduates' of the transition to the labour market." Master's thesis, University of Copenhagen.

Nilsen, Vetle, and Giovanni Anelli. 2016. "Knowledge transfer at CERN." *Technological Forecasting and Social Change, 112*, 113–120.

Nordberg, Markus, Alexandra Campbell, and Alain Verbeke. 2003. "Using customer relationships to acquire technological innovation: A value-chain analysis of supplier contracts with scientific research institutions." *Journal of Business Research, 56*(9), 711–719.

Norwegian Association of Higher Education Institutions. 2014. *A Norwegian Research Infrastructure Resource Model: A Methodology for Declaring the Costs and Pricing the Use of Research Infrastructure in Externally Funded Projects at Universities and Colleges.* http://www.uhr.no/documents/a_norwegian_research_infrastructure_resource_model_270214.pdf. Access date: March 20, 2019.

O'Brien, Dave. 2010. *Measuring the Value of Culture*: *A Report to the Department for Culture Media and Sport.* Economic and Social Research Council. https://assets.publishing.service.gov.uk/government/uploads/system/uploads/attachment_data/file/77933/measuring-the-value-culture-report.pdf. Access date: March 20, 2019.

O'Byrne, John, Alberto Mendez, Manjula Sharma, Les Kirkup, and Dale Scott. 2008. "Physics graduates in the workforce: Does physics education help?" School of Physics, Sydney University and Department of Physics & Advanced Materials, University of Technology Sydney. http://www.physics.usyd.edu.au/super/ALTC/publications/AIP-Graduates(Paper).pdf. Access date: March 20, 2019.

O'Leary, Nigel C., and Peter J. Sloane. 2005. "The return to a university education in Great Britain." *National Institute Economic Review, 193*(1), 75–89.

Organisation of Economic Co-operation and Development (OECD). 2006. *Quasi option Value in Cost Benefit Analysis and the Environment: Recent Developments.* Paris: OECD Publishing.

Organisation of Economic Co-operation and Development (OECD). 2009. *OECD Patent Statistics Manual.* Paris: OECD Publishing. https://www.oecd-ilibrary.org/docserver/9789264056442-en.pdf?expires=1553095039&id=id&accname=guest&checksum=2A96CF022AC1292883C342067FDC094C. Access date: March 20, 2019.

Organisation of Economic Co-operation and Development (OECD). 2010. *Valuing Lives Saved from Environmental, Transport and Health Policies: A Meta-analysis of Stated Preference Studies.* Working Party on National Environmental Policies. Paris. http://www.oecd.org/officialdocuments /publicdisplaydocumentpdf/?doclanguage=en&cote=env/epoc/wpnep(2008)10/final. Access date: May 27, 2019.

Organisation of Economic Co-operation and Development (OECD). 2012. *The Value of Statistical Life: A Meta-Analysis.* In Working Party on National Environmental Policies. Paris: OECD Publishing.

Organisation of Economic Co-operation and Development (OECD). 2014a. *International Distributed Research Infrastructure: Issues and Options.* Paris: OECD Publishing. https://www.oecd.org /sti/sci-tech/international-distributed-research-infrastructures.pdf.

Organisation of Economic Co-operation and Development (OECD). 2014b. *Report on the Impacts of Large Research Infrastructures on Economic Innovation and on Society.* Case studies at CERN, Global Science Forum report. Paris: OECD Publishing.

Organisation of Economic Co-operation and Development (OECD). 2015a. *Government at a Glance 2015.* Paris: OECD Publishing. https://www.oecd-ilibrary.org/docserver/gov_glance-2015 -en.pdf?expires=1553095449&id=id&accname=guest&checksum=5F90863AFF5DD7208D2928 F2C192455A. Access date: March 20, 2019.

Organisation of Economic Co-operation and Development (OECD). 2015b. *Measuring Societal Dimension in STI.* Paris: OECD Publishing. http://www.oecd.org/sti/inno/Scientometrics_OECD _Summary_final.pdf. Access date: March 20, 2019.

Organisation of Economic Co-operation and Development (OECD). 2017. Public Procurement for Innovation *Good Practices and Strategies.* Paris: OECD Publishing. http://www.oecd.org /gov/public-procurement-for-innovation-9789264265820-en.htm. Access date: March 20, 2019.

Organisation of Economic Co-operation and Development (OECD). 2019. *Reference Framework for Assessing the Scientific and Socio-economic Impact of Research Infrastructure.* https://www .oecd-ilibrary.org/docserver/3ffee43b-en.pdf?expires=1558088424&id=id&accname=guest&che cksum=52416A0B75DB7D53AF2A1686DE1ACBA. Access date: March 20, 2019.

Oskarsson, Ingvi, and Alexander Schläpfer. 2008. "The performance of spin-off companies at the Swiss Federal Institute of Technology Zurich." Thesis for the Masters in Finance Program (MSc Finance) at London Business School. ETH-Transfer.

Ostrom, Elinor. 1996. "Crossing the great divide: Coproduction, synergy, and development." *World Development, 24*(6), 1073–1087.

PAERIP. 2013. "Socio-economic impact of African-European Research Infrastructure cooperation." Woburn House, London. http://www.paerip.eu/sites/www.paerip.org/files/2%20-%20RI%20socio -economic%20impact%20-%20London%2017-10-2012.pdf. Access date: March 20, 2019.

Pakes, Ariel. 1986. "Patents as options: Some estimates of the value of holding European patent stocks." *Econometrica, 54*(4), 755–784.

Pakes, Ariel, and Mark Schankerman. 1979. "The rate of obsolescence of knowledge, research gestation lags, and the private rate of return to research resources." In Zvi Griliches (ed.), *R & D, Patents, and Productivity* (chap. 4). National Bureau of Economic Research (NBER), Chicago: University of Chicago Press.

Pancotti, Chiara, Julie Pellegrin, and Silvia Vignetti. 2014. *Appraisal of Research Infrastructures: Approaches, Methods, and Practical Implications.* No. 2014–13.

Pancotti, Chiara, Giuseppe Battistoni, Mario Genco, Maria Livraga, Paola Mella, Sandro Rossi, and Silvia Vignetti. 2015. *The Socio-economic Impact of the National Hadrontherapy Centre for Cancer Treatment (CNAO): Applying a Cost-Benefit Analysis Analytical Framework.* DEMM Working Paper 2015–05.

Panofsky, W. 1997. "The evolution of particle accelerators and colliders." *Beam Line, 27*(1):36–44.

Pearce, David W. 1993. *Economic Values and the Natural World.* London: Earthscan.

Pearce, David, Giles Atkinson, and Susana Mourato. 2006. *Cost-Benefit Analysis and the Environment: Recent Developments.* Paris: OECD Publishing.

Pearl, Robert. 2017. "Why patent protection in the drug industry is out of control." *Forbes,* January 19.

Peplow, Mark. 2016. "UK graphene inquiry reveals commercial struggles: Concerns about the University of Manchester's National Graphene Institute reflect a broader decline in industrial research and development." *Nature.* https://www.nature.com/news/uk-graphene-inquiry-reveals -commercial-struggles-1.19840. Access date: March 21, 2019.

Pero, Mickael. 2013. "RRI Governance in Research Infrastructures." *ResAGorA.* http://www.res -agora.eu/. Access date: March 21, 2019.

Phillips, W., R. Lamming, J. Bessant, and H. Noke. 2006. "Discontinuous innovation and supply relationships: Strategic alliances," *R&D Management, 36*(4), 451–461.

Picot, Arnold, Massimo Florio, Nico Grove, and Johann Kranz (eds.). 2015. *The Economics of Infrastructure Provisioning: The Changing Role of the State.* Cambridge, MA: MIT Press.

Pigou, Arthur Cecil. 1912. *Wealth and Welfare.* London: Palgrave Macmillan,

Pinski, Gabriel, and Francis Narin. 1976. "Citation influence for journal aggregates of scientific publications: Theory, with application to the literature of physics." *Information Processing & Management, 12*(5), 297–312.

Poincaré, Henri. 1905. *La Valeur de la Science.* Paris: Flammarion.

Polachek, Solomon W. 2007. *Earnings over the Lifecycle: The Mincer Earnings Function and Its Applications.* IZA Discussion Paper No. 3181. http://ftp.iza.org/dp3181.pdf. Access date: March 20, 2019.

Poor, P. Joan, and Jamie M. Smith. 2004. "Travel cost analysis of a cultural heritage site: The case of historic St. Mary's City of Maryland."*Journal of Cultural Economics, 28*(3), 217–229.

Popular Science. 2011. "Big Science: The 10 most ambitious experiments in the universe today." https://www.popsci.com/science/article/2011-07/supersized-10-most-awe-inspiring-projects -universe. Access date: March 20, 2019.

Porter, Sandra. 2012. *How Much Does It Cost to Get a Scientific Paper?* http://scienceblogs.com /digitalbio/2012/01/09/how-much-does-it-cost-to-get-a/. Access date: March 20, 2019.

Portney, Paul R. 1994. "The contingent valuation debate: Why economists should care." *Journal of Economic Perspectives, 8*(4), 3–17.

Posner, Richard A. 2004. *Catastrophe: Risk and Response.* New York: Oxford University Press.

Pouliquen, Louis Y. 1970. *Risk Analysis in Project Appraisal.* Baltimore: Johns Hopkins University.

Powe, N. A., G. D. Garrod, P. L. McMahon, and K. G. Willis. 2004. "Assessing customer preferences for water supply options using mixed methodology choice experiments." *Water Policy, 6*(5), 427–441.

Powell, Walter W., and Kaisa Snellman. 2004. "The knowledge economy." *Annual Review of Sociology, 30,* 199–220.

PricewaterhouseCoopers. 2007. *The Costs and Benefits of UK World Heritage Site Status.* A literature review for the Department for Culture, Media and Sport. https://assets.publishing.service.gov .uk/government/uploads/system/uploads/attachment_data/file/78452/PwC_fullreport.pdf. Access date: March 20, 2019.

PricewaterhouseCoopers. 2016. *Study to Examine the Socioeconomic Impact of Copernicus in the EU Report on the Copernicus Downstream Sector and User Benefits.* European Commission. https://publications.europa.eu/en/publication-detail/-/publication/97a5cf70-aa5f-11e6-aab7 -01aa75ed71a1. Access date: March 20, 2019.

Priemus, Hugo, Bent Flyvbjerg, and Bert van Wee. 2008. *Decision-Making on Mega-Projects: Cost-Benefit Analysis, Planning, and Innovation.* Cheltenham, UK: Edward Elgar Publishing.

Psacharopoulos, George. 1987. "The cost–benefit model." In G. Psacharopoulos (ed.), *Economics of Education: Research and Studies* (pp. 342–347). Oxford: Pergamon Press.

Psacharopoulos, George. 1994. "Returns to investment in education: A global update." *World Development, 22*(9): 1325–1343.

Psacharopoulos, George. 1995. *The Profitability of Investment in Education: Concepts and Methods.* World Bank Human Capital Development and Operations Policy Working Paper No. 63.

Psacharopoulos, George, 2009. *Returns to Investment in Higher Education: A European Survey.* European Commission, Brussels.

Psacharopoulos, George, and Harry Anthony Patrinos. 2004. "Returns to investment in education: A further update." *Education Economics 12*(2): 111–134.

Punj, Girish. 2015. "The relationship between consumer characteristics and willingness to pay for general online content: Implications for content providers considering subscription-based business models." *Marketing Letters, 26*(2), 175–186.

Quinet, Emile. 2007, "Cost benefit analysis of transport projects in france." In M. Florio (ed.), *Cost Benefit Analysis and Incentives in Evaluation* (pp. 164–188). Cheltenham, UK: Edward Elgar Publishing.

Quinet, Emile. 2013. *Cost Benefit Assessment of Public Investments: Final Report Summary and Recommendations.* http://www.strategie.gouv.fr/sites/strategie.gouv.fr/files/atoms/files/cgsp -calcul_socioeconomique_english4.pdf. Access date: March 20, 2019.

Radicchi, Filippo, Santo Fortunato, and Claudio Castellano. 2008. "Universality of citation distributions: Toward an objective measure of scientific impact." *Proceedings of the National Academy of Sciences, 105*(45), 17268–17272.

Reutlinger, Shlomo. 1970. *Techniques for Project Appraisal under Uncertainty.* World Bank Staff Occasional Paper No. 10, Baltimore: Johns Hopkins University Press.

Rice, Dorothy P. 1967. "Estimating the cost of illness." *American Journal of Public Health and the Nations Health, 57*(3), 424–440.

Rice, Dorothy P., Thomas A. Hodgson, and Andrea N. Kopstein. 1985. "The economic costs of illness: A replication and update." *Health Care Financing Review, 7*(1), 61.

Riordan, M., L. Hoddeson, and A. W. Kolb. 2015. *The Rise and Fall of the Superconducting Super Collider.* Chicago: University of Chicago Press.

Robert, Christian P., and George Casella. 2004. *Monte Carlo Methods.* New York: John Wiley & Sons.

Rodrigues, Susan, Russel Tytler, Linda Darby, Peter Hubber, David Symington, and Jane Edwards. 2007. "The usefulness of a science degree: The 'lost voices' of science trained professionals." *International Journal of Science Education, 29*(11), 1411–1433.

Romer, Paul M. 1990. "Endogenous technological change." *Journal of Political Economy, 98*(5, Part 2), S71–S102.

Rong, Zhao, Xiaokai Wu, and Philipp Boeing. 2017. "The effect of institutional ownership on firm innovation: Evidence from Chinese listed firms." *Research Policy, 46*(9), 1533–1551.

Rørstad, Kristoffer, and Dag W. Aksnes. 2015. "Publication rate expressed by age, gender and academic position—A large-scale analysis of Norwegian academic staff." *Journal of Informetrics, 9*(2), 317–333.

Rosen, Sherwin. 1992. "Distinguished fellow: Mincering labor economics." *Journal of Economic Perspectives, 6*(2), 157–170.

Rosenberg, Nathan, and Richard R. Nelson. 1994. "American universities and technical advance in industry." *Research Policy, 23*(3), 323–348.

Rosenblatt, Michael. 2017. "The large pharmaceutical company perspective." *New England Journal of Medicine, 376*(1), 52–60.

Roson, Roberto. 2000. "Social cost pricing when public transport is an option value." *Innovation: The European Journal of Social Science Research, 13*(1), 81–94.

Rossi, Lucio, and E. Todesco. 2011. *Conceptual Design of 20 T Dipoles for High-Energy LHC.* arXiv preprint arXiv:1108.1619.

Rossi, Sandro. 2015. "The National Centre for Oncological Hadrontherapy (CNAO): Status and perspectives." *Physica Medica, 31*(4), 333–351.

Rothwell, Roy. 1994. "Towards the fifth-generation innovation process." *International Marketing Review, 11*(1), 7–31.

Rowe, Gene. 2007. "A guide to Delphi." *Foresight, 8*, 11–16.

Royal Society. 2018. *A Snapshot of UK Research Infrastructures*. UCL Public Policy. https://royalsociety.org/~/media/policy/Publications/2018/snapshot-uk-research-infrastructures.pdf. Access date: March 20, 2019.

Ruijgrok, E. C. M. 2006. "The three economic values of cultural heritage: A case study in the Netherlands." *Journal of Cultural Heritage, 7*(3), 206–213.

Sakakibara, Mariko. 2010. "An empirical analysis of pricing in patent licensing contracts." *Industrial and Corporate Change, 19*, 927–945.

Salina, G. 2006. "Dalla ricerca di base al trasferimento tecnologico: impatto dell'attività scientifica dell'istituto nazionale di fisica nucleare sull'industria italiana." *Rivista di cultura e politica scientifica*. No. 2/2006.

Salle, M. Caroline, Scott D. Watkins, and Alex L. Rosaen. 2011. "The Economic Impact of Fermi National Accelerator Laboratory." Anderson Economic Group, Chicago.

Salter, Ammon J., and Ben R. Martin. 2001. "The economic benefits of publicly funded basic research: A critical review." *Research Policy, 30*(3), 509–532.

Samuelson, Paul. 1954. "The pure theory of public expenditure." *Review of Economics and Statistics, 36*(4), 387–389.

Sanchez, Y., J. R. Penrod, X. L. Qiu, J. Romley, J. Thornton Snider, and T. Philipson. 2012. "The option value of innovative treatments in the context of chronic myeloid leukemia." *American Journal Management Care, 18*(11 Suppl), S265–S271.

Sanderson, T., G. Hertzler, T. Capon, and P. Hayman. 2016. "A real options analysis of Australian wheat production under climate change." *Australian Journal of Agricultural and Resource Economics, 60*(1), 79–96.

Sattout, Elsa J., Salma N. Talhouk, and Peter D. S. Caligari. 2007. "Economic value of cedar relics in Lebanon: An application of contingent valuation method for conservation." *Ecological Economics, 61*(2–3), 315–322.

Sauermann Henry, and Chiara Franzoni. 2015. "Crowd science user contribution patterns and their implications." *Proceedings of the National Academy of Sciences of the United States of America, 112*(3), 679–684.

Schankerman, Mark. 1998. "How valuable is patent protection? Estimates by technology field." *RAND Journal of Economics, 29*(1), 77–107.

Schankerman, Mark, and Ariel Pakes. 1986. "Estimates of the value of patent rights in European countries during the post-1950 period." *Economic Journal, 96*(384), 1052–1076.

Scherer, Frederic M. 1965. "Firm size, market structure, opportunity, and the output of patented inventions." *American Economic Review, 55*(5), 1097–1125.

Schmied, Helwig. 1977. "A study of economic utility resulting from CERN contracts." *IEEE Transactions on Engineering Management, 4*, 125–138.

Schmied, Helwig. 1982. "Results of attempts to quantify the secondary economic effects generated by big research centers." *IEEE Transactions on Engineering Management, 4*, 154–165.

Schmookler, Jacob. 1966. *Invention and Economic Growth*. Cambridge, MA: Harvard University Press.

Schopper, Herwig. 2009. *LEP-The Lord of the Collider Rings at CERN 1980–2000: The Making, Operation and Legacy of the World's Largest Scientific Instrument*. Springer Science & Business Media.

Schopper, Herwig. 2016. "Some remarks concerning the cost/benefit analysis applied to LHC at CERN." *Technological Forecasting and Social Change, 112*, 54–64.

Schultz, Theodore W. 1961. "Investment in human capital." *American Economic Review, 51*(1), 1–17.

Schuster, R., H. K. Cordell, and Brad Phillips. 2005. "Understanding the cultural, existence, and bequest value of wilderness." *International Journal of Wilderness*, *11*(3), 22–25.

Schweitzer, L., S. Lyons, L. K. J. Kuron, and E. S. W. Ng. 2014. "The gender gap in pre-career salary expectations: a test of five explanations." *Career Development International*, *19*(4), 404–425.

Science and Technology Facilities Council (STFC). 2010. *New Light on Science: The Social & Economic Impact of the Daresbury Synchrotron Radiation Source, (1981–2008)*. https://stfc.ukri .org/stfc/cache/file/4304D848-4E42-468A-89984CE70C5CB565.pdf. Access date: March 20, 2019.

Scientific American. 2015. "Top 10 emerging technologies of 2015." https://www.scientificamerican .com/article/top-10-emerging-technologies-of-20151/. Access date: March 20, 2019.

Scott, David Meerman, and Richard Jurek. 2014. *Marketing the Moon: The Selling of the Apollo Lunar Program*. Cambridge, MA: MIT Press.

Sell, Axel. 1991. *Project Evaluation: An Integrated Financial and Economic Analysis*. Avebury.

Seppä, Tuukka J., and Tomi Laamanen. 2001. "Valuation of venture capital investments: empirical evidence." *R&D Management*, *31*(2), 215–230.

Serjeant, Stephen. 2017. *Successes in Open Data and Citizen Science: ASTERICS and the Open Science Laboratory*. UNOOSA workshop on the Open Universe Initiative, November. http://www .unoosa.org/documents/pdf/psa/activities/2017/OpenUniverse/slides/Presentation41.pdf. Access date: March 20, 2019.

Serrano, Carlos J. 2008. *The Dynamics of the Transfer and Renewal of Patents*. NBER Working Papers 13938, National Bureau of Economic Research (NBER).

Shah, Vivek, and James Keefe. 2010. "Free software: Uses of free software and its implications in the software industry." *Business Quest*, 1–16. https://www.westga.edu/~bquest/2010/software10 .pdf. Access date: March 20, 2019.

Shane, Scott Andrew. 2004. *Academic Entrepreneurship: University Spinoffs and Wealth Creation*. Cheltenham, UK: Edward Elgar Publishing.

Sharma, Manjula, et al. 2007. "What does a physics undergraduate education give you? A perspective from Australian physics." *European Journal of Physics*, *29*(1), 59.

Shelley, Kristina J. 1994. "More job openings—even more new entrants: The outlook for college graduates, 1992–2005." *Occupational Outlook Quarterly*, *38*(2), 5–9.

Sidiropoulos, Antonis, Dimitrios Katsaros, and Yannis Manolopoulos. 2007. "Generalized Hirsch h-index for disclosing latent facts in citation networks." *Scientometrics*, *72*(2), 253–280.

Simkin, Mikhail V., and Vwani P. Roychowdhury. 2007. "A mathematical theory of citing." *Journal of the Association for Information Science and Technology*, *58*(11), 1661–1673.

Simmonds, P., E. Kraemer-Mbula, A. Horvath, J. Stroyan, and E, Zuijdam. 2013. *Big Science and Innovation*. Brighton, UK: Technopolis Group.

Sneed, Katherine A., and Daniel KN Johnson. 2009, "Selling ideas: The determinants of patent value in an auction environment." *R&D Management*, *39*(1), 87–94.

Snowball, Jeanette D. 2007. *Measuring the Value of Culture: Methods and Examples in Cultural Economics*. Springer Science & Business Media.

Solow, Robert M. 1956. "A contribution to the theory of economic growth." *Quarterly Journal of Economics*, *70*(1), 65–94.

Solow, Robert M. 1997. *Learning from "Learning by Doing": Lessons for Economic Growth*. Stanford, CA: Stanford University Press.

Sorg, Cindy F., and John B. Loomis. 1984. *Empirical Estimates of Amenity Forest Values: A Comparative Review*. General Technical Report, Rocky Mountain Forest and Range Experiment Station, US Department of Agriculture, Forest Service RM–107.

Space Telescope Science Institute (STScI). 2016. *Enhancing STScI's Astronomical Data Science Capabilities over the Next Five Years*, Science Definition Team Report. March 15. https://archive .stsci.edu/reports/BigDataSDTReport_Final.pdf. Access date: March 20, 2019.

Squicciarini, Mariagrazia, Hélène Dernis, and Chiara Criscuolo. 2013. *Measuring Patent Quality: Indicators of Technological and Economic Value.* OECD Science, Technology and Industry Working Papers, 2013/03. Paris: OECD Publishing.

SQW Consulting. 2004. *Costs and Business Models in Scientific Research Publishing.* A report commissioned by the Wellcome Trust. https://wellcome.ac.uk/sites/default/files/wtd003184_0 .pdf. Access date: March 20, 2019.

SQW Consulting. 2008. *Review of Economic Impacts Relating to the Location of Large-Scale Science Facilities in the UK: Final Report.*

Stahlecker, Thomas, and Henning Kroll. 2013. *Policies to Build Research Infrastructures in Europe: Following Traditions or Building New Momentum?* No. R4/2013. Working Papers Firms and Region, 2013.

Stange, Kevin M. 2012. "An empirical investigation of the option value of college enrollment."*American Economic Journal: Applied Economics*, *4*(1), 49–84.

Stephan, Paula E. 1996. "The economics of science." *Journal of Economic Literature, 34*(3), 1199–1235.

Stephens, Z. D., et al. 2015. "Big data: Astronomical or genomical?" *PLoS Biology*, *13*(7), e1002195.

Stern, Nicholas. 2015. *Why Are We Waiting? The Logic, Urgency, and Promise of Tackling Climate Change.* Cambridge, MA: MIT Press.

Steuer, Ralph E. 1986. *Multiple Criteria Optimization: Theory, Computation, and Applications.* Wiley.

Stiglitz, Joseph E. 1999. "Knowledge as a global public good." *Global Public Goods: International Cooperation in the 21st Century*, *308*, 308–25.

Stokey, Nancy L. 1988. "Learning by doing and the introduction of new goods." *Journal of Political Economy*, *96*(4), 701–717.

Strachey, Lytton. 1948. *Eminent Victorians.* Oxford: Oxford University Press.

Strandburg, Katherine J., Brett M. Frischmann, and Michael J. Madison (eds.). 2017. *Governing Medical Knowledge Commons.* New York: Cambridge University Press.

Strandburg, K. J., B. M. Frischmann, and M. J. Madison. 2017. "The knowledge commons framework." In K. J. Strandburg, B. M. Frischmann, and M. J. Madison (eds.), *Governing Medical Knowledge Commons* (pp. 9–19). New York: Cambridge University Press.

Strazzera, Elisabetta, Elisabetta Cherchi, and Silvia Ferrini. 2010. "Assessment of regeneration projects in urban areas of environmental interest: a stated choice approach to estimate use and quasi-option values." *Environment and Planning A*, *42*(2), 452–468.

Sun, Jianjun, Chao Min, and Jiang Li. 2016. "A vector for measuring obsolescence of scientific articles." *Scientometrics*, *107*(2), 745–757.

Swan, Trevor W. 1956 "Economic growth and capital accumulation." *Economic Record*, *32*(2), 334–361.

Swerdlow, Robert, Dorothee Teichmann, and Tim Young. 2016. *Economic Analysis of Research Infrastructure Projects in the Programming Period 2014–2020.* JASPERS Working Paper, Luxembourg.

Tassa, Alessandra. 2019. "The socio-economic value of satellite earth observations: huge, yet to be measured." *Journal of Economic Policy Reform*, 1–15. Forthcoming. https://doi.org/10.1080 /17487870.2019.1601565.

Tauri Group. 2013. *NASA Socio-Economic Impacts.* https://www.nasa.gov/sites/default/files/files /SEINSI.pdf. Access date: March 20, 2019.

Technopolis. 2010a. *BBMRI: An Evaluation Strategy for Socio-economic Impact Assessment.* http://www.iss.it/binary/eric/cont/Technopolis_5MB.pdf. Access date: March 20, 2019.

Technopolis. 2010b. *Impacts of European RTOs—A Study of Social and Economic Impacts of Research and Technology Organisations.* http://www.earto.eu/fileadmin/content/03_Publications /TechnopolisReportFinalANDCorrected.pdf. Access date: March 20, 2019.

Technopolis. 2011. *The Role and Added Value of Large-Scale Research Facilities: Final Report.* http://www.technopolis-group.com/wp-content/uploads/2011/02/1379_Report_Large-scale _Research_Facilities_EN1.pdf. Access date: March 20, 2019.

Technopolis. 2015. *Guide to Impact Assessment Research Infrastructures.* http://www .technopolisgroup.com/wpcontent/uploads/2015/04/2015_Technopolis_Group_guide_to_impact _assessment_of_research_infrastructures.pdf. Access date: March 20, 2019.

Teichmann, Dorothee, and C. Schempp. 2013. *Calculation of GHG Emissions of Waste Management Projects.* JASPERS Staff Working Papers, Luxembourg.

Teichmann, Dorothee, and Robert Swerdlow. 2016. *JASPERS Working Paper Economic Analysis of Research Infrastructure Projects in the Programming Period 2014–2020.* http://www .jaspersnetwork.org/download/attachments/21168696/3.%20JASPERS%20guidance%20and%20 worked%20example.pdf?version=1&modificationDate=1465313906000&api=v2. Access date: March 20, 2019.

Terleckyj, Nestor. 1974. *Effects of R&D on the Productivity Growth of Industries: An Exploratory Study.* Report No. 140, Washington, DC: National Planning Association.

Terleckyj, Nestor. 1980. "Direct and indirect effects of industrial research and development on the productivity growth of industries." In John W. Kendrick and Beatrice N. Vaccara (eds.), *New Developments in Productivity Measurement* (pp. 357–386). Chicago: University of Chicago Press.

Theobald, E. J., et al. 2015. "Global change and local solutions: Tapping the unrealized potential of citizen science for biodiversity research." *Biological Conservation, 181,* 236–244.

Thomas, H., A. Marian, A. Chervyakov, S. Stückrad, D. Salmieri, and C. Rubbia. 2016. "Superconducting transmission lines–Sustainable electric energy transfer with higher public acceptance?" *Renewable and Sustainable Energy Reviews, 55,* 59–72.

Thompson, Peter. 2010. "Learning by doing." In B. Hall and N. Rosenberg (eds.), *Handbook of the Economics of Innovation* (Vol. 1, pp. 429–476). Oxford: North-Holland.

Throsby, David. 1999. "Cultural capital." *Journal of Cultural Economics 23*(1–2), 3–12.

Tietenberg, Thomas H., and Lynne Lewis. 2009. *Environmental and Natural Resource Economics,* 8th ed. Routledge.

Togridou, Anatoli, Tasos Hovardas, and John D. Pantis. 2006. "Determinants of visitors' willingness to pay for the National Marine Park of Zakynthos, Greece." *Ecological Economics, 60*(1), 308–319.

Tohmo, Timo. 2004. "Economic value of a local museum: Factors of willingness-to-pay." *Journal of Socio-Economics, 33*(2), 229–240.

Tonin, Mirco, and Michael Vlassopoulos. 2010. "Disentangling the sources of pro-socially motivated effort: A field experiment." *Journal of Public Economics, 94*(11–12), 1086–1092.

Torrance, George W. 1986 "Measurement of health state utilities for economic appraisal: A review." *Journal of Health Economics, 5*(1), 1–30.

Trajtenberg, Manuel, Rebecca Henderson, and Adam Jaffe. 1997. "University versus corporate patents: A window on the basicness of invention." *Economics of Innovation and New Technology, 5*(1), 19–50.

Tresch, Richard W. 2008. *Public Sector Economics.* London: Palgrave McMillan.

Tripp, Simon, and Martin Grueber. 2011. *Economic Impact of the Human Genome Project.* Battelle. https://www.battelle.org/docs/default-source/misc/battelle-2011-misc-economic-impact -human-genome-project.pdf. Access date: March 20, 2019.

Trow, Martin. 2007. "Reflections on the transition from elite to mass to universal access: Forms and phases of higher education in modern societies since WWII." In *International Handbook of Higher Education* (pp. 243–280). Dordrecht, Netherlands: Springer.

Tsujii, H., T. Kamada, T. Shirai, K. Noda, H. Tsuji, and K. Karasawa (eds.). 2014. *Carbon-Ion Radiotherapy Principles, Practices, and Treatment Planning.* Tokyo: Springer.

Tuan, Tran Huu, and Stale Navrud. 2006 "Capturing the benefits of preserving cultural heritage." *Journal of Cultural Heritage, 9*(3), 326–337.

Turner, R. Kerry. 1999. "The place of economic values in environmental valuation." In I. J. Bateman and K. G. Willis (eds.), *Valuing Environmental Preferences: Theory and Practice of the Contingent Valuation Method in the US, EU, and Developing Countries* (pp. 17–41). Oxford: Oxford University Press.

UNESCO. 2009. *UNESCO Framework for Cultural Statistics*. https://unesdoc.unesco.org/ark: /48223/pf0000191061. Access date: March 20, 2019.

Uninett Sigma2. 2016. *A Contribution Model for Funding of the National E-Infrastructure*. https://www.sigma2.no/sites/default/files/media/Dokumenter/a_contribution_model_for_funding _of_the_national_e-infrastructure_ver_d.pdf. Access date: March 20, 2019.

Unnervik, Anders. 2009. "Lessons in Big Science management and contracting." In L. R. Evans (ed.), *The Large Hadron Collider: A Marvel of Technology* (pp. 38–56). Lausanne, Switzerland: EPFL Press.

US Department of Energy (DoE). 2017a. *Annual Report on the State of DOE National Laboratories*. US Department of Energy. https://www.energy.gov/sites/prod/files/2017/02/f34/DOE%20 State%20of%20the%20National%20Labs%20Report%2002132017.pdf.

US Department of Energy (DoE). 2017b. *FY 2018 Congressional Budget Request, Budget in Brief.* Office of Chief Financial Officer, US Department of Energy, DOE/CF-0134.

US Green Book. 1950. *Proposed Practices for Economic Analysis of River Basin Projects*. Prepared by the Subcommittee on Benefits and Costs, https://planning.erdc.dren.mil/toolbox/library /Guidance/Green%20Book%201950%20complete2.pdf. Access date: March 20, 2019.

Valero, Anna, and John Van Reenen. 2016. *The Economic Impact of Universities: Evidence from across the Globe*. No. 22501. National Bureau of Economic Research (NBER).

Van Maanen, John Eastin, and Edgar Henry Schein. 1977. "Toward a theory of organizational socialization." In Barry Staw (ed.), *Annual Review of Research in Organizational Behaviour* (Vol. 1, pp. 209–264). New York: JIP Press.

Van Noorden, Richard. 2013. "The true cost of science publishing." *Nature*, *495*(7442), 426–429.

Venables, Anthony. 2007. "Evaluating urban transport improvements: Cost-benefit analysis in the presence of agglomeration and income taxation." *Journal of Transport Economics and Policy*, *41*, 173–188.

Venter J. Craig, et al. 2001. "The sequence of the human genome." *Science*, *291*(5507), 1304–1351.

Vermeulen, Niki, John N. Parker, and Bart Penders. 2013. "Understanding life together: A brief history of collaboration in biology." *Endeavour*, *37*(3), 162–171.

Vesely, Éva-Terézia. 2007. "Green for green: The perceived value of a quantitative change in the urban tree estate of New Zealand." *Ecological Economics*, *63*(2–3), 605–615.

Viscusi, W. Kip. 2014. "The value of individual and societal risks to life and health." In M. J. Machina and W. K. Viscusi (eds.), *Handbook of the Economics of Risk and Uncertainty* (pp. 385–446). Oxford: Elsevier.

Viscusi, W. Kip. 2018. "Best estimate selection bias in the value of a statistical life." *Journal of Benefit-Cost Analysis*, *9*(2), 205–246.

Viscusi, W. Kip, and Joseph E. Aldy. 2003. "The value of a statistical life: A critical review of market estimates throughout the world." *Journal of Risk and Uncertainty*, *27*(1), 5–76.

Vock, Marlene, Willemijn Van Dolen, and Ko De Ruyter. 2013. "Understanding willingness to pay for social network sites." *Journal of Service Research*, *16*(3), 311–325.

Von der Fehr, Nils-Henrik, Leonardo Meeus, Isabel Azevedo, Xian He, Luis Olmos, and Jean-Michel Glachant. 2013. *Cost Benefit Analysis in the Context of the Energy Infrastructure Package*. THINK Report. https://www.eui.eu/Projects/THINK/Documents/Thinktopic/THINKTopic10.pdf. Access date: March 20, 2019.

Von Hippel, Eric.1986. "Lead users: a source of novel product concepts." *Management Science*, *32*(7), 791–805.

Von Hippel, Eric. 2005. *Democratizing Innovation*. Cambridge, MA: MIT Press.

Vose, David. 2008. *Risk Analysis: A Quantitative Guide*. Chichester, UK: John Wiley and Sons.

Walsh, Richard G., John B. Loomis, and Richard A. Gillman. 1984. "Valuing option, existence, and bequest demands for wilderness." *Land Economics*, *60*(1), 14–29.

Wang, C. L., Y. Zhang, L. R. Ye, and D. D. Nguyen. 2005. "Subscription to fee-based online services: What makes consumer pay for online content?" *Journal of Electronic Commerce Research*, *6*(4), 304.

Wang, Yifang. 2018. *The Future of High-Energy Physics and China's Role*. Presentation at the Department of Physics, University of Milan, May 8.

Wardman, M. R., P. Chintakayala, and G. C. de Jong. 2016, "European wide meta-analysis of values of travel time." *Transportation Research A*, *94*, 93–111.

Webb, D., and E. White. 2009. *Economic Impact of the John Innes Centre*. (Edinburgh). https://www.jic.ac.uk/research-impact/. Access date: March 21, 2019.

Weikard, Hans-Peter. 2005. *Why Non-use Values Should Not Be Used*. No. 22. Mansholt Graduate School, Hollandseweg, NL.

Weimer, David. 2017. *Behavioural Economics for Cost-Benefit Analysis*. Cambridge: Cambridge University Press.

Weinberg, Alvin. M. 1961. "Impact of large-scale science on the United States." *Science*, *134*(3473), 161–164.

Weinberg, Steven. 2012. "The crisis of Big Science." *New York Review of Books*, *59*(8), 1–4.

Weisbrod, Burton A. 1964. "Collective-consumption services of individual-consumption goods." *Quarterly Journal of Economics*, *78*(3), 471–477.

Weitzman, Martin L. 2001. "Gamma discounting." *American Economic Review*, *91*(1), 260–271.

Weitzman, Martin L. 2007. "A review of the Stern Review on the economics of climate change." *Journal of Economic Literature*, *45*(3), 703–724.

Westland, J. Christopher. 2010. "Critical mass and willingness to pay for social networks." *Electronic Commerce Research and Applications*, *9*(1), 6–19.

Westwick, Peter J. 2003. *The National Labs: Science in an American System, 1947–1974*. Cambridge, MA: Harvard University Press.

Wetterstrand, Kris A. 2013. *DNA Sequencing Costs: Data from the NHGRI Genome Sequencing Program (GSP)*. https://www.genome.gov/sequencingcostsdata/.

Whitehead, John C., and Glenn Blomquist. 1991. "A link between behaviour, information, and existence value." *Leisure Sciences*, *13*(2), 97–109.

Wilkie, Tom. 1993. *Perilous Knowledge: The Human Genome Project and Its Implications*. Berkeley: University of California Press.

Wilkinson, Mark, et al. 2016. "The FAIR guiding principles for scientific data management and stewardship." *Scientific Data*, *3*, article number 160018.

Willis, Kenneth George. 1994. "Paying for heritage: What price for Durham Cathedral?" *Journal of Environmental Planning and Management*, *37*(3), 267–278.

Wing, Matthew. 2018. *Particle Physics Experiments Based on the AWAKE Acceleration Scheme*. arXiv:1810.12254.

Woodhall, Maureen.1992. *Cost-Benefit Analysis in Educational Planning*. Paris: UNESCO, International Institute for Educational Planning.

Wooldridge, Jeffrey. 2010. *Econometric Analysis of Cross Section and Panel Data*. Cambridge, MA: MIT Press.

World Bank. 2003. *Building Safer Cities: The Future of Disaster Risk*. http://documents.worldbank.org/curated/en/584631468779951316/Building-safer-cities-the-future-of-disaster-risk. Access date: March 20, 2019.

World Health Organization (WHO). 2006. *Guidelines for Conducting Cost-Benefit Analysis of Household Energy and Health Interventions*. WHO Publication, Geneva.

Wright, Mike, and Ken Robbie. 1996. "Venture capitalists, unquoted equity investment appraisal and the role of accounting information." *Accounting and Business Research*, *26*(2), 153–168.

Young, O., L. King, and H. Schroeder. 2008. *Institutions and Environmental Change*. Cambridge, MA: MIT Press.

Zellner, Arnold, and Henri Theil. 1962. "Three-stage least squares: Simultaneous estimation of simultaneous equations." *Econometrica: Journal of the Econometric Society*: 54–78.

Zerbe, Richard O., and Dwight Dively. (1994). *Benefit-Cost Analysis in Theory and Practice*. HarperCollins Series in Economics. New York: HarperCollins College Publishers.

Zuniga, Pluvia, and Paulo Correa. 2013. *Technology Transfer from Public Research Organizations: Concepts, Markets, and Institutional Failures*. World Bank.

Index